METHODS IN MOLECULAR BIOLOGY

Series Editor:
John M. Walker
School of Life and Medical Sciences
University of Hertfordshire
Hatfield, Hertfordshire, AL10 9AB, UK

For further volumes:
http://www.springer.com/series/7651

Fungal Genomics

Methods and Protocols

Second Edition

Edited by

Ronald P. de Vries

Fungal Physiology, Westerdijk Fungal Biodiversity Institute, Utrecht, The Netherlands

Adrian Tsang

Centre for Structural and Functional Genomics, Concordia University, Montreal, QC, Canada

Igor V. Grigoriev

US Department of Energy Joint Genome Institute, Walnut Creek, CA, USA

Humana Press

Editors
Ronald P. de Vries
Fungal Physiology, Westerdijk Fungal
Biodiversity Institute
Utrecht, The Netherlands

Igor V. Grigoriev
US Department of Energy
Joint Genome Institute
Walnut Creek, CA, USA

Adrian Tsang
Centre for Structural and Functional Genomics
Concordia University
Montreal, QC, Canada

ISSN 1064-3745 ISSN 1940-6029 (electronic)
Methods in Molecular Biology
ISBN 978-1-4939-7803-8 ISBN 978-1-4939-7804-5 (eBook)
https://doi.org/10.1007/978-1-4939-7804-5

Library of Congress Control Number: 2018941834

Printed on acid-free paper

This Humana Press imprint is published by the registered company Springer Science+Business Media, LLC part of Springer Nature.
The registered company address is: 233 Spring Street, New York, NY 10013, U.S.A.

Preface

The development of genomics has had a major impact on many fields of research and the fungal field is no exception. The ability to compare species at the genomic level has significantly enhanced our understanding of fungal taxonomy, evolution, physiology, and cell biology and allowed us to trace the origin of genes within the fungal kingdom.

Transcriptomics and proteomics have increased our understanding of the response of fungi to various culture conditions and environmental situations, thus allowing us to go beyond an inventory of the genes fungi possess to a functional analysis of the relevance of such genes. They have also helped in discovering the function of genes that previously were considered to encode unknown proteins, and as such significantly deepened our understanding of fungal biology. Comparative transcriptomics and proteomics of fungal species revealed that sometimes species with a highly similar genome content use this in very different ways, while species with more diverse content give an overall similar response. A major factor in high-quality genomics, transcriptomics, and proteomics is the quality of the reference genome that is used, as this differs strongly among fungal genomes.

Fungal metabolomics is still in its infancy, although major advances have been made in recent years. The challenge in this field is the generation of reference databases of compounds that can be produced by fungi. This has already received much attention in some fields (e.g., secondary metabolism), but less so in others.

Despite fungal genomics having reached a high level of maturity, comparison of studies to each other is often challenging due to the diversity of methods that is being used. The need for more standardized approaches and better reporting on the details of the methodology has become widely recognized and has inspired several consortia to move toward this.

This book aims to contribute to the development of fungal genomics by presenting a set of protocols that are widely applicable in fungal genomics and related biotechnologies that for the most part have already been embraced by part of the fungal research community. The protocols are not limited to the experimental part of genomics, but also cover analysis and processing of data.

We are very grateful to all the authors of the chapters who together enable this book to cover nearly all aspects currently addressed in fungal genomics, and we hope that the book will serve as a reference across the fungal research community.

Utrecht, The Netherlands — *Ronald P. de Vries*
Montreal, QC, Canada — *Adrian Tsang*
Walnut Creek, CA, USA — *Igor V. Grigoriev*

Contents

Part II Data Processing and Analysis

Contributors

MARK ARENTSHORST • *Molecular Microbiology and Biotechnology, Institute of Biology Leiden, Leiden University, Leiden, The Netherlands*

ANNIE BELLEMARE • *Centre for Structural and Functional Genomics, Concordia University, Montreal, QC, Canada*

ISABELLE BENOIT-GELBER • *Department of Biology, Centre for Structural and Functional Genomics, Concordia University, Montreal, QC, Canada*

MARC BUÉE • *Institut National de la Recherche Agronomique, UMR1136 INRA-Université de Lorraine Interactions Arbres/Microorganismes, Laboratoire d'Excellence ARBRE, Champenoux, France*

CINDY CHEN • *United States Department of Energy Joint Genome Institute, Walnut Creek, CA, USA*

ALICIA CLUM • *United States Department of Energy Joint Genome Institute, Walnut Creek, CA, USA*

CHRISTOPHER DAUM • *United States Department of Energy Joint Genome Institute, Walnut Creek, CA, USA*

RONALD P. DE VRIES • *Fungal Physiology, Westerdijk Fungal Biodiversity Institute, Utrecht, The Netherlands*

MARCOS RAFAEL DI FALCO • *Centre for Structural and Functional Genomics, Concordia University, Montreal, QC, Canada*

LAURE FAUCHERY • *Institut National de la Recherche Agronomique, UMR1136 INRA-Université de Lorraine Interactions Arbres/Microorganismes, Laboratoire d'Excellence ARBRE, Champenoux, France*

AILEEN R. FERRARO • *Department of Microbiology, University of Georgia, Athens, GA, USA*

R. J. FORSTER • *Lethbridge Research and Development Centre, Agriculture and Agri-Food Canada, Lethbridge, AB, Canada*

IGOR V. GRIGORIEV • *United States Department of Energy Joint Genome Institute, Walnut Creek, CA, USA*

R. J. GRUNINGER • *Lethbridge Research and Development Centre, Agriculture and Agri-Food Canada, Lethbridge, AB, Canada*

SAJEET HARIDAS • *United States Department of Energy Joint Genome Institute, Walnut Creek, CA, USA*

GUIFEN HE • *United States Department of Energy Joint Genome Institute, Walnut Creek, CA, USA*

HEINO M. HEYMAN • *Earth and Biological Sciences Directorate, Pacific Northwest National Laboratory, Richland, WA, USA*

KRISTIINA HILDÉN • *Department of Microbiology, University of Helsinki, Helsinki, Finland*

JAKOB BLÆSBJERG HOOF • *Department of Biotechnology and Biomedicine, Technical University of Denmark, Kgs.Lyngby, Denmark*

TRICIA JOHN • *Centre for Structural and Functional Genomics, Concordia University, Montreal, QC, Canada*
YOUNG-MO KIM • *Earth and Biological Sciences Directorate, Pacific Northwest National Laboratory, Richland, WA, USA*
ANNEGRET KOHLER • *Institut National de la Recherche Agronomique, UMR1136 INRA-Université de Lorraine Interactions Arbres/Microorganismes, Laboratoire d'Excellence ARBRE, Champenoux, France*
RITA C. KUO • *United States Department of Energy Joint Genome Institute, Walnut Creek, CA, USA*
ZACHARY A. LEWIS • *Department of Microbiology, University of Georgia, Athens, GA, USA*
ANNA LIPZEN • *United States Department of Energy Joint Genome Institute, Walnut Creek, CA, USA*
RONNIE LUBBERS • *Fungal Physiology, Westerdijk Fungal Biodiversity Institute, Utrecht, The Netherlands*
MIIA R. MÄKELÄ • *Department of Microbiology, University of Helsinki, Helsinki, Finland; Fungal Physiology, Westerdijk Fungal Biodiversity Institute, Utrecht, The Netherlands*
SANDRINE MARQUETEAU • *Centre for Structural and Functional Genomics, Concordia University, Montreal, QC, Canada*
JOEL MARTIN • *United States Department of Energy Joint Genome Institute, Walnut Creek, CA, USA*
T. A. MCALLISTER • *Lethbridge Research and Development Centre, Agriculture and Agri-Food Canada, Lethbridge, AB, Canada*
ERIN MCDONNELL • *Centre for Structural and Functional Genomics, Concordia University, Montreal, QC, Canada*
STEPHEN J. MONDO • *United States Department of Energy Joint Genome Institute, Walnut Creek, CA, USA*
UFFE H. MORTENSEN • *Department of Biotechnology and Biomedicine, Technical University of Denmark, Kgs.Lyngby, Denmark*
CHRISTINA S. NØDVIG • *Department of Biotechnology and Biomedicine, Technical University of Denmark, Kgs.Lyngby, Denmark*
LASZLO NAGY • *Institute of Biochemistry, Biological Research Centre of the Hungarian Academy of Sciences, Szeged, Hungary*
MICHELLE A. O'MALLEY • *Department of Chemical Engineering, University of California, Santa Barbara, CA, USA*
RONAN O'MALLEY • *United States Department of Energy Joint Genome Institute, Walnut Creek, CA, USA*
ALEKSANDRINA PATYSHAKULIYEVA • *Fungal Physiology, Westerdijk Fungal Biodiversity Institute, Utrecht, The Netherlands*
XUEFENG PENG • *Department of Chemical Engineering, University of California, Santa Barbara, CA, USA*
ARTHUR F. J. RAM • *Molecular Microbiology and Biotechnology, Institute of Biology Leiden, Leiden University, Leiden, The Netherlands*
IAN REID • *Centre for Structural and Functional Genomics, Concordia University, Montreal, QC, Canada*
ROBERT RILEY • *United States Department of Energy Joint Genome Institute, Walnut Creek, CA, USA*

Asaf Salamov • *United States Department of Energy Joint Genome Institute, Walnut Creek, CA, USA*
Wendy Schackwitz • *United States Department of Energy Joint Genome Institute, Walnut Creek, CA, USA*
Vasanth R. Singan • *United States Department of Energy Joint Genome Institute, Walnut Creek, CA, USA*
Letian Song • *Centre for Structural and Functional Genomics, Concordia University, Montreal, QC, Canada*
Kimchi Strasser • *Centre for Structural and Functional Genomics, Concordia University, Montreal, QC, Canada*
Candice L. Swift • *Department of Chemical Engineering, University of California, Santa Barbara, CA, USA*
Michael K. Theodorou • *Animal Production, Welfare and Veterinary Sciences, Harper Adams University, Newport, Shropshire, UK*
Adrian Tsang • *Centre for Structural and Functional Genomics, Concordia University, Montreal, QC, Canada*
Stéphane Uroz • *Institut National de la Recherche Agronomique, UMR1136 INRA-Université de Lorraine Interactions Arbres/Microorganismes, Laboratoire d'Excellence ARBRE, Champenoux, France*
Ad Wiebenga • *Fungal Physiology, Westerdijk Fungal Biodiversity Institute, Utrecht, The Netherlands*
Sherry Wu • *Centre for Structural and Functional Genomics, Concordia University, Montreal, QC, Canada*
Yuko Yoshinaga • *United States Department of Energy Joint Genome Institute, Walnut Creek, CA, USA*

Chapter 1

Introduction: Overview of Fungal Genomics

Ronald P. de Vries, Igor V. Grigoriev, and Adrian Tsang

Abstract

Genome sequences and postgenomic tools have had a major impact on fungal research. When the first fungal genome sequences became available it became clear how much more complex fungal biology was than had been previously assumed. Since then an increasing number of genomes have become available enabling detailed comparative studies, especially when combined with postgenomic tools such as transcriptomics and proteomics. This chapter provides an overview and current state of fungal genomics.

Key words Fungi, Genomes, Transcriptomics, Proteomics, Metabolomics

1 The Influence of Genomics on Fungal Research

The availability of genomics and postgenomics technologies has had a major impact of fungal research as it has in many other biological research fields. The first fungal genome to be published was that of the yeast *Saccharomyces cerevisiae* [1], while the first genome of a filamentous fungus was that of *Neurospora crassa* [2]. While initially the number of fungal genomes increased very slowly, especially compared to bacteria, in recent years large genome sequencing programs have resulted in more than 1000 fungal genome sequences (see below).

These genomes have provided a much more detailed look into various aspects of fungal biology and applications, but have also raised many new questions. The number of genes without known function in fungal genomes is still significant (on average between 30 and 50% of all genes), and while in silica studies rapidly increase the number of putative genes in many gene or PFAM families, the experimental confirmation of function is falling behind.

Comparative genomics have revealed the high diversity in the fungal kingdom at the genomic level, such as with respect to plant pathogenicity [3–5], their ability to degrade plant biomass [6–8], as well as stress response [9]. Genomic diversity is found at all levels of fungal taxonomy, even within genera such as recently described for *Aspergillus* [10].

Ronald P. de Vries et al. (eds.), *Fungal Genomics: Methods and Protocols*, Methods in Molecular Biology, vol. 1775, https://doi.org/10.1007/978-1-4939-7804-5_1, © Springer Science+Business Media, LLC, part of Springer Nature 2018

The expansion of fungal genomes has also resulted in an increasing number of fungal transcriptome and proteome studies covering many aspects of fungal biology. Most of these studies cover only one or a few species and are usually highly specific in their conditions, making it difficult to compare different proteome or transcriptome datasets. While a way to solve this is to perform a novel study in which the different species and mutants are compared under identical conditions, as was done for the role of the (hemi)cellulolytic regulator XlnR/Xyr1/Xlr1 in five species [11], this is not a feasible option for many studies. Recently, a study comparing available transcriptome datasets of basidiomycete fungi revealed that with sufficient care and taking into account the experimental variation of the original studies, it is possible to identify core sets of genes that are part of the common response of fungi to a certain condition [12]. Reuse of transcriptome and proteome data is currently still limited compared to reuse of genome data, but will likely increase as methods that take experimental variation into account are further developed.

2 Current Status of Fungal Genomics

Over 4000 whole genome sequencing (WGS) projects are registered in GOLD database [13], and numerous additional WGS and resequencing projects are in progress in various labs around the world. WGS results in assemblies with predicted genes and annotations (*see* Chapters 13 and 15), while resequencing usually involves lower depth sequencing and produces single-nucleotide polymorphisms (SNPs) and structural rearrangements based on mapping reads to a reference genome but with a limited ability to identify lineage specific genes (Chapter 18). According to GOLD complete, with no gaps, or draft genome assemblies are available for ~1500 fungal species, i.e., ~15% of known species [14], and are heavily biased. The majority (~60%) of sequenced species are represented by a single strain, while, for example, 220 of strains of Saccharomyces cerevisiae have already been sequenced by 40 different institutes, and over 500 strains are in progress [15]. Moreover, 90% of sequenced genomes are in Dikarya leaving only 10% for early diverging fungi despite their diversity and unique properties [15]. To address the latter bias, the US Department of Energy Joint Genome Institute (JGI)'s 1000 Fungal Genomes Project has been open to the entire research community to nominate species for sequencing and provides at least one genome per family [16].

This ocean of genomics data requires integration and assessment of data consistency for proper interpretation. SGD [17] and CGI [18] are examples of specialized databases for Saccharomycetes and Candida, while FungiDB [19], CFGP [20], Ensembl Fungi [21], NCBI [22], and Mycocosm [16] represent a broader diversity of fungi. Mycocosm not only offers one of the largest collections of fungal genomes but builds whole genome phylogenies, helps identify gaps in the fungal tree of life [23], and fill them out with new species to be sequenced within the 1000 Fungal Genomes Project.

Besides the phylogenetic context (*see* Chapter 20), the success of genome interpretation and in particular of comparative genomics requires knowledge of how the data been produced (types of sequencing) and processed (types of data mapping/assembly/annotations). Different sequencing platforms (*see* Chapter 4) have their own pros and cons, different requirements for DNA quality and quality (Chapter 2), and the corresponding algorithms for assembly (Chapter 13). Annotation approaches vary dramatically as reviewed in Chapter 15. Frequently, omics (e.g., transcriptomics and proteomics) data are also produced to improve gene annotations, study gene behavior under different conditions, and integrate these data into metabolic models or gene network analysis (Chapters 8–10).

3 The Value of Gold-Standard Genomes

When a genome is published or released publicly, there is a general perception by the public and the research community that the published genome sequence is complete. Moreover, most researchers assume that the genes identified in the genomes to be correct and comprehensive. Unfortunately except for a few manually curated genomes [24–28], there is a vast disconnect between the actual and perceived state of completeness and accuracy in eukaryotic genomics.

In the early days of genomics, Nobel laureate Sydney Brenner often advocated the need to achieve CAP (Complete, Accurate, and Permanent) criteria for genome sequencing. The current sequencing technologies afford deep coverage with relatively long sequence reads at reasonable cost. With appropriate correction tools, the sequences can be highly accurate and, in all likelihood, the entire genome sequence is represented. However, high-quality assembly of the sequence reads is required to produce a complete genome where each chromosome is assembled into one contiguous fragment (contigs) with both ends decorated by telomere repeats. Such complete genomes are difficult to achieve with older generations of sequencing technology and assembly methods. Hence, most fungal genomes that are publicly accessible have been assembled to hundreds or thousands of contigs even though most fungi contain fewer than ten chromosomes. Whole genomes that have assembled to fewer than a hundred contigs are considered very high quality, and only two fungal species, *Myceliophthora thermophila* and *Thielavia terrestris*, have been reported to have completely finished genomes [29]. Typically centromeres, which are often rich in AT residues and highly repetitive, and telomeres are not assembled. These repetitive regions in fungal genomes can be spanned using appropriate sequencing and assembly methods

(Chapters 3 and 12). Thus complete and accurate fungal genomes are achievable. As for permanency, a stable repository where genome sequence can be accessed and retrieved is required. For the past years, Mycocosm maintains a comprehensive set of annotated fungal genomes in easily accessible format.

Genome annotation involves two sequential processes: structural annotation to find genes and functional annotation to assign function to each predicted gene. The algorithms to find genes in fungal genomes have improved dramatically in recent years. The new algorithms make use of the extensive data on transcripts provided by RNA-seq to predict gene models [30–32]. Chapter 16 compares the sensitivity and selectivity of these gene-prediction programs and found them to predict accurately 86–92% of a reference set of manually curated gene models. Functional assignment to genes is based on experimental evidence, sequence similarity to characterized proteins, and domains and motifs found in the gene products. Chapter 14 describes the electronic pipeline used in functional annotation, and Chapter 15 details the procedures in manual assignment of function.

Incomplete genome sequences can lead to the loss of genome information. The most recent gene-calling programs still result in about 10% of the genes either not called or with defective structure. Moreover, they can overcall 10% or more genes than there are in the genome (Chapter 16). These issues impact adversely our understanding of genome organization and function, and the interpretation of results from genome evolution and comparison. Few eukaryotic genomes have achieved gold-standard status where structure and function of gene models are manually curated [24–28]. For filamentous fungi, attempts were made to create a gold-standard genome resource for Aspergilli [33, 34]. The fungal research community is motivated to generate gold-standard genome resources. Several research groups are collaborating to develop gold-standard genome resources for flagship fungi including *Aspergillus niger*, A. nidulans, *Trichoderma reesei*, and *Phanerochaete chrysosporium*. The first filamentous genome that can claim to achieve gold standard is A. niger strain NRRL3, see www.fungalgenomics.ca.

4 Methodology Related to Fungal Genomics

Methods related to fungal genomics are highly diverse, in particular due to the variations needed to obtain high-quality genomic DNA, RNA, or protein samples from different species. However, over the years several more robust methods for sample generation as well as sophisticated methods for genome/transcriptome/proteome generation and analysis have been developed. This book includes protocols for many of these methodologies to facilitate genomics research in the fungal community.

References

1. Goffeau A, Barrell BG, Bussey H, Davis RW, Dujon B, Feldmann H, Galibert F, Hoheisel JD, Jacq C, Johnston M, Louis EJ, Mewes HW, Murakami Y, Philippsen P, Tettelin H, Oliver SG (1996) Life with 6000 genes. Science 274:546–547
2. Galagan JE, Calvo SE, Borkovich KA, Selker EU, Read ND, Jaffe D, FitzHugh W, Ma LJ, Smirnov S, Purcell S, Rehman B, Elkins T, Engels R, Wang S, Nielsen CB, Butler J, Endrizzi M, Qui D, Ianakiev P, Bell-Pedersen D, Nelson MA, Werner-Washburne M, Selitrennikoff CP, Kinsey JA, Braun EL, Zelter A, Schulte U, Kothe GO, Jedd G, Mewes W, Staben C, Marcotte E, Greenberg D, Roy A, Foley K, Naylor J, Stange-Thomann N, Barrett R, Gnerre S, Kamal M, Kamvysselis M, Mauceli E, Bielke C, Rudd S, Frishman D, Krystofova S, Rasmussen C, Metzenberg RL, Perkins DD, Kroken S, Cogoni C, Macino G, Catcheside D, Li W, Pratt RJ, Osmani SA, DeSouza CP, Glass L, Orbach MJ, Berglund JA, Voelker R, Yarden O, Plamann M, Seiler S, Dunlap J, Radford A, Aramayo R, Natvig DO, Alex LA, Mannhaupt G, Ebbole DJ, Freitag M, Paulsen I, Sachs MS, Lander ES, Nusbaum C, Birren B (2003) The genome sequence of the filamentous fungus *Neurospora crassa*. Nature 422(6934):859–868
3. Amselem J, Cuomo CA, van Kan JA, Viaud M, Benito EP, Couloux A, Coutinho PM, de Vries RP, Dyer PS, Fillinger S, Fournier E, Gout L, Hahn M, Kohn L, Lapalu N, Plummer KM, Pradier JM, Quevillon E, Sharon A, Simon A, ten Have A, Tudzynski B, Tudzynski P, Wincker P, Andrew M, Anthouard V, Beever RE, Beffa R, Benoit I, Bouzid O, Brault B, Chen Z, Choquer M, Collemare J, Cotton P, Danchin EG, Da Silva C, Gautier A, Giraud C, Giraud T, Gonzalez C, Grossetete S, Guldener U, Henrissat B, Howlett BJ, Kodira C, Kretschmer M, Lappartient A, Leroch M, Levis C, Mauceli E, Neuveglise C, Oeser B, Pearson M, Poulain J, Poussereau N, Quesneville H, Rascle C, Schumacher J, Segurens B, Sexton A, Silva E, Sirven C, Soanes DM, Talbot NJ, Templeton M, Yandava C, Yarden O, Zeng Q, Rollins JA, Lebrun MH, Dickman M (2011) Genomic analysis of the necrotrophic fungal pathogens *Sclerotinia sclerotiorum* and *Botrytis cinerea*. PLoS Genet 7(8):e1002230. https://doi.org/10.1371/journal.pgen.1002230
4. de Wit PJ, van der Burgt A, Okmen B, Stergiopoulos I, Abd-Elsalam KA, Aerts AL, Bahkali AH, Beenen HG, Chettri P, Cox MP, Datema E, de Vries RP, Dhillon B, Ganley AR, Griffiths SA, Guo Y, Hamelin RC, Henrissat B, Kabir MS, Jashni MK, Kema G, Klaubauf S, Lapidus A, Levasseur A, Lindquist E, Mehrabi R, Ohm RA, Owen TJ, Salamov A, Schwelm A, Schijlen E, Sun H, van den Burg HA, van Ham RC, Zhang S, Goodwin SB, Grigoriev IV, Collemare J, Bradshaw RE (2012) The Genomes of the fungal plant pathogens *Cladosporium fulvum* and *Dothistroma septosporum* reveal adaptation to different hosts and lifestyles but also signatures of common ancestry. PLoS Genet 8(11):e1003088. https://doi.org/10.1371/journal.pgen.1003088
5. Ohm RA, Feau N, Henrissat B, Schoch CL, Horwitz BA, Barry KW, Condon BJ, Copeland AC, Dhillon B, Glaser F, Hesse CN, Kosti I, Labutti K, Lindquist EA, Lucas S, Salamov AA, Bradshaw RE, Ciuffetti L, Hamelin RC, Kema GH, Lawrence C, Scott JA, Spatafora JW, Turgeon BG, de Wit PJ, Zhong S, Goodwin SB, Grigoriev IV (2012) Diverse lifestyles and strategies of plant pathogenesis encoded in the genomes of eighteen dothideomycetes fungi. PLoS Pathog 8(12):e1003037. https://doi.org/10.1371/journal.ppat.1003037
6. Floudas D, Binder M, Riley R, Barry K, Blanchette RA, Henrissat B, Martinez AT, Otillar R, Spatafora JW, Yadav JS, Aerts A, Benoit I, Boyd A, Carlson A, Copeland A, Coutinho PM, de Vries RP, Ferreira P, Findley K, Foster B, Gaskell J, Glotzer D, Gorecki P, Heitman J, Hesse C, Hori C, Igarashi K, Jurgens JA, Kallen N, Kersten P, Kohler A, Kues U, Kumar TK, Kuo A, LaButti K, Larrondo LF, Lindquist E, Ling A, Lombard V, Lucas S, Lundell T, Martin R, McLaughlin DJ, Morgenstern I, Morin E, Murat C, Nagy LG, Nolan M, Ohm RA, Patyshakuliyeva A, Rokas A, Ruiz-Duenas FJ, Sabat G, Salamov A, Samejima M, Schmutz J, Slot JC, St John F, Stenlid J, Sun H, Sun S, Syed K, Tsang A, Wiebenga A, Young D, Pisabarro A, Eastwood DC, Martin F, Cullen D, Grigoriev IV, Hibbett DS (2012) The Paleozoic origin of enzymatic lignin decomposition reconstructed from 31 fungal genomes. Science 336(6089):1715–1719. https://doi.org/10.1126/science.1221748
7. Peng M, Dilokpirnol A, Makela MR, Hilden K, Bervoets S, Riley R, Grigoriev IV, Hainaut M, Henrissat B, de Vries RP, Granchi Z (2017) The draft genome sequence of the ascomycete fungus *Penicillium subrubescens* reveals a highly enriched content of plant biomass related CAZymes compared to related fungi. J Biotechnol 246:1–3
8. Rytioja J, Hildén K, Yuzon J, Hatakka A, de Vries RP, Mäkelä MR (2014) Plant-polysaccharide-degrading enzymes from Basidiomycetes. Microbiol Mol Biol Rev 78(4):614–649. https://doi.org/10.1128/MMBR.00035-14

9. Acton E, Lee AHY, Zhao PJ, Flibotte S, Neira M, Sinha S, Chiang J, Flaherty P, Nislow C, Giaever G (2017) Comparative functional genomic screens of three yeast deletion collections reveal unexpected effects of genotype in response to diverse stress. Open Biol 7(6). https://doi.org/10.1098/rsob.160330
10. de Vries RP, Riley R, Wiebenga A, Aguilar-Osorio G, Amillis S, Uchima CA, Anderluh G, Asadollahi M, Askin M, Barry K, Battaglia E, Bayram O, Benocci T, Braus-Stromeyer SA, Caldana C, Canovas D, Cerqueira GC, Chen F, Chen W, Choi C, Clum A, Dos Santos RA, Damasio AR, Diallinas G, Emri T, Fekete E, Flipphi M, Freyberg S, Gallo A, Gournas C, Habgood R, Hainaut M, Harispe ML, Henrissat B, Hilden KS, Hope R, Hossain A, Karabika E, Karaffa L, Karanyi Z, Krasevec N, Kuo A, Kusch H, LaButti K, Lagendijk EL, Lapidus A, Levasseur A, Lindquist E, Lipzen A, Logrieco AF, MacCabe A, Makela MR, Malavazi I, Melin P, Meyer V, Mielnichuk N, Miskei M, Molnar AP, Mule G, Ngan CY, Orejas M, Orosz E, Ouedraogo JP, Overkamp KM, Park HS, Perrone G, Piumi F, Punt PJ, Ram AF, Ramon A, Rauscher S, Record E, Riano-Pachon DM, Robert V, Rohrig J, Ruller R, Salamov A, Salih NS, Samson RA, Sandor E, Sanguinetti M, Schutze T, Sepcic K, Shelest E, Sherlock G, Sophianopoulou V, Squina FM, Sun H, Susca A, Todd RB, Tsang A, Unkles SE, van de Wiele N, van Rossen-Uffink D, Oliveira JV, Vesth TC, Visser J, Yu JH, Zhou M, Andersen MR, Archer DB, Baker SE, Benoit I, Brakhage AA, Braus GH, Fischer R, Frisvad JC, Goldman GH, Houbraken J, Oakley B, Pocsi I, Scazzocchio C, Seiboth B, vanKuyk PA, Wortman J, Dyer PS, Grigoriev IV (2017) Comparative genomics reveals high biological diversity and specific adaptations in the industrially and medically important fungal genus *Aspergillus*. Genome Biol 18(1):28. https://doi.org/10.1186/s13059-017-1151-0
11. Klaubauf S, Narang HM, Post H, Zhou M, Brunner K, Mach-Aigner AR, Mach RL, Heck AJ, Altelaar AF, de Vries RP (2014) Similar is not the same: differences in the function of the (hemi-)cellulolytic regulator XlnR (Xlr1/Xyr1) in filamentous fungi. Fungal Genet Biol 72:73–81. https://doi.org/10.1016/j.fgb.2014.07.007
12. Peng M, Aguilar-Pontes MV, Hainaut M, Henrissat B, Hildén K, Mäkelä MR, de Vries RP (2017) Comparative analysis of basidiomycete transcriptomes reveals a core set of expressed genes encoding plant biomass degrading enzymes. Fungal Genet Biol. https://doi.org/10.1016/j.fgb.2017.08.001
13. Mukherjee S, Stamatis D, Bertsch J, Ovchinnikova G, Verezemska O, Isbandi M, Thomas AD, Ali R, Sharma K, Kyrpides NC, Reddy TB (2017) Genomes OnLine Database (GOLD) v.6: data updates and feature enhancements. Nucleic Acids Res 45(D1):D446–D456. https://doi.org/10.1093/nar/gkw992
14. Rossman AY (1995) A strategy for an all-taxa inventory of fungal biodiversity. In: Peng C, Chou CH (eds) Biodiversity and terrestrial ecosystems. Academia Sinica, Taiwan, pp 169–194
15. Hittinger CT, Rokas A, Bai FY, Boekhout T, Goncalves P, Jeffries TW, Kominek J, Lachance MA, Libkind D, Rosa CA, Sampaio JP, Kurtzman CP (2015) Genomics and the making of yeast biodiversity. Curr Opin Genet Dev 35:100–109. https://doi.org/10.1016/j.gde.2015.10.008
16. Grigoriev IV, Nikitin R, Haridas S, Kuo A, Ohm R, Otillar R, Riley R, Salamov A, Zhao X, Korzeniewski F, Smirnova T, Nordberg H, Dubchak I, Shabalov I (2014) MycoCosm portal: gearing up for 1000 fungal genomes. Nucleic Acids Res 42(Database issue):D699–D704. https://doi.org/10.1093/nar/gkt1183
17. Sheppard TK, Hitz BC, Engel SR, Song G, Balakrishnan R, Binkley G, Costanzo MC, Dalusag KS, Demeter J, Hellerstedt ST, Karra K, Nash RS, Paskov KM, Skrzypek MS, Weng S, Wong ED, Cherry JM (2016) The Saccharomyces Genome Database Variant Viewer. Nucleic Acids Res 44(D1):D698–D702. https://doi.org/10.1093/nar/gkv1250
18. Skrzypek MS, Binkley J, Binkley G, Miyasato SR, Simison M, Sherlock G (2017) The Candida Genome Database (CGD): incorporation of Assembly 22, systematic identifiers and visualization of high throughput sequencing data. Nucleic Acids Res 45(D1):D592–D596. https://doi.org/10.1093/nar/gkw924
19. Stajich JE, Harris T, Brunk BP, Brestelli J, Fischer S, Harb OS, Kissinger JC, Li W, Nayak V, Pinney DF, Stoeckert CJ Jr, Roos DS (2012) FungiDB: an integrated functional genomics database for fungi. Nucleic Acids Res 40(Database issue):D675–D681. https://doi.org/10.1093/nar/gkr918
20. Park J, Park B, Jung K, Jang S, Yu K, Choi J, Kong S, Park J, Kim S, Kim H, Kim S, Kim JF, Blair JE, Lee K, Kang S, Lee YH (2008) CFGP: a web-based, comparative fungal genomics platform. Nucleic Acids Res 36(Database issue):D562–D571. https://doi.org/10.1093/nar/gkm758
21. Kersey PJ, Allen JE, Armean I, Boddu S, Bolt BJ, Carvalho-Silva D, Christensen M, Davis P, Falin LJ, Grabmueller C, Humphrey J, Kerhornou A, Khobova J, Aranganathan NK, Langridge N, Lowy E, McDowall MD, Maheswari U, Nuhn M, Ong CK, Overduin B, Paulini M, Pedro H, Perry E, Spudich

G, Tapanari E, Walts B, Williams G, Tello-Ruiz M, Stein J, Wei S, Ware D, Bolser DM, Howe KL, Kulesha E, Lawson D, Maslen G, Staines DM (2016) Ensembl Genomes 2016: more genomes, more complexity. Nucleic Acids Res 44(D1):D574–D580. https://doi.org/10.1093/nar/gkv1209

22. Robbertse B, Tatusova T (2011) Fungal genome resources at NCBI. Mycology 2(3): 142–160

23. Spatafora JW, Aime MC, Grigoriev IV, Martin F, Stajich JE, Blackwell M (2017) The Fungal Tree of Life: from molecular systematics to genome-scale phylogenies. Microbiol Spectr 5(5). https://doi.org/10.1128/microbiolspec.FUNK-0053-2016

24. Fey P, Gaudet P, Curk T, Zupan B, Just EM, Basu S, Merchant SN, Bushmanova YA, Shaulsky G, Kibbe WA, Chisholm RL (2009) dictyBase—a Dictyostelium bioinformatics resource update. Nucleic Acids Res 37(Database issue):D515–D519. https://doi.org/10.1093/nar/gkn844

25. Harrow J, Frankish A, Gonzalez JM, Tapanari E, Diekhans M, Kokocinski F, Aken BL, Barrell D, Zadissa A, Searle S, Barnes I, Bignell A, Boychenko V, Hunt T, Kay M, Mukherjee G, Rajan J, Despacio-Reyes G, Saunders G, Steward C, Harte R, Lin M, Howald C, Tanzer A, Derrien T, Chrast J, Walters N, Balasubramanian S, Pei B, Tress M, Rodriguez JM, Ezkurdia I, van Baren J, Brent M, Haussler D, Kellis M, Valencia A, Reymond A, Gerstein M, Guigo R, Hubbard TJ (2012) GENCODE: the reference human genome annotation for The ENCODE Project. Genome Res 22(9):1760–1774. https://doi.org/10.1101/gr.135350.111

26. Cherry JM, Hong EL, Amundsen C, Balakrishnan R, Binkley G, Chan ET, Christie KR, Costanzo MC, Dwight SS, Engel SR, Fisk DG, Hirschman JE, Hitz BC, Karra K, Krieger CJ, Miyasato SR, Nash RS, Park J, Skrzypek MS, Simison M, Weng S, Wong ED (2012) Saccharomyces Genome Database: the genomics resource of budding yeast. Nucleic Acids Res 40(Database issue):D700–D705. https://doi.org/10.1093/nar/gkr1029

27. Howe KL, Bolt BJ, Shafie M, Kersey P, Berriman M (2017) WormBase ParaSite - a comprehensive resource for helminth genomics. Mol Biochem Parasitol 215:2–10. https://doi.org/10.1016/j.molbiopara.2016.11.005

28. Cheng CY, Krishnakumar V, Chan AP, Thibaud-Nissen F, Schobel S, Town CD (2017) Araport11: a complete reannotation of the Arabidopsis thaliana reference genome. Plant J 89(4):789–804. https://doi.org/10.1111/tpj.13415

29. Berka RM, Grigoriev IV, Otillar R, Salamov A, Grimwood J, Reid I, Ishmael N, John T, Darmond C, Moisan MC, Henrissat B, Coutinho PM, Lombard V, Natvig DO, Lindquist E, Schmutz J, Lucas S, Harris P, Powlowski J, Bellemare A, Taylor D, Butler G, de Vries RP, Allijn IE, van den Brink J, Ushinsky S, Storms R, Powell AJ, Paulsen IT, Elbourne LD, Baker SE, Magnuson J, Laboissiere S, Clutterbuck AJ, Martinez D, Wogulis M, de Leon AL, Rey MW, Tsang A (2011) Comparative genomic analysis of the thermophilic biomass-degrading fungi *Myceliophthora thermophila* and *Thielavia terrestris*. Nat Biotechnol 29(10):922–927. https://doi.org/10.1038/nbt.1976

30. Reid I, O'Toole N, Zabaneh O, Nourzadeh R, Dahdouli M, Abdellateef M, Gordon PM, Soh J, Butler G, Sensen CW, Tsang A (2014) SnowyOwl: accurate prediction of fungal genes by using RNA-Seq and homology information to select among ab initio models. BMC Bioinformatics 15:229. https://doi.org/10.1186/1471-2105-15-229

31. Testa AC, Hane JK, Ellwood SR, Oliver RP (2015) CodingQuarry: highly accurate hidden Markov model gene prediction in fungal genomes using RNA-seq transcripts. BMC Genomics 16:170. https://doi.org/10.1186/s12864-015-1344-4

32. Hoff KJ, Lange S, Lomsadze A, Borodovsky M, Stanke M (2016) BRAKER1: unsupervised RNA-Seq-based genome annotation with GeneMark-ET and AUGUSTUS. Bioinformatics 32(5):767–769. https://doi.org/10.1093/bioinformatics/btv661

33. Arnaud MB, Chibucos MC, Costanzo MC, Crabtree J, Inglis DO, Lotia A, Orvis J, Shah P, Skrzypek MS, Binkley G, Miyasato SR, Wortman JR, Sherlock G (2010) The Aspergillus Genome Database, a curated comparative genomics resource for gene, protein and sequence information for the Aspergillus research community. Nucleic Acids Res 38(Database issue):D420–D427. https://doi.org/10.1093/nar/gkp751

34. Cerqueira GC, Arnaud MB, Inglis DO, Skrzypek MS, Binkley G, Simison M, Miyasato SR, Binkley J, Orvis J, Shah P, Wymore F, Sherlock G, Wortman JR (2014) The Aspergillus Genome Database: multispecies curation and incorporation of RNA-Seq data to improve structural gene annotations. Nucleic Acids Res 42(Database issue):D705–D710. https://doi.org/10.1093/nar/gkt1029

Part I

Technical Aspects of Genomics

Chapter 2

Fungal Genomic DNA Extraction Methods for Rapid Genotyping and Genome Sequencing

Annie Bellemare, Tricia John, and Sandrine Marqueteau

Abstract

Isolation of fungal genomic DNA of high quality is required for a number of downstream biotechnology-derived applications such as genome sequencing, microarrays, and digital PCR technologies, to only name a few. In most cases, not only a high molecular weight DNA of superior grade is required but also large quantities. On the other hand, a number of laboratory experiments, such as polymerase chain reaction (PCR) for medical diagnostic or for genotyping, have to be conducted in a limited amount of time and can provide complete results with the use of lower quality DNA. We describe here two different fungal DNA extraction approaches, which are applicable to a wide range of fungal species.

First, we adapted a DNA extraction method for PCR-based genotyping which allows analysis of single to hundreds of colonies simultaneously. Cells are disrupted in the presence of sodium dodecyl sulfate and Proteinase K which are then removed by precipitation and centrifugation. The cleared lysate is used for PCR reaction.

Secondly, we describe a method to obtain genome sequencing quality grade DNA from fungal liquid cultures. Mycelia are harvested by either filtration or centrifugation. Cells are mechanically disrupted by liquid nitrogen grinding, followed by genomic DNA extraction using the QIAGEN's DNeasy® Plant Kit. The quality and quantity of genomic DNA is monitored by fluorometry.

Key words Fungal genomic DNA extraction, Colony PCR, Mycelia, DNA quantification, Spectrophotometer, Agarose gel electrophoresis, Fluorometry

1 Introduction

When it comes to chromosomal or genomic DNA isolation from fungi, published reports offer varied options [1, 2]. Many of the described protocols provide DNA quantity and quality that serve for a specific downstream application. These methods include direct colony PCR for diagnostic of food contaminants [3, 4], DNA isolation from spores for infectious disease detection [5, 6], determination of genetic engineering content [7], and metagenomic studies from soil microorganisms [8]. Most downstream operations can be performed using partially purified fungal cell

Ronald P. de Vries et al. (eds.), *Fungal Genomics: Methods and Protocols*, Methods in Molecular Biology, vol. 1775, https://doi.org/10.1007/978-1-4939-7804-5_2,

extracts, as long as inhibitors of DNA synthesis are removed. For these applications, we adapted the direct colony PCR method described by Alshahni et al. [3]. The generated DNA is of good enough quality so that PCR fragments as large as 11 kb can be amplified.

For systems analysis of fungal species, genome sequencing represents an integral part. High-quality genome assembly requires genomic DNA of superior quality as well as sufficient quantity to obtain a precise quantification of the material [9]. The most challenging step in the isolation of fungal DNA is to be able to disrupt the cell wall while preventing the DNA from shearing. Because of the fungal cell wall strength, a mechanical method such as mill grinding [6], liquid nitrogen grinding or glass beads cell disruption [10] has to be used for cell lysis. This step is often followed by a chemical treatment with sodium dodecyl sulfate (SDS), Triton, Tween 80 or cetyltrimethylammonium bromide (CTAB) [11].

Commercial kits are usually preferred for genomic DNA purification as they often provide an environment free of harmful chemicals in comparison with phenol–chloroform–isoamyl alcohol methods. The reagents cost of commercial kits are generally higher than most traditional methods but they are less labor-intensive. They also allow removal of common DNA preparation contaminants such as guanidine, pigments, and calcium ions.

Quality of genomic DNA preparation can be assessed by agarose gel electrophoresis. The visual inspection of DNA following agarose gel electrophoresis will allow the user to confirm the integrity of high molecular weight DNA. Presence of multiple bands or smearing is a sign of DNA degradation. Contaminating RNA or proteins can also be detected.

If the downstream step is to be used for genome sequencing, up to 20 μg of genomic DNA are required depending on the sequencing method, at a concentration of 2–100 ng/μL. Sequencing centers often recommend the use of fluorescent nucleic acid stain for quantifying double-stranded DNA in solution. The method described here emphasizes the critical steps for obtaining fungal genomic DNA of high quality. The protocol can be applied to a wide range of fungal species, preferably from liquid shaking culture, as nonsporulating conditions will minimize production of potential contaminants such as pigments. Finally, besides commercial kits for DNA isolation and quantitation, the procedure requires few laboratory instruments and reagents. We have used the described method for the preparation of genomic DNA from over 50 evolutionarily distant fungal species to generate high-quality genome assemblies.

2 Materials

All procedures are performed using aseptic consumables and reagents. Used materials should be placed in a biohazard bag and autoclaved before discarding. When handling liquid nitrogen, wear proper personal protection equipment such as insulated gloves and face shield. Wear gloves during the entire process. Consult Material Safety Data Sheets (MSDS) for all chemicals and reagents and discard them in proper containers under the chemical fume hood when advised. All solutions should be prepared with double distilled water (ddH_2O).

2.1 Buffers

1. Lysis buffer for colony PCR. To 70 mL ddH_2O, add 2 mL of 1 M Tris–HCl, pH 8.0, 1 mL of 500 mM EDTA, and 10 mL of 4 M NaCl. Add 3 mL of 10% SDS solution and fill to 100 mL with ddH_2O before autoclaving the solution. Final concentrations are as follows: 20 mM Tris–HCl, pH 8.0, 5 mM EDTA, 400 mM NaCl, 0.3% SDS (*see* **Note 1**). Add proteinase K to 200 μg/mL just before using and only to the amount of lysis buffer that will be used.
2. Tris buffer. To 70 mL ddH_2O, add 1 mL of 1 M Tris–HCl, pH 8.0. Adjust pH to 8.0 if necessary and add ddH_2O up to 100 mL before autoclaving the solution. Final concentration is 10 mM Tris–HCl, pH 8.0. Store at room temperature.
3. TBE Buffer. Dissolve 10.8 g Tris and 5.5 g boric acid in 900 mL ddH_2O. Add 4 mL 0.5 M EDTA and adjust the volume to 1 L. Final concentrations are as follows: 89 mM Tris base, 89 mM boric acid, 2 mM EDTA. Store at room temperature.
4. Agarose gel loading dye (6×). Dissolve 50 mg Bromophenol Blue, 50 mg Xylene Cyanol in 10 mL ddH_2O. Add 6 mL glycerol and mix well. Add ddH_2O up to 20 mL. Final concentrations are 0.25% Bromophenol Blue, 0.25% Xylene Cyanol, 30% glycerol.

2.2 Reagents

1. DNeasy® Plant Mini Kit (QIAGEN). Buffer AP1 should be preheated to 65 °C before use to remove any precipitate that may have formed. Otherwise, store at room temperature. Buffers AW1 and AW2 are supplied as concentrates. Before using for the first time, add the appropriate amount of 95% ethanol as indicated on the bottle to obtain a working solution. Store at room temperature.

3 Methods

3.1 Preparation of Cell Extract from Mycelia for PCR-Based Analyses

1. Fungal cultures are performed in sterile 96-well flat bottomed plates by using a sterile moistened toothpick to transfer a small amount of spores or mycelia (if nonsporulating) into each well containing 150 μL of the culture media (*see* **Notes 2** and **3**).
2. When the mycelial mats have formed but the sporulation has not started, use a sterile toothpick to transfer the mats into a clean 96-well PCR plate and centrifuge at 2400 × *g* for 10 min at room temperature (*see* **Note 4**).
3. Use a single- or multichannel pipettor to remove extracellular fluids that had carried over with the mats. At this point, the mat can be stored at −80 °C until use.
4. If using frozen mycelia, make sure to warm the plate to room temperature before adding the lysis buffer that contains SDS. Add 25 μL lysis buffer containing the proteinase K at a concentration of 200 μg/mL to each well. Use multichannel pipet tips to resuspend the mat in the buffer and to dislodge any bubbles.
5. Carefully seal the PCR plate using a sticky foil mat and incubate for 1 h at 55 °C in a prewarmed heating block. Place a metal block on top of the incubating plate to keep the heat uniform and to prevent evaporation.
6. At the end of proteinase K digestion step, centrifuge the plate at 2400 × *g* for 2 min to collect any condensation, unseal the plate and add 50 μL Tris buffer.
7. Seal the plate with sticky foil mat, centrifuge at 2400 × *g* for 2 min and incubate at 95 °C for 15 min. Transfer on ice to precipitate SDS. Centrifuge at 2400 × *g* for 10 min to pellet SDS. Transfer 30 μL of the cell lysate into a clean 96-well PCR plate. Avoid transferring any spore as it will inhibit PCR (*see* **Note 5**).
8. Use 2 μL of cleared cell lysate per 50 μL PCR reaction.
9. Store the lysate at −20 °C (*see* **Note 6**). To use, thaw the frozen lysate on ice. Centrifuge at 2400 × *g* for 30 min. Transfer 2 μL for PCR directly to the PCR reaction mixture.

3.2 Harvesting Mycelia from 100 mL Liquid Shaking Cultures via Vacuum Filtration for Genomic DNA Extraction

1. After wiping the working bench with 70% isopropanol, light a Bunsen burner. Set up a vacuum pump with a flask and a Büchner funnel. Flame tweezers by dipping into a container with ethanol and gently passing through the flame. Let it cool for a few seconds. Use the flamed tweezers to place one miracloth layer on the Büchner funnel and start the vacuum pump.
2. Remove the cover the fungal culture flask and quickly flame the opening. Pour the mycelia over the miracloth filtration

unit. Once all the culture media passes through the filter, wash the mycelia with four volumes of ddH_2O. Break the vacuum and stop the pump.

3. Using flamed tweezers, place the miracloth with the mycelia on top of another miracloth with the mycelia in between the two miracloths. Sandwich them with paper towels. Press dry.
4. Transfer mycelia into sterile 50 mL Falcon tube and record wet weight, Quick freeze the tube in liquid nitrogen and store at −80 °C until genomic DNA extraction will be performed or proceed directly to Subheading 3.4. Clean working bench with sporicidal cleaning agent followed by 70% isopropanol.

3.3 Harvesting Mycelia from Liquid Shaking Cultures via Centrifugation for Genomic DNA Extraction

1. In a biosafety cabinet, spray the surface of the culture flask with 70% isopropanol. Remove the cover of the culture flask and pour the contents into one or multiple 50-mL centrifuge tubes. Proceed with centrifugation at 3200 × *g* for 15 min at 4 °C. Bring the centrifuged cultures back to the biosafety cabinet and carefully remove the supernatant by pipetting it out of the tube.
2. Wash the mycelium with an equivalent volume of ddH_2O and invert to mix. Centrifuge again at 3200 × *g* for 15 min at 4 °C. After centrifugation, carefully remove the supernatant with a pipette. Repeat wash and centrifugation one more time and remove the supernatant by pipetting out of the tube.
3. Quick freeze the tube in liquid nitrogen and store at −80 °C until genomic DNA extraction will be performed or proceed directly to Subheading 3.4. Clean working bench with sporicidal cleaning agent followed by 70% isopropanol.

3.4 Grinding Mycelia in Liquid Nitrogen

1. Place sterile mortar and pestle in a Styrofoam box. Pour ~100 mL of liquid nitrogen on the mortar and pestle to chill them (*see* **Note 7**).
2. After the nitrogen has nearly evaporated completely, use a sterile spatula to add the previously harvested mycelia into the mortar. Start grinding using the cold pestle. Pour more liquid nitrogen as needed to help finely grind the mycelial mass. Grind the biomass to a talc-like fine powder.
3. Prechill 1.5-mL Eppendorf tubes in liquid nitrogen. Aliquot approximately 100 mg of ground mycelium into the prechilled tubes. Store at −80 °C or proceed with genomic DNA extraction (Subheading 3.5).

3.5 Genomic DNA Extraction Using QIAGEN's DNeasy® Plant Kit (Mini)

If genomic DNA is to be used for sequencing, a minimum of 12 tubes of mycelium should be processed at a time. For detailed information, the QIAGEN DNeasy® Plant kit handbook should be consulted.

1. Preheat buffer AP1 to 65 °C.
2. Add 400 μL buffer AP1 and 4 μL RNase A (stock 100 mg/mL) to frozen ground mycelia in each 1.5-mL tube. Buffer AP1 contains SDS, which would facilitate cell lysis. Do not premix buffer AP1 and RNase A. Vortex at maximum speed until no more clumps are visible. Incubate 10 min at 65 °C and mix by inverting the tube 2–3 times during incubation to help complete cell lysis.
3. Proceed with the salt precipitation of proteins and polysaccharides by adding 130 μL of buffer P3 to the cell lysate. Incubate for 5 min on ice.
4. Centrifuge cell lysate at 20,000 × *g* for 5 min at room temperature (*see* **Note 8**).
5. Pipet lysate into a QIAshredder Mini spin column placed in a 2-mL collection tube for the complete removal of cell debris. Centrifuge at 20,000 × *g* for 2 min at room temperature.
6. Transfer the flow-through (which contains the DNA) to a new tube without disturbing the pellet and record the volume. Add 1.5 volume of buffer AW1 and mix immediately by pipetting. It is possible to observe a precipitate forming at this step.
7. Transfer 650 μL of the mixture onto a DNeasy Mini spin column placed in a 2 mL collection tube, including the precipitate if there is one. At this step, the DNA will bind to the column. Centrifuge at 6000 × *g* for 1 min at room temperature and discard the flow-through. Repeat the centrifugation step to remove the residual liquid in the column completely.
8. Place the DNeasy Mini spin column on a new collection tube and add 500 μL of buffer AW2 to wash the column. Centrifuge at 6000 × *g* for 1 min and discard the flow-through. Add another 500 μL of buffer AW2 to the column and centrifuge at 20,000 × *g* for 2 min.
9. Transfer the DNeasy Mini spin column on a 1.5 mL tube and let the column dry for 2–5 min. Any remaining ethanol in solution AW2 may interfere with the subsequent DNA elution step.
10. For Elution 1, pipet 100 μL of buffer AE directly onto the DNeasy membrane and incubate for 5 min. Centrifuge at 6000 × *g* for 1 min at room temperature to elute DNA.
11. For Elution 2, repeat **step 10**. Keep Elution 1 separate from Elution 2.
12. If you start with 12 tubes of mycelia, pool Elution 1 by groups of four in a 1.5 mL tube. You will have three tubes containing 400 μL each of Elution 1. Repeat this step with the 12 Elution 2 tubes.
13. Precipitate DNA by adding 0.1 volume of 3 M sodium acetate (40 μL) and 2 volumes of 95% cold ethanol (80 μL) to each tube. Mix and incubate overnight at −20 °C.

3.6 DNA Cleanup

1. In this section, all centrifugations are performed at 4 °C and tubes are kept on ice at all times.
2. Centrifuge tubes from **step 13** of Subheading 3.5 at 16000 × *g* for 30 min. Carefully remove the supernatant by decanting and wash the pellet by gently adding 750 μL of cold 70% ethanol. Avoid disturbing the pellet. Centrifuge immediately at 16000 × *g* for 5 min. Discard the supernatant very carefully by decanting.
3. Centrifuge for a few seconds to collect the residual ethanol and remove it with a pipettor. Repeat the rapid centrifugation and liquid removal step.
4. Resuspend the pellet in 30 μL of TE buffer or nuclease-free H_2O (or any other non DEPC-treated nuclease-free water), depending on the downstream application requirements. Keep the tube on ice for 30 min to fully hydrate the DNA.
5. Pool Elution 1 tubes. Use 10 μL of eluent to rinse the inside of the other two Elution 1 tubes and add to the pooled DNA. Do the same thing with Elution 2 DNA.
6. Remove 10 μL of Elution 1 and transfer to a new tube. This aliquot will be used for quality control. Freeze the remaining DNA at −20 °C.

3.7 Genomic DNA Quality Control Using Gel Electrophoresis

Intactness of high molecular weight genomic DNA is confirmed by visual inspection following agarose gel electrophoresis. Always thaw DNA on ice.

1. Weigh 1 g of agarose and mix with 50 mL of 1× TBE buffer in a 250-mL Erlenmeyer flask. Microwave until boiling point and swirl to mix. Let the mixture cool until the flask can be held comfortably in hands for 5 s and add 1 μL of 10 mg/mL Ethidium Bromide (EtBr) solution. Swirl to mix and cast the gel using the appropriate comb. Let it solidify for at least 20 min at room temperature. Place the casting tray in the electrophoresis chamber containing enough 1× TBE buffer to cover the gel and remove the comb.
2. Pipet 1 μL of gel loading dye (6×) into three 0.2 mL tubes. In the first and second tube, add 4 μL of TE buffer pH 8.0. In the first tube, add 1 μL of ready-to-use DNA ladder. In the second tube, add 1 μL of thawed genomic DNA preparation from Subheading 3.6. In the third tube, add 5 μL of genomic DNA. Mix each tube gently and centrifuge for a few seconds to collect all of the solution at the bottom of the tubes.
3. Using a different pipet tip for each sample, load the 6 μL of DNA ladder and genomic DNA preparations into designated well of the prepared gel. Run electrophoresis at 100 V for 1 h.
4. Carefully transport the gel (in a plastic container) to the Imaging Station. Capture an image under UV light exposure.

5. Genomic DNA should look like a tight band of molecular weight higher than 23 kb with minimal smearing which would be a sign of DNA degradation. Contaminating RNA will be detected in the low molecular weight region of the gel. It is also possible to find undesirable proteins and polysaccharides impurities near the gel wells, which can have adverse effect on DNA library construction.

3.8 Double-Stranded DNA (dsDNA) Quantitation Using Fluorometry

Quant-iT™ PicoGreen dsDNA kit is used to determine DNA concentration. The advantage of using double-stranded DNA labeling is that the presence of other nucleic acids and proteins in the sample will not interfere with the quantification result. Quant-iT™ PicoGreen dsDNA kit is selective for dsDNA and allows detection of DNA concentration as low as 25 pg/mL. Follow instructions from the kit.

3.9 Assessment of Nucleic Acid Purity

1. Set the spectrophotometer at 260 nm wavelength. In a 1-cm path length quartz cuvette, add 800 μL of nuclease-free water and carefully wipe the cuvette to remove any dust. Use it as a blank to set the spectrophotometric reading to zero. In the same cuvette, add 2 μL of genomic DNA solution, cover the cuvette, and mix by inverting several times. Wipe the cuvettes sides and read the optical density at 260 nm. DNA concentration is calculated as follows:

   Optical density of 1.0 at 260 nm corresponds to a dsDNA concentration of 50 μg/mL (1 cm path length cuvette).

   $$\text{dsDNA concentration} = 50\ \mu\text{g/mL} \times \text{OD} \times \text{dilution factor}$$

2. Set the spectrophotometer at 280 nm wavelength. Repeat procedure described in **step 1** and record the optical density at 280 nm. Calculate the optical density ratio 260/280 nm (*see* **Note 9**).
3. Set the spectrophotometer at 230 nm wavelength. Repeat procedure described in **step 1** and record the optical density at 230 nm. Calculate the optical density ratio 260/230 nm (*see* **Note 10**).

4 Notes

1. *Taq* polymerase is inhibited by SDS concentration higher than 0.01%, and therefore, it is not recommended to use higher concentration than 0.3% in the lysis buffer [12].
2. The method also suits the analysis of a low number of colonies. In this case, PCR strip tubes can be used to culture transformants instead of 96-well PCR plates.
3. The use of any rich fungal culture media such as potato dextrose broth (PDB) is preferred for rapid growth. In most cases, overnight incubation is sufficient.

4. When culture growth is poor, it may be difficult to grab mycelium with a toothpick or to transfer small amounts of mycelia onto the dry surface of the PCR plate well. In this case, aspirate media avoiding mycelia. Then, use a toothpick to transfer the now visible hyphae into a PCR well containing 25 μL lysis buffer.
5. If samples still contain spores and debris, add 75 μL of ddH_2O to each well, seal the plate, and incubate on ice for 10 min. Centrifuge at 2400 × *g* for 30 min at 4 °C. The cold temperature will precipitate the SDS and trap some of the debris in the pellet. Carefully aspirate 25 μL of the cleared supernatant and use 2 μL for PCR.
6. Note that crude lysates may be unstable in storage and PCR reaction may not work properly after freeze–thaw cycles. The PCR reaction works most reliably when lysate is used on the same day that lysis was performed.
7. During grinding, it is of crucial importance to never allow the samples to thaw as it may affect the quality of the DNA.
8. A cell debris white pellet will form and some particles will float. In this case, it is better to add an extra centrifugation step for 5 min at 20,000 × *g*. Then, proceed with **step 5** of Subheading 3.5.
9. A 260/280 nm absorbance ratio should be in the range of 1.8–2.0 for DNA. A value below 1.8 may indicate a protein contamination.
10. A 260/230 nm absorbance ratio should be in the range of 2.0–2.2 for DNA. A value below 1.8 may indicate a salt or solvent contamination.

Acknowledgment

This work was supported by Genome Canada and Génome Québec.

References

1. Fredricks DN, Smith C, Meier A (2005) Comparison of six DNA extraction methods for recovery of fungal DNA as assessed by quantitative PCR. J Clin Microbiol 43(10):5122–5128
2. Tan SC, Yiap BC (2009) DNA, RNA, and protein extraction: the past and the present. J Biomed Biotechnol 2009:574398
3. Alshahni MM et al (2009) Direct colony PCR of several medically important fungi using Ampdirect plus. Jpn J Infect Dis 62(2):164–167
4. Umesha S, Manukumar HM, Raghava S (2016) A rapid method for isolation of genomic DNA from food-borne fungal pathogens. 3 Biotech 6(2):123
5. Khot PD, Fredricks DN (2009) PCR-based diagnosis of human fungal infections. Expert Rev Anti-Infect Ther 7(10):1201–1221
6. Black JA, Foarde KK (2007) Comparison of four different methods for extraction of Stachybotrys chartarum spore DNA and verification by real-time PCR. J Microbiol Methods 70(1):75–81

7. Demeke T, Jenkins GR (2010) Influence of DNA extraction methods, PCR inhibitors and quantification methods on real-time PCR assay of biotechnology-derived traits. Anal Bioanal Chem 396(6):1977–1990
8. Liles MR et al (2008) Recovery, purification, and cloning of high-molecular-weight DNA from soil microorganisms. Appl Environ Microbiol 74(10):3302–3305
9. Robin JD et al (2016) Comparison of DNA quantification methods for next generation sequencing. Sci Rep 6:24067
10. Taskova RM et al (2006) A comparison of cell wall disruption techniques for the isolation of intracellular metabolites from Pleurotus and Lepista sp. Z Naturforsch C 61(5–6): 347–350
11. Zachová I et al (2003) Detection of aflatoxigenic fungi in feeds using the PCR method. Folia Microbiol 48(6):817–821
12. Weyant RS, Edmonds P, Swaminathan B (1990) Effect of ionic and nonionic detergents on the Taq polymerase. BioTechniques 9(3):308–309

Chapter 3

Purification of Fungal High Molecular Weight Genomic DNA from Environmental Samples

Laure Fauchery, Stéphane Uroz, Marc Buée, and Annegret Kohler

Abstract

Sequencing of a high number of fungal genomes has become possible due to the development of next generation sequencing techniques (NGS). The most recent developments aim to sequence single-molecule long-reads in order to improve genome assemblies, but consequently needs higher quality (minimum >20 kbp) DNA as starting material. However, environmental-derived samples from soil, wood, or litter often contain phenolic compounds, pigments, and other molecules that can be inhibitors for reactions during sequencing library construction. In this chapter, we propose an optimized protocol allowing the preparation of high quality and long fragment DNA from different samples (mycelium, fruiting body, soil) compatible with the current sequencing requirements.

Key words CTAB-based DNA extraction, Sucrose density gradient ultracentrifugation, Fruiting body

1 Introduction

Fungi represent one of the most important microbial guilds occurring in our environment. From aquatic to terrestrial environments, they can be found everywhere. Around 100,000 species are described by taxonomists and distributed in seven phyla [1]. Fungi are capable of releasing crucial nutrients through their ability to degrade organic matter and to weather minerals. Beside these functions, many fungal species are associated in symbiotic interactions with plants and trees, contributing through these interactions to the growth and health of their plant partners. The analysis of the functional and taxonomic diversity of fungi is also justified by their ability to produce enzymes and biomolecules with industrial and medical interests. In this sense, the sequencing of fungal genomes opened a new area of research, giving access to the full catalog of genes carried by fungi and an unprecedented view of their functional potential.

The fungal genomic revolution began in the 1990s. The first fungal genome sequence from the budding yeast *Saccharomyces*

Ronald P. de Vries et al. (eds.), *Fungal Genomics: Methods and Protocols*, Methods in Molecular Biology, vol. 1775, https://doi.org/10.1007/978-1-4939-7804-5_3, © Springer Science+Business Media, LLC, part of Springer Nature 2018

cerevisiae was published in 1996 [2], followed by the first filamentous fungus, *Neurospora crassa* in 2003 [3]. Then the first sequenced basidiomycete was in 2004 *Phanerochaete chrysosporium* [4], *a white rot fungus*, capable of efficiently depolymerizing and mineralizing lignin. Since then, many fungal genomes from different taxa and from fungi with different lifestyles have become available. Sequencing of a high number of fungal genomes has become possible due to the development of next generation sequencing techniques (NGS, many reviews, e.g., [5]) that replaced the previously used, time-consuming, and much more expensive Sanger sequencing [6]. The most recent developments (e.g., PacBio, http://www.pacb.com/) aim to sequence single-molecule long-reads in order to improve genome assemblies, especially important for genomes with a high-repeat content. These sequence resources are notably concentrated and analyzed on the Mycocosm database (http://genome.jgi.doe.gov/programs/fungi/index.jsf). This fungal genomics database was developed by the US Department of Energy Joint Genome Institute to support integration, analysis and dissemination of fungal genome sequences [7]. It contains currently (June 2017) 772 fungal genomes. Global initiatives such as the 1000 fungal genome project (http://1000.fungalgenomes.org/home/) aim to fill the gaps in the fungal tree of life and to sequence more ecologically important species.

In the frame of the 1000 fungal genome project, fungi have to be collected in their natural habitats, in form of fruiting bodies or for metagenome sequencing projects directly from their environment without culturing step or fruiting bodies. However, many of the samples considered such as soil, wood, and litter often contain phenolic compounds, pigments, and other molecules that can be inhibitors for reactions during sequencing library construction. Another important feature is the development of long-read sequencing technologies. These methods provide high quality sequences, but require also high quality (minimum >20 kbp) DNA as starting material. In this context, it is crucial to use an adapted extraction/purification protocol to allow the sequencing of environmental DNA without reducing the DNA quality and in particular the fragment size needed for long-read sequencing. In this chapter, we propose an optimized protocol permitting to generate high quality and long fragment DNA from different samples (mycelium, fruiting body, soil) compatible with the current sequencing requirements (Fig. 1).

2 Materials

2.1 Collecting and Preparing of Fungal Material

1. Hagem medium (per 1 L; [8]): 0.5 g NH_4Cl, 0.5 g KH_2PO_4, 0.5 g $(NH_4)_2HPO_4$, 0.5 g $MgSO_4{\cdot}7H_2O$, 5 g glucose D+, 50 μg thiamine–HCl, 100 μL Kanieltra 6 FE (Yara, Nanterre,

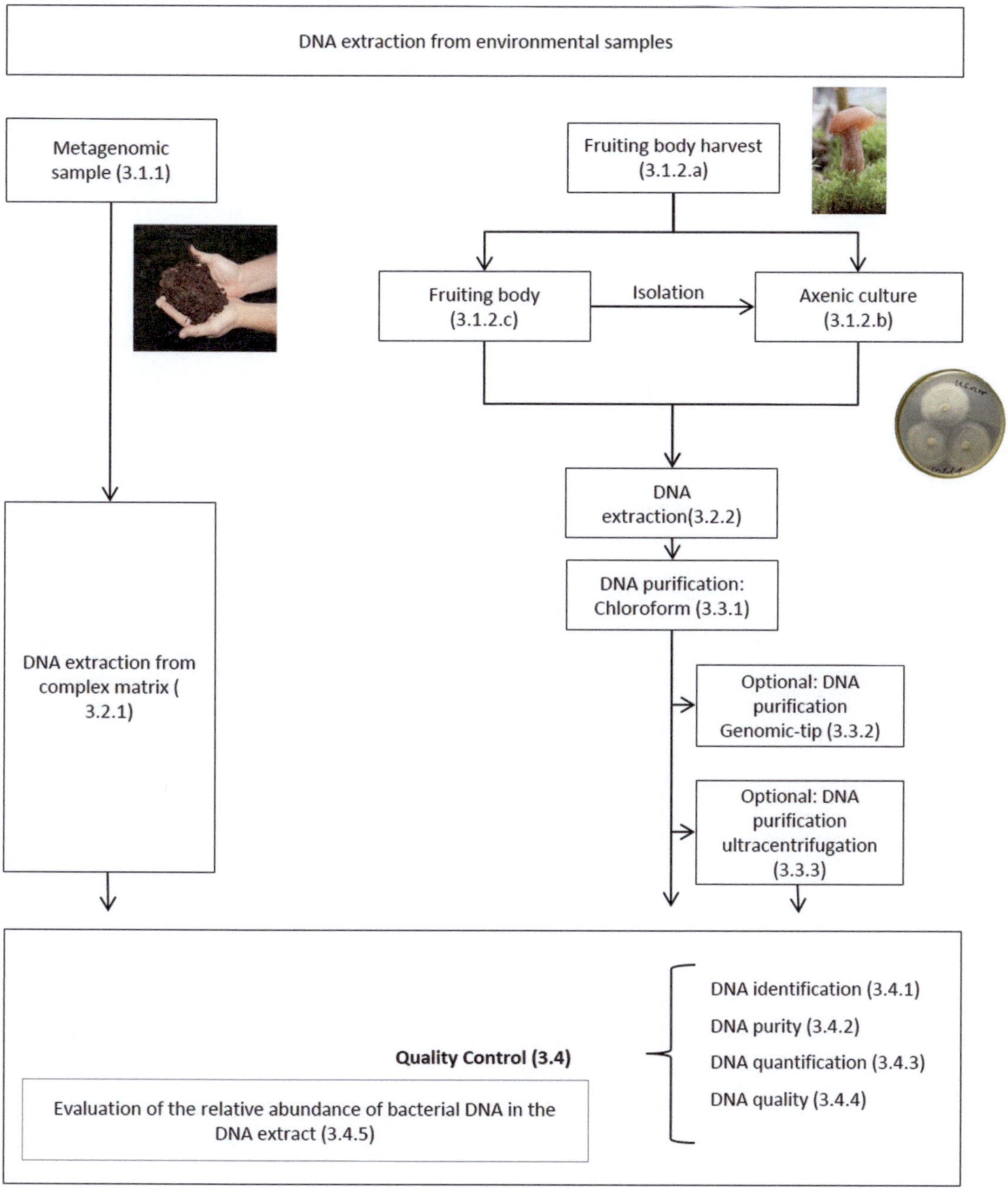

Fig. 1 Workflow summarizing DNA extraction procedure and quality control for environmental samples of different origin. In parentheses the link to the respective protocol is given

France) in distilled water. Adjust the pH to 5.5 and complete to 1 L final volume and add 15 g agar. Sterilize by autoclaving. Add benomyl, chloramphenicol, and rifampicin at 40, 100, and 100 μg/mL final concentration. Benomyl has to be solubilized in ethanol. Chloramphenicol has to be solubilized in water and rifampicin in methanol (*see* **Note 1**).

2. Extract-N-Amp™ Plant PCR Kit (Sigma-Aldrich, Saint-Louis, MO).
3. Cellophane membranes (Hutchinson, Paris, France) preparation: Trim the cellophane membranes to the size of the petri dish used (90 mm). Put the cellophane membranes 20 min into boiling distilled water containing EDTA (1 g/L) in order to permeabilize the membrane. Rinse the membranes four times in a big container with distilled water. Autoclave the cellophane membranes two times. Place one cellophane membrane on the surface of the agar medium of each petri dish.
4. MNC medium (per 1 L; [9]): 1 g KH_2PO_4, 0.5 g $MgSO_4 \cdot 7H_2O$, 0.5 mL $ZnSO_4$ 0.2%, 0.5 g NH_4-tartrate, 0.5 mL citrate Fe 1%, 0.5 mL thiamine (final concentration 0.05 μg/mL), 0.23 g casein hydrolysate, 0.5 g yeast extract, 10 g glucose, 15 g agar (without for liquid medium), qsp 1 L water distilled. Sterilize by autoclaving.
5. Vacuum pump and filter 12 μm.
6. Incubator.

2.2 DNA Extraction from Fungal Material

For this step, prepare all solutions with ultrapure water (UltraPure™ DNase/RNase-Free Distilled Water, ThermoScientific, Waltham, MA) and use molecular biology grade reagents.

1. PowerMax Soil DNA Isolation Kit, now DNeasy PowerMax Soil Kit (Qiagen, Courtaboeuf, France).
2. The components of the lysis buffer have to be prepared separately, and mixed together just before use (modified from [10]). Solution A: 0.35 M sorbitol; 0.1 M Tris–HCl pH 9; 5 mM EDTA (pH 8). Solution B: 0.2 M Tris–HCl pH 9; 50 mM EDTA pH 9; 2 M NaCl; 2% w/v CTAB. Solution C: 5% v/v *N*-lauroylsarcosine sodium salt (SIGMA L5125, Sigma-Aldrich, Saint-Louis, MO). Solution D: Proteinase K (recombinant) PCR grade at 20 mg/mL (ThermoScientific, Waltham, MA).
3. Potassium acetate KAc (Sigma-Aldrich, Saint-Louis, MO) at 5 M. Adjust the pH at 7.
4. Chloroform–isoamyl alcohol 24:1 v/v, suitable for nucleic acid purification (Sigma-Aldrich, Saint-Louis, MO).
5. RNase A, DNase and Protease-free at 10 mg/mL (ThermoScientific, Waltham, MA).
6. Sodium acetate (NaAc) solution 3 M.
7. Isopropanol: Propan-2-ol 100% VWR (International, Radnor, PA).
8. Ethanol: Ethanol absolute 99.9% VWR (International, Radnor, PA).

9. Ethanol 70%: Mix 70 mL ethanol absolute and 30 mL ultrapure water (UltraPure™ DNase/RNase-Free Distilled Water, ThermoScientific, Waltham, MA) in a 50 mL falcon and vortex briefly.
10. TE buffer (Tris–EDTA buffer): 10 mM Tris–HCl, 0.1 mM EDTA pH 7.8 (Sigma-Aldrich, Saint-Louis, MO) (*see* **Note 2**).
11. Qiagen Genomic-tip 500/G (Qiagen, Hilden, Germany; *see* **Note 3**).
12. Ultracentrifuge, a swinging buckets rotor, adapted tubes and a fraction recovery system (*see* **Note 4**).
13. Sucrose gradient solution preparation: A series of sucrose solutions has to be prepared in distilled water as follows: 40% sucrose solution: 4 g sucrose qsp 10 mL MQ water and mix well. Proceed In the same manner for 30%, 20%, 10%, and 5% sucrose solutions but add respectively 3 g, 2 g, 1 g, and 0.5 g sucrose (*see* **Note 5**).

2.3 Quality Control

1. NanoDrop-1000 spectrophotometer (ThermoScientific, Waltham, MA).
2. Qubit 2.0 (Thermo Fisher Scientific, Waltham, MA).
3. Ladder: DNA Molecular Weight Marker II—0.12–23.1 kbp-(Roche, Basel, Switzerland).
4. Electrophoresis buffer (Tris–Borate–EDTA; TBE): 89 mM Tris, 89 mM boric acid, 2 mM EDTA disodium salt, pH 8).
5. Molecular Biology Grade Agarose (Eurogentec, Liege, Belgium).
6. Electrophoresis equipment and DNA staining (e.g., Ethidium Bromide).
7. Ultraviolet light (UV) transilluminator.
8. iQ SybrGreen Supermix (Bio-Rad, Hercules, CA).
9. Polymerase Chain Reaction System (PCR).
10. Real-Time PCR System (qPCR).

3 Methods

3.1 Collection and Preparation of Fungal Material

Fungal DNA can be obtained from different sources such as complex matrix (soil, wood, litter, roots, etc.) or from fruiting bodies or axenic cultures. From fruiting bodies or pure culture the extracted DNA is used to characterize the fungal species by marker gene amplification (ITS: [11]) and subsequent Sanger sequencing. For DNA extracted from complex matrix (such as soil and wood) or nonsterile fruiting bodies (potentially mixed with other fungal species and/or other organisms), a fungal quantification based on qPCR is recommended to determine the amount of nonfungal DNA in the sample (ratio fungi/bacteria).

3.1.1 Fungal Material: Complex Matrix

Collect the environmental samples in 50 mL falcons (usually 5 g is enough) and store at −80 °C.

3.1.2 Fungal Material: Extraction from Fruit Bodies

(a) Fruiting body harvest

1. Collect the fresh fruiting body using sterile material (glove, scalpel, tubes). If possible prefer to harvest one single carpophore (one genotype) with enough biomass for DNA extraction. If not possible, collect several carpophores from the same location to avoid mixing of several genotypes (*see* **Note 6**). Prefer to use younger carpophores without insects or other visuals affections (like molds or rotting area).
2. At this step, the fruiting body material is split in three aliquots: (a) about 10 mg of fresh material conserved at −20 °C for genotyping (*see* Subheading 3.4.1), (b) about 10 mg of fresh material to generate a pure culture (*see* Subheading 3.1.2b; *see* **Note 7**), and (c) the remaining material for extraction of high quality genomic DNA directly (*see* Subheading 3.1.2c).
3. After DNA extraction, verify the fungal species by using adapted primers (ITS, [11]) before proceeding to further genomic DNA extractions.

(b) Establishment of axenic culture (preferable)

This approach is time consuming but more efficient in terms of DNA yield. It results in genomic DNA samples without any contamination and allows further investigations of the sequenced fungal strain.

To obtain pure cultures from field-collected carpophores, prepare Hagem agar medium (Subheading 2.1) or an appropriate medium for your target species, supplemented with antibiotics to avoid bacterial contamination.

1. Cut a small piece (10–50 mg) of tissue from the inner part of a fruiting body using a sterile scalpel blade under a clean bench, place it on Hagem agar plates and incubate at 25 °C (temperature has to be adapted according to the requirements of your fungal species).
2. After growth of the mycelium and visual inspection of its purity, cut a piece of 5 mg mycelium in the periphery of the colony and place it in a 1.5 mL tube for ITS genotyping. The remaining culture is used as starting culture for further steps.
3. The 5 mg mycelium (from **step 2**) are used for DNA extraction (*see* Subheading 3.4.1) and subsequent ITS Sanger

sequencing to verify the identity of the isolated fungus (*see* Subheading 3.4.1).

4. Cut and transfer mycelium containing agar blocks from the starting cultures on ten medium agar plates with cellophane membranes (*see* Subheading 2.1), and incubate them at the appropriate temperature (usually 23–24 °C for most of the fungal species) to allow the fungus to grow.
5. Grow more fungus in liquid culture (MNC medium, or other according to your species requirements, *see* Subheading 2.1) or alternatively on agar medium with cellophane in order to obtain enough fresh fungal material (*see* **Note 8**).
6. Depending on the fungus, the necessary starting material for sufficient quantity and quality of genomic DNA for sequencing can strongly vary. In general with 2 g of mycelium about 10 μg high quality genomic DNA can be obtained, but for difficult-to extract fungi up to 5 g of mycelium can be needed. To harvest the mycelium from liquid culture, use a vacuum nozzle with filter (12 μm) under gentle vacuum and wash the mycelium twice with ultrapure water. To collect the fungus, use a sterile scalpel, place the mycelium in 50 mL falcons and store the tube at −80 °C. To harvest the mycelium from agar plates covered with cellophane membranes (*see* **Note 9**), use a sterile scalpel, transfer the mycelium in 50 mL falcons and store the tube at −80 °C.

(c) Sampling from fruiting bodies

1. With a sterile scalpel and gloves, remove the exterior parts of the mushroom to eliminate most of the contaminants.
2. With a sterile and clean scalpel cut the remaining fungal tissue in small pieces and transfer them in 50 mL falcons. Store tubes at −80 °C.

3.2 DNA Extraction

3.2.1 DNA Extraction from Complex Matrix

The quality of genomic DNA from complex matrices can be altered by a multitude of organic compounds like humic acids, polysaccharides, lipids, polyphenols, tannins, or other inhibitors. It is therefore important to adapt the DNA extraction protocol to the expected organic contaminants from each matrix. For soil samples the PowerMax Soil DNA Isolation kit can be used directly. For other matrices such as wood, litter or roots, it is necessary to carry out a mortar/pestle grinding step in liquid nitrogen first.

1. For grinding in liquid nitrogen, place about 2 g of material in a mortar precooled with liquid nitrogen and grind it with the pestle. Add liquid nitrogen if necessary to avoid unfreezing.
2. Use only 1 g of starting material from difficult-to extract matrices to avoid overloading and saturation of the DNA columns from the PowerMax Soil DNA Isolation Kit.

3. Perform DNA extraction and purification with the PowerMax Soil DNA Isolation Kit according to the manufacturer's recommendations.

3.2.2 DNA Extraction from Pure Culture or Fruiting Bodies

1. Grind the fungal material in liquid nitrogen using a precooled mortar and pestle until a fine powder is obtained (*see* **Note 10**).
2. Distribute 500 mg fungal powder into each of four 50 mL falcons and store the rest powder in 50 mL falcons at −80 °C.
3. Prepare lysis buffer for DNA extraction the same day: Mix 6.5 mL of solution A; 6.5 mL of solution B prewarmed at 65 °C; 2.6 mL; solution C; 1.75 mL PVP 0.1% w/v; and solution D: 125 μL of proteinase K (20 mg/mL).
4. Resuspend the 500 mg fungal powder in 17.5 mL of lysis buffer and immediately mix by moderate vortexing (*see* **Note 11**).
5. Perform lysis for 30 min at 65 °C and invert the falcon tubes gently after 10, 20, and 30 min (*see* **Note 12**).
6. Add 5.75 mL of KAc (5 M, pH 7.5) and mix by inverting the tube. Incubate for 30 min on ice (*see* **Note 13**). Centrifuge for 20 min at 5000 × *g* at 4 °C. Transfer the supernatant into a new 50 mL falcon tube.

3.3 DNA Purification

3.3.1 DNA Purification Using Chloroform–Isoamyl Alcohol

Addition of chloroform–isoamyl alcohol allows the separation of nucleic acids from proteins, various contaminants and cell debris by formation of two phases: the upper aqueous phase with contains the nucleic acids and the lower organic phase with the junk. Isoamyl alcohol is added along with chloroform (in a ratio of 24 parts chloroform to 1 part isoamyl alcohol) to reduce foaming and to stabilize the interphase (proteins) between the aqueous and the organic phase.

1. For each of the four tubes (Subheading 3.2.2), add 1 volume of chloroform–isoamyl alcohol (24:1 v/v) to the supernatant and gently invert the tubes for mixing, until the solution became milky (usually 30 times).
2. After 10 min of centrifugation at 4000 × *g* at 4 °C, carefully transfer the aqueous DNA containing phase to new 50 mL falcon tubes (*see* **Note 14**).
3. Repeat **steps 1** and **2** up to six times depending on the fungus. In general two times is sufficient to eliminate the majority of the contaminants.
4. Add RNase A (100 μL of 10 mg/mL; Sigma) and incubate for 90–120 min at 37 °C.
5. Add 1/10 volume of 3 M NaAc solution to facilitate the precipitation of nucleic acids and mix gently by inverting. Then, add 1 volume of room temperature 100% isopropanol and mix by gently inverting the two times.

6. Centrifuge for 30 min at 10,000 × *g* at 4 °C. Discard the supernatant by inverting the tube.
7. Wash the DNA pellet with 2 mL of 70% ice-cold ethanol and centrifuge for 5 min at 10,000 × *g* at 4 °C.
8. Carefully pipet off the supernatant and dry the pellet for 5 min at room temperature.
9. Resuspend the pellet of each of the four tubes (Subheading 3.2.2) in 500 μL TE buffer (10 mmol/L Tris–HCl, 0.1 mmol/L EDTA pH 7.8) at 65 °C. Avoid pipetting.

The DNA purification step described above may in many cases be sufficient to obtain the DNA quality and quantity necessary for genome sequencing. Therefore, purity, quantity, and quality of the DNA should be controlled at this point (*see* Subheading 3.4). Some sequencing methods like PacBio (Pacific Biosciences) require exclusively large size DNA (more than 23 kb), while other sequencing methods as Illumina (Illumina Inc.) accept smaller DNA fragments.

If the quality criteria (Subheading 3.4) are not reached, further purification becomes necessary. If the electrophoresis gel of your DNA reveals the presence of non-DNA residues like RNA, proteins, polyphenols, or polysaccharides, use Qiagen genomic-tip columns (*see* Subheading 3.3.2) according to the manufacturer's recommendations for further cleanup. They eliminate more efficiently non-DNA residues than chloroform. If the electrophoresis gel identifies in addition partial DNA degradation (presence of a smear below 23 kb), proceed with sucrose density gradient ultracentrifugation (*see* Subheading 3.3.3). The density gradient ultracentrifugation requires large amount of DNA (preferably 100 μg of starting DNA), but this technique allows, in addition to the elimination of non-DNA residues, the selection of the longest DNA fragments from your DNA pool.

3.3.2 DNA Purification Using Qiagen Genomic-Tip Columns (Optional)

1. Add 2 mL QBT buffer to each DNA extract and combine the four extractions.
2. Place the Qiagen Genomic-tip 500/G column in a new 50 mL falcon using the provided support.
3. Equilibrate the Qiagen Genomic-tip 500/G column with 10 mL of QBT buffer. Allow the QIAGEN Genomic-tip column to drain completely.
4. Load the DNA-QBT mix onto the column. Wait until complete infiltration.
5. Wash the genomic-tip columns twice with 15 mL of QC buffer. Place the Genomic-tip column onto a new 50 mL flacon.
6. Elute DNA with 15 mL of prewarmed (50 °C) QF buffer.

7. Add 1/10 volume of 3 M NaAc solution and mix by inverting the tube. Add 1 volume of room temperature 100% isopropanol, invert gently to mix two times and incubate 5 min at room temperature.
8. Centrifuge for 30 min at 10,000 × *g* at 4 °C. Discard the supernatant by inverting the tube.
9. Wash the DNA pellet with 2 mL of 70% ice-cold ethanol and centrifuge for 5 min at 10,000 × *g* at 4 °C. Carefully discard the supernatant by pipetting and dry the pellet for 5 min at room temperature.
10. Resuspend the pellet by flickering, in 500 μL TE buffer at 65 °C and store at −80 °C.

3.3.3 Sucrose Density Gradient Ultracentrifugation (Optional)

A DNA fragment size selection in addition to the DNA clean-up is possible by using sucrose gradient ultracentrifugation. A Sucrose density gradient is produced by gently overlaying sucrose solutions in varying concentrations.

1. A sucrose gradient is obtained by gently and successively depositing 2.5 mL of each sucrose solution (40%, 30%, 20%, 10%, and 5%) in adapted tubes.
2. To stabilize sucrose gradient, the tubes have to be incubated at 4 °C overnight.
3. Load gently 100 μg of genomic DNA on the top of the sucrose gradient of each tube.
4. Equilibrating the tubes is important and critical for ultracentrifugation.
5. Centrifuge for 18 h at 133,907 × *g* at 15 °C.
6. Once the ultracentrifugation run is completed, the content of each tube is collected using a fraction recovery system (Beckman). To do so, the bottom of each tube is pierced using a capillary needle and successive fractions of about 100 μL (8 drops) are collected in 1.5 mL tubes.
7. The DNA of each fraction is precipitated by adding 1/10 volume NaCl 3 M and 2 volume 100% ice-cold ethanol overnight at −20 °C.
8. Centrifuge the tubes for 30 min at 10,000 × *g*.
9. Wash the DNA pellets with 1 mL of 70% ice-cold ethanol and centrifuge for 5 min at 10,000 × *g* at 4 °C. Discard the supernatant by inverting the tube and dry the pellet for 10 min at room temperature.
10. Resuspend the pellet by flickering, in 20 μL TE buffer at 65 °C and store at −80 °C.
11. Load 1 μL of each DNA fraction together with an appropriate DNA ladder on a 0.8% agarose gel. Run gel for 90 min at

~70 V in 1× TBE buffer. If a different electrophoresis set-up is used, make sure that the genomic DNA bands have minimum run ≥2 cm down from the wells and that a separation of the ladder is apparent. Finally combine all fractions showing DNA bands above 23 kb.

3.4 Quality Control

3.4.1 Verification of the Fungal Species

Perform a DNA extraction with Extract-N-Amp™ Plant PCR Kit (Sigma-Aldrich, Saint-Louis, MO), amplify the ITS fragment from the DNA template and sequence the fragment by Sanger Sequencing in order to verify the identity of the fungus.

1. Use 5–10 mg of fungal material (mycelium, fruiting body) in 1.5 mL tubes.
2. Add 100 μL Extraction Solution and grind quickly with a small pestle.
3. Incubate at 95 °C for 10 min.
4. Add 100 μL Dilution Solution.
5. Set up the PCR mix: 10 μL REDExtract-N-Amp PCR ReadyMix, 0.4 μL primer ITS1f (5′-CTTGGTCATTTA GAGGAAGTAA-3′) at 10 μM, 0.4 μL primer ITS4 (5′-TCC TCCGCTTATTGATATGC-3′) at 10 μM [11], 8.2 μL ultrapure water, and 1 μL of previously extracted DNA. Perform four PCR reactions using a serial dilution of the DNA template (pure to 1/1000 dilution in ultrapure water).
6. Run PCR in a thermocycler, using the following program: denaturation 3 min at 94 °C, 30 cycles (denaturation: 45 s at 94 °C, annealing: 45 s at 55 °C, elongation: 1 min at 72 °C), final elongation time 10 min at 72 °C and hold at 4 °C.
7. Load 2 μL of PCR product on a 1.5% agarose gel in 100 mL 1% TBE. Run gel for 40 min at ~90 V in 1× TBE buffer. Incubate the gel for 10 min in Ethidium Bromide solution and destain it for 10 min in water. Visualize the PCR fragments by UV transillumination. One single band indicates the presence of a single fungus in the DNA extract.
8. Perform Sanger sequencing and confirm the fungal species by blasting the obtained ITS sequence to an appropriate database (e.g., UNITE (https://unite.ut.ee/).

3.4.2 DNA Purity and Concentration

Perform first a blank with TE buffer using a NanoDrop-1000 spectrophotometer and measure then the DNA sample. The A260nm/A280nm ratio should be around 1.8 and A260nm/A230nm ratio between 1.8 and 2.2 for pure DNA.

3.4.3 DNA Quantification

To quantify DNA properly, use a Qubit 2.0 Fluorometer with dsDNA BR Assay Kit according to the manufacturer's recommendations. Perform measurements with 3 μL DNA sample.

3.4.4 DNA Quality

1. Dilute DNA samples to a concentration of about 12 ng/μL and load 5 μL on a 0.8% agarose gel in 100 μL 1% TBE. Load 2 μL of ladder in another well.
2. Run gel for 90 min at ~70 V in 1× TBE buffer. If a different electrophoresis set-up is used, make sure that the genomic DNA bands have ran ≥2 cm down from the wells and a separation of the ladder is apparent.
3. Stain the gel for10 min in Ethidium Bromide (EtBr at 0.5 μg/mL final concentration) solution and transfer it then into water for 10 min for destaining. Visualize the DNA fragments by UV transillumination.

The genomic DNA should be visible as a clear band without smear above the 23 kb band of the ladder. Bands or smears between loading well and the 23 kb ladder band usually indicate presence of impurities (proteins, polysaccharides or pigments that can inhibit the construction of the library). If a smear is visible below the DNA band, it means that the genomic DNA is partially degraded or RNAse treatment was inefficient. In this case, it is recommended to perform an additional purification with Qiagen genomic tips columns (*see* Subheading 3.3.2) or to add a sucrose gradient ultracentrifugation (*see* Subheading 3.3.3). For genomic DNA of complex matrices (*see* Subheading 3.2.1) a band above 23 kb with a smear is possible and indicates the presence of a mix of organisms with different genome sizes.

3.4.5 Evaluation of the Relative Abundance of Bacterial DNA in the DNA Extract

When genomic DNA was isolated directly from fruiting bodies or complex matrices (*see* Subheadings 3.1.2c and 3.1.1), a qPCR run should be performed to determine the relative purity of the fungal DNA. It is possible that an overrepresentation of bacteria in the starting material can result in genomic DNA of good quality but contain only a small portion of fungal-derived DNA.

To determine if there is bacterial contamination, the extracted DNA is used as template to amplify the 16S rRNA region for bacteria [12] and the 18S rRNA region for fungi [13].

Total bacterial and fungal communities can be quantified from total DNA by using 16S and 18S rRNA gene-specific primers (10 μM each; 968F/1401R, and FR1/FF390r, respectively, *see* **Note 15**) and SybrGreen detection (iQ SybrGreen Supermix, Bio-Rad) according to the manufacturer's recommendations. A four-step (45 cycles: 20 s at 95 °C, 30 s at the specific annealing temperature, 20 s at 72 °C and fluorescence acquisition at 82 °C) amplification protocol was performed using previously described protocols with annealing temperatures of 56 °C for bacteria and 50 °C for fungi by qPCR. Absolute quantifications were performed using serial dilutions of standard plasmids containing bacterial 16S rDNA or fungal 18S rDNA inserts (from 10^9 to 10^2 gene copies/μL). DNA suitable for fungal genome sequencing should have a ratio copy number of 18S rRNA/ 16S rRNA ≤15.

4 Notes

1. Benomyl has no effect on Basidiomycetes and many Mucorales. Avoid using it with Ascomycetes.
2. Ultrapure water can be used alternatively to TE Buffer, but long time storage of DNA is better in TE.
3. QBT QC QF buffers don't have to be purchased, the chemical composition is provided by Qiagen in the kit manual.
4. In our case equipment from Beckman was used (an Ultracentrifuge Optima XP80 model, a SW32 rotor, Polyallomer tubes [ref. 337986] and a fraction recovery system [ref. 270-331580]).
5. To dissolve sucrose, especially the higher concentrations, distilled water has to be heated before adding sucrose.
6. Carefully choose the fruiting bodies since if possible only one genotype should be used for genomic DNA sequencing. Prefer sampling of fruiting bodies as close from each other as possible. When only distant fruiting bodies are available, collect them in separate tubes and perform genotyping first.
7. If possible invest some time in the isolation of pure mycelium from fruiting bodies and perform DNA extractions from axenic cultures (*see* Subheading 3.1.2b). DNA extracted directly from fruiting bodies is often of lower quality and concentration. Isolation of mycelium is also recommended from small fruiting bodies, for the case that biomass is not sufficient.
8. Some fungi grow better in liquid culture, while other prefer agar medium. The method has to be adapted depending on your fungus.
9. The use of cellophane membranes makes it possible to harvest mycelium without agar and medium. Rests of culture medium and agar can inhibit the efficiency of the DNA extraction.
10. During this step it is important to keep the biological material frozen and to use precooled accessories.
11. The frozen fungal powder has to be transferred rapidly into the lysis buffer to avoid unfreezing. Immediately mix the sample by moderate vortexing until the suspension becomes less viscous. The formation of spume indicates the detachment of DNA from polysaccharides. For most fungi 10 s vortexing is sufficient.
12. After this step it is important to avoid to vortex the sample. Prefer to carefully pipette instead, in order to not break down the DNA molecules and to not alter the quality of the genomic DNA.

13. This step allows the precipitation of polysaccharides, and the low temperature facilitates their elimination.
14. Harvest the aqueous phase (upper phase) by avoiding the interface. This step allows separation of nucleic acids from proteins and other contaminants.
15. Primer sequences for qPCR:
    - 16S: 968F 5′-AACGCGAAGAACCTTAC-3′; 1401R 5′-CGGTGTGTACAAGACCC-3′.
    - 18S: FR1 5′-ANCCATTCAATCGGTANT-3′; FF390r 5′-CGATAACGAACGAGACCT-3′.

Acknowledgments

Our research was financed by the “Institut National de la Recherche Agronomique” (INRA), the Région Grand Est and the lab of excellence ARBRE (ANR-11-LABX-0002-01). The protocols were developed at the INRA Grand Est-Nancy Ecogenomics facilities. Part of our research was sponsored by the Genomic Science Program of the US Department of Energy, Office of Science, Biological and Environmental Research (under contract DE-AC05-00OR22725) and the US Department of Energy (DOE) Joint Genome Institute (JGI; Office of Science of the US Department of Energy). We would like to thank Francis Martin for helpful discussions and input on this chapter.

References

1. Kirk PM, Cannon PF, Minter DW, Stalpers J (2008) Dictionary of the fungi. CABI, Wallingford
2. Goffeau A, Barrell BG, Bussey H, Davis RW, Dujon B, Feldmann H, Galibert F, Hoheisel JD, Jacq C, Johnston M, Louis EJ, Mewes HW, Murakami Y, Philippsen P, Tettelin H, Oliver SG (1996) Life with 6000 genes. Science 274:546
3. Galagan JE, Calvo SE, Borkovich KA, Selker EU, Read ND, Jaffe D, FitzHugh W, Ma LJ, Smirnov S, Purcell S, Rehman B, Elkins T, Engels R, Wang S, Nielsen CB, Butler J, Endrizzi M, Qui D, Ianakiev P, Bell-Pedersen D, Nelson MA, Werner-Washburne M, Selitrennikoff CP, Kinsey JA, Braun EL, Zelter A, Schulte U, Kothe GO, Jedd G, Mewes W, Staben C, Marcotte E, Greenberg D, Roy A, Foley K, Naylor J, Stange-Thomann N, Barrett R, Gnerre S, Kamal M, Kamvysselis M, Mauceli E, Bielke C, Rudd S, Frishman D, Krystofova S, Rasmussen C, Metzenberg RL, Perkins DD, Kroken S, Cogoni C, Macino G, Catcheside D, Li W, Pratt RJ, Osmani SA, DeSouza CP, Glass L, Orbach MJ, Berglund JA, Voelker R, Yarden O, Plamann M, Seiler S, Dunlap J, Radford A, Aramayo R, Natvig DO, Alex LA, Mannhaupt G, Ebbole DJ, Freitag M, Paulsen I, Sachs MS, Lander ES, Nusbaum C, Birren B (2003) The genome sequence of the filamentous fungus Neurospora crassa. Nature 422:857–868
4. Martinez D, Larrondo LF, Putnam N, Gelpke MD, Huang K, Chapman J, Helfenbein KG, Ramaiya P, Detter JC, Larimer F, Coutinho PM, Henrissat B, Berka R, Cullen D, Rokhsar D (2004) Genome sequence of the lignocellulose degrading fungus Phanerochaete chrysosporium strain RP78. Nat Biotechnol 22:695–700
5. Goodwin S, McPherson JD, McCombie WR (2016) Coming of age: ten years of next-generation sequencing technologies. Nat Rev Genet 17(6):333–351

6. Sanger F, Nicklen S, Coulson AR (1977) DNA sequencing with chain-terminating inhibitors. Proc Natl Acad Sci U S A 74(12):5463–5467
7. Grigoriev IV, Nikitin R, Haridas S, Kuo A, Ohm R, Otillar R, Riley R, Salamov A, Zhao X, Korzeniewski F, Smirnova T, Nordberg H, Dubchak I, Shabalov I (2014) MycoCosm portal: gearing up for 1000 fungal genomes. Nucleic Acids Res 42(1):D699–D704
8. Laiho O (1970) Paxillus involutus as a mycorrhizal symbiont of forest trees. Acta Forestalia Fennica 106:1–73
9. Yamad A, Katsuya K (1995) Mycorrhizal association of isolates from sporocarps and ectomycorrizal with *Pinus densiflora* seedings. Mycroscience 36:315–323
10. Fulton PJ, Chungwongse J, Tanksley SS (1995) Microprep protocol for extraction of DNA from tomato and other herbaceous plants. Plant Mol Biol Report 13:207–209
11. Manter DK, Vivanco JM (2007) Use of the ITS primers, ITS1F and ITS4, to characterize fungal abundance and diversity in mixed-template samples by qPCR and length heterogeneity analysis. J Microbiol Methods 71:7–14
12. Cébron A, Norini MP, Beguiristain T, Leyval C (2008) Real-time PCR quantification of PAH-ring hydroxylating dioxygenase (PAH-RHDα) genes from Gram positive and Gram negative bacteria in soil and sediment samples. J Microbiol Methods 73:148–159
13. Chemidlin Prévost-Bouré N, Christen R, Dequiedt S, Mougel C, Lelièvre M, Jolivet C, Shahbazkia HR, Guillou L, Arrouays D, Ranjard L (2011) Validation and application of a PCR primer set to quantify fungal communities in the soil environment by real-time quantitative PCR. PLoS One 6(9):e24166

Chapter 4

Genome Sequencing

Yuko Yoshinaga, Christopher Daum, Guifen He, and Ronan O'Malley

Abstract

Strategies for sequencing fungal genomes on next-generation sequencing (NGS) platforms depend on the characteristics of the genome of the targeted species, quantity and quality of the genomic DNA, and cost considerations. Massively parallel sequencing with sequencing by synthesis (SBS) approach by Illumina produces terabases of short read sequences (i.e., ~300 bp) in a time and cost-effective manner, though the read length can limit the assembly particularly in repetitive regions. The single molecule, real-time (SMRT) sequencing approach by Pacific Biosciences (PacBio) produces longer reads (i.e., ~12,500 bp) which can facilitate de novo assembly of genomes that contain long repetitive sequences, though due to the lower-throughput of this platform achieving the coverage needed for assembly is more expensive than by SBS. Additionally, the Illumina SBS platforms can handle low quantity/quality of genomic DNA materials, while the SMRT system requires undamaged long DNA fragments as input to ensure that high-quality data is produced. Both platforms are discussed in this chapter including key decision-making points.

Key words Next-generation sequencing (NGS), Illumina, HiSeq 2500, Sequencing by synthesis (SBS), Pacific Biosciences, RS II, Single molecule real-time (SMRT) sequencing

1 Introduction

The short-read approach by Illumina makes large-scale whole-genome sequencing accessible and practical for individual research labs. Illumina's sequencing by synthesis (SBS) [1] utilizes fluorescently labeled reversible-terminator nucleotides, on clonally amplified DNA templates immobilized on the surface of a glass flow cell. The Illumina HiSeq 2500 has set the standard for high-throughput massively parallel sequencing.

Pacific Biosciences (PacBio) has developed the single molecule, real-time (SMRT) approach for sequencing multikilobase length reads [2]. In the SMRT approach, DNA polymerase molecules bound to a DNA template are attached to the bottom of 50 nm-wide wells termed zero-mode waveguides (ZMWs). Each polymerase then carries out second strand DNA synthesis in the presence of γ-phosphate fluorescently labeled nucleotides. The width of the ZMW is such that light cannot propagate through the waveguide,

Ronald P. de Vries et al. (eds.), *Fungal Genomics: Methods and Protocols*, Methods in Molecular Biology, vol. 1775, https://doi.org/10.1007/978-1-4939-7804-5_4,

but energy can penetrate a short distance and excite the fluorophores attached to those nucleotides in the vicinity of the polymerase bound to the bottom of the well. As each of the four bases is labeled with a unique fluorophore a pulse of fluorescence is detected in real time corresponding to the incorporated base.

Three important differences of the Illumina and PacBio platforms are read length, cost, and input DNA quality and quantity requirements. Illumina SBS approach generates relatively short reads, and in this chapter we primarily discuss 300 bp reads which result from paired end (PE) reads of 150 bp from each end. The Illumina short reads are very valuable for counting experiments (i.e., RNA-seq) and for discovery of sequence differences within or between related species by mapping reads from an individual to a reference genome, a strategy known as resequencing. Though the Illumina read length can be used for de novo genome assembly, the reads are too short to span most repetitive regions which can result in many breaks in the assembly. The PacBio SMRT approach on the other hand generates kilo-base size (described here is above 10 kb) reads which allow them to span most repetitive regions making them very well suited for de novo genome assembly. However, one important advantage of the Illumina system is that it can produce 25-times more bases than PacBio in the same time frame at an eighth of the cost per base. The Illumina system also allows for lower input amounts, as low as 100 ng of DNA for PCR-free library, while PacBio requires >1 μg of high molecular weight DNA as input. Finally, the PacBio requires long undamaged DNA fragments for sequencing while the Illumina platform is more forgiving in terms of DNA quality. Therefore, while in general PacBio sequencing results in higher quality of assembly especially for diploid repetitive genomes, choice of sequencing platform will also depend on the quality and quantity of the DNA that can be extracted from a particular fungal species and on cost considerations.

All NGS platforms require library preparation to attach specific adapters at the both ends of fragmented DNA pieces. Fragmentation process involves physical shearing by sonication (LE220 and S2, 100 bp–5 kb size, Covaris), hydrodynamic shearing using g-TUBE (6–20 kb sizes, Covaris), 26-gauge needles (20–40 kb sizes), or Megaruptor (2–75 kb sizes, Diagenode), or by enzymatic reactions (Hyper Plus, Kapa Biosystems; NEBNext dsDNA fragmentase, New England Biolabs; Nextera, Illumina). The ends of fragments are repaired and ligated with blunt-end adapters (PacBio), or A-tailed and ligated with T-tailed adapters (Illumina). PacBio utilizes looped adapters combined with exonuclease treatment (SMRTbell libraries) to eliminate DNA fragments lacking adapters.

Size selections of the DNA are required for both platforms to eliminate unwanted shorter fragments, which will tend to preferentially sequence and skew the average read lengths to shorter sizes. A less stringent size selection can be performed by using a polyethylene glycol (PEG) and sodium chloride solution to preferentially bind longer DNA onto a paramagnetic bead. The concentration of PEG determines the size of the DNA that binds to beads, and increasing or decreasing the PEG concentration can be used to allow for a specific size cutoff [3]. The PEG-based bead method is used to target library templates up to 20 kb and is used in both Illumina and PacBio preparation methods. Strict size selection to target library templates above 20 kb requires automated (e.g., BluePippin, Sage Science) or manual electrophoresis-based size selection, and verification of the sizes using pulsed-field gel electrophoresis (PFGE), Pippin Pulse (Sage Science), or Fragment Analyzer gDNA kit which extend the library preparation process to over 3 days. The strict size selection method for large templates is used in the PacBio preparation method to take advantage of the long reads generated by this sequencing platform.

A combination of Illumina short PE reads and Illumina long-mate pairs (LMP) [4] can be used to overcome the challenges of assembling long repetitive sequence. Briefly, LMPs are created by circularization of size-selected fragments by CRE-loxP reaction [5]. Circular DNA is fragmented by sonication, and the biotinylated fragments are pulled down to make Illumina libraries. The final libraries consist of short fragments made up of two DNA segments that were originally separated by several kilobases. While combining LMP with PE reads is an effective strategy for genome assembly, the LMP library preparation process is time, labor, and sample intensive. Because the PacBio platform produces better assemblies at a lower-cost and higher-throughput it has largely replaced the LMP with PE strategy. Therefore, we will discuss here using the Illumina HiSeq 2500 sequencing platform with PCR-free 300-bp libraries for standard genome assembly from low DNA quality and quantity inputs, and the PacBio RSII sequencing platform with a range of sizes (10-kb, >10-kb, >20-kb, and 30-kb) for de novo assembly of highly repetitive genomes starting from high DNA quality and quantity input.

2 Materials

Prepare all solutions using nuclease-free water (such as Ambion, AM9938) and molecular biology grade reagents. Prepare and store all reagents at room temperature (unless indicated otherwise). You would require standard laboratory equipment such as pipettes, vortex, centrifuges, thermal cycler, and consumables of tips and tubes.

2.1 Quality Assessment of Genomic DNA

1. QuantiFluor™ dsDNA System (Promega, E2670) or Qubit dsDNA HA and Broad Range Kit (Life Technologies, Q32854 and Q32853).
2. Microplate reader (such as BioTek, Synergy H1) or Qubit fluorometer 2.0 (Life Technologies, Q32866).

3. High Sensitivity Large Fragment 50 Kb Analysis Kit (AATI, DNF-464).
4. AATI Fragment Analyzer (*see* **Note 1**).
5. NanoDrop Spectrophotometer (Thermo Scientific, ND-1000).

2.2 Illumina 300-bp Fragment Library

1. Covaris microTube (Covaris, 520052 or 520045) for Sonicator (Covaris, LE220).
2. Indexed TruSeq adapters adjusted to 18 μM.
3. Kapa Library Preparation Kit (Kapa Biosystems, KK8201 or KK8208).
4. AMPure XP beads (Agencourt, A63880 or A63882).
5. Buffer EB (Qiagen, 19086).
6. 75% ethanol (prepare freshly for the date of use).
7. 96-well Magnetic Particle Concentrator (Edge Biosystems, 57624).
8. High Sensitivity Kit (Agilent, 5067-4626) or High Sensitivity NGS Fragment Analysis Kit (AATI, DNF-474).
9. BioAnalyzer 2100 (Agilent) or Fragment Analyzer (AATI).

2.3 PacBio SMRTbell Library Above 10 kb

1. g-TUBE (Covaris, 520104) system with Eppendorf Mini Spin Plus (VWR, 47727-636) centrifuge or Hydro Tubes (Diagenode, E07010002) with Hydropore-long (for 10–75 kb, Diagenode, C30010018) on Megaruptor (Diagenode).
2. PacBio SMRTbell Template Prep Kit (Pacific Bioscience, 100-259-100).
3. AMPure PB Beads (Pacific Biosciences, 100-265-900).
4. 70% ethanol (prepare freshly for the date of use).
5. Eppendorf ThermoMixer® C (Thermo Fisher, 05-412-503).
6. 96-well Magnetic Particle Concentrator (Edge Biosystems, 57624).
7. High Sensitivity Large Fragment 50 Kb Analysis Kit (AATI, DNF-464).
8. AATI Fragment Analyzer (*see* **Note 1**).
9. Qubit High Sensitivity Assay Kit (Life Technologies, Q32854) on Qubit fluorometer 2.0 (Life Technologies, Q32866).
10. 0.75% Agarose, 4–20 kb High Pass, S1. 10/pkg (Sage Science, BLF7510) or 0.75% Agarose, 30–40 kb High Pass, U1.10/pkg Sage Science, BUF7510).
11. BluePippin (Sage Science, BLU0001).
12. 96-well Magnetic Particle Concentrator (Edge Biosystems, 57624).

2.4 qPCR Quantification of Illumina Libraries

1. Kapa SYBR Fast Illumina Library Quantification Kit, catalog # KK4854.
2. Roche LightCycler 480.

2.5 Illumina HiSeq 2500 Sequencing

1. Illumina HiSeq PE Cluster Kit v4, catalog # PE-401-4001 (contains a HiSeq flow cell, flow cell manifold, and reagents for clustering).
2. Illumina HiSeq SBS Kit v4 (250 cycles), catalog # FC-401-4003.

2.6 Pacific Biosciences RS II Sequencing

The RS II's automated sequencing workflow uses ready-to-load SMRT sequencing kits from PacBio that contain the necessary reagents for sequencing prepared SMRTbell template libraries on SMRT Cells.

1. PacBio RS II DNA/Polymerase Binding Kit P6, catalog # 100-372-700.
2. PacBio RS II DNA Sequencing Reagent 4.0 v2, catalog # 100-612-400.
3. SMRT Cell 8Pac (8 Cells) v3, catalog # 100-171-800.

3 Methods

3.1 Quality Assessment of Genomic DNA

DNA sample quality assessment is focused on three areas to (1) determine nucleic acid concentration, (2) detect degradation and unusual profiles in sample traces, and (3) check absorbance readings to report on sample purity. For a single tube sample assays, Qubit fluorimeter system 2.0 (Life Technologies) is used following the manufacturer's protocol with 2 μL of each sample with 2-point standard curve for broad-range (0 and 1000 ng/μL) or high sensitivity assay (0 and 100 ng/μL). For samples in a microplate, the DNA concentration is determined using the QuantiFluor™ dsDNA System (Promega) and microplate reader. First, the DNA sample is serial-diluted in 10 mM Tris–HCl 1 mM EDTA (pH 8.0) and the concentration is then determined by comparison to a 4-point standard curve. For both the Illumina and PacBio platforms the DNA purity is initially checked by spectrometer (i.e., NanoDrop) absorbance ratios. Absorbance ratios between 1.8 and 2.0 for 260 nm/280 nm, and 2.0 and 2.2 for 260 nm/230 nm are recommended, as enzyme activity required for library construction can be sensitive to sample contaminants.

As DNA degradation will impact the quality and read length of PacBio libraries we perform additional quality control steps on DNA samples destined for this platform. The Fragment Analyzer (AATI) using the High Sensitivity Large Fragment 50 Kb Analysis Kit (AATI, DNF-464) is recommended for evaluating DNA degradation prior to 10 kb and >20 kb library constructions. Due to

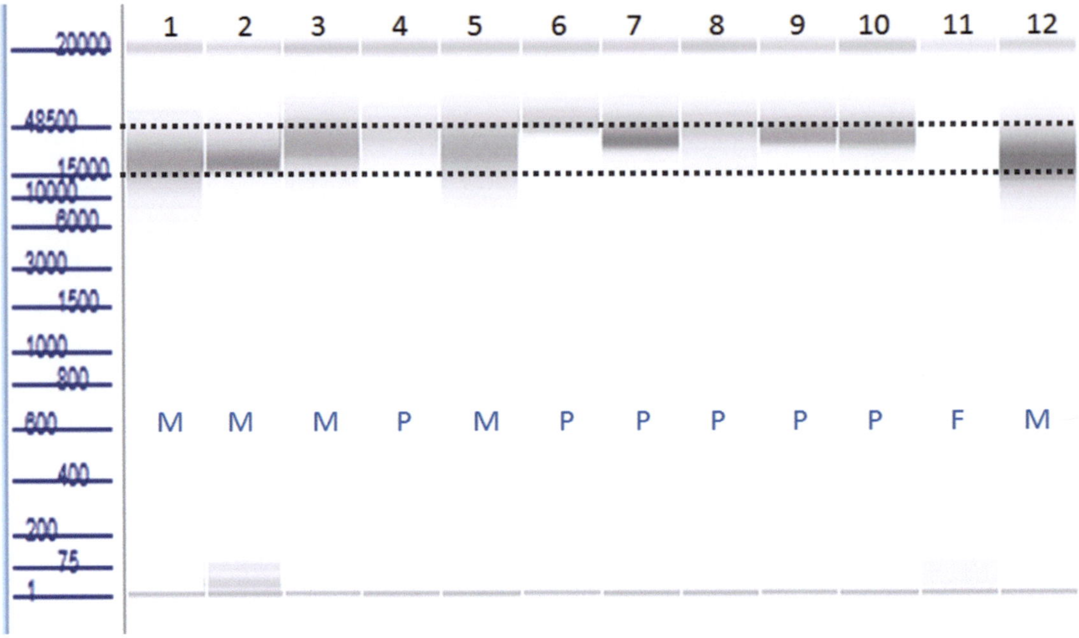

Fig. 1 An example of genomic DNA quality assessment on Fragment Analyzer with High Sensitivity Large Fragment 50 Kb Analysis Kit (AATI, DNF-464) showing *P*: pass, *M*: marginal, and *F*: failed qualities

the high-sensitivity of this platform only 1 ng is need as input. Alternatively, 0.7% agarose gel in 1×TAE buffer (40 mM Tris, 20 mM acetic acid, and 1 mM EDTA) with 50 ng of samples can be used. We recommend running the gel for 90 min at 90 V for standard 10 kb PacBio libraries. For >20 kb libraries a pulsed-field gel electrophoresis (PFGE) or Pippin Pulse system (Sage Science) is recommended. We evaluate the quality of the sample as: (1) "pass" if majority of DNA is above 48-kb ladder band, (2) "marginal" if majority of DNA is between 15 and 48-kb ladder bands, and (3) "fail" if majority of DNA is below 15-kb ladder band (Fig. 1). Any sample that stays in the well of the gel may indicate that the sample contains high molecular weight DNA, though this can potentially be due to other contaminants. The assessment of degradation of DNA is important to determine the shearing method. Although only "marginal" to "pass" quality samples are usable for 10 kb or above PacBio library construction, lower cutoff size could be used to make libraries from "fail" quality DNA without shearing. The DNA quality issue is discussed with shearing methods at the PacBio library creation.

3.2 Illumina 300-bp Fragment PCR-Free Library

1. DNA Shearing

   100 ng of genomic DNA in 50 μL volume is transferred into microTube (Covaris) and sheared by sonication using Covaris LE220 (Duty Cycle: 15%, PIP: 450, Cycles per Burst: 200, Time per run: 120 s, Temperature: 7 °C).

2. Double Size Selection

   Transfer the sheared DNA into 1.5-mL or an 8-strip tube (*see* **Note 2**), add 50 μL of 10 mM Tris–HCl, 1 mM EDTA (pH 8.0) buffer into the sample, mix well and add 60 μL of AMPure XP beads (0.6× volume of beads). Mix well and incubate at room temperature for 5 min. Place on the magnet until the supernatant becomes clear. Transfer the supernatant into a new tube. It is important to take all the supernatant but not beads (this step will remove DNA fragments above 500-bp size which are bound to the beads). Add 30 μL beads into the transferred supernatant (total 0.9× beads). Mix well and incubate at room temperature for 5 min. Place on magnet until the supernatant becomes clear. Discard the supernatant (this step will remove below 150-bp size). Keeping the tube on magnet, add 200 μL of 75% ethanol. Incubate for 30 s and discard the ethanol. Repeat this step twice. Place samples on a thermal cycler with lid open, and incubate at 37 °C until residual ethanol has evaporated (which is 2–3 min). Resuspend the beads in 53 μL of Buffer EB. Mix well and incubate at room temperature for 1 min. Place on the magnet until the supernatant becomes clear. Transfer the supernatant into a new tube.

3. Size Check

   Take 1 μL of sample to check library profile and concentration on Bioanalyzer using High Sensitivity Kit (Fig. 2) or Fragment Analyzer with NGS kit to ensure that the peak size at around 300 bp which will be the insert size of the library.

4. End repair

   To 50 μL of size-selected DNA, add 26 μL water, 9 μL of 10× End Repair Buffer, and 5 μL of End Repair Enzyme to make 90 μL reaction. Mix well and incubate at 30 °C for 30 min on a thermal cycler. After the reaction, add 126 μL of AMPure XP beads (1.4× beads) and incubate at room temperature for 5 min. Place on magnet until the supernatant becomes clear. Discard the supernatant. Keeping the tube on the magnet, add 200 μL of 75% ethanol. Incubate for 30 s and discard the ethanol. Repeat this step twice. Place samples on a thermal cycler with lid open, and incubate at 37 °C until residual ethanol has evaporated (for 2–3 min). Resuspend the beads in 17.5 μL of Buffer EB. Mix well and incubate at room temperature for 1 min. Place on the magnet until the supernatant becomes clear. Transfer the supernatant into a new tube.

5. A-Tailing and Adapter Ligation

   To 15 μL of end-repaired fragments, add 9 μL of water, 3 μL of 10× A-Tailing Buffer, and 3 μL of A-Tailing Enzyme to make 30 μL reaction. Mix well and incubate at 30 °C for 30 min, 70 °C for 5 min, then cool down to 4 °C on a thermal cycler. Proceed to adaptor ligation no more than 10 min after

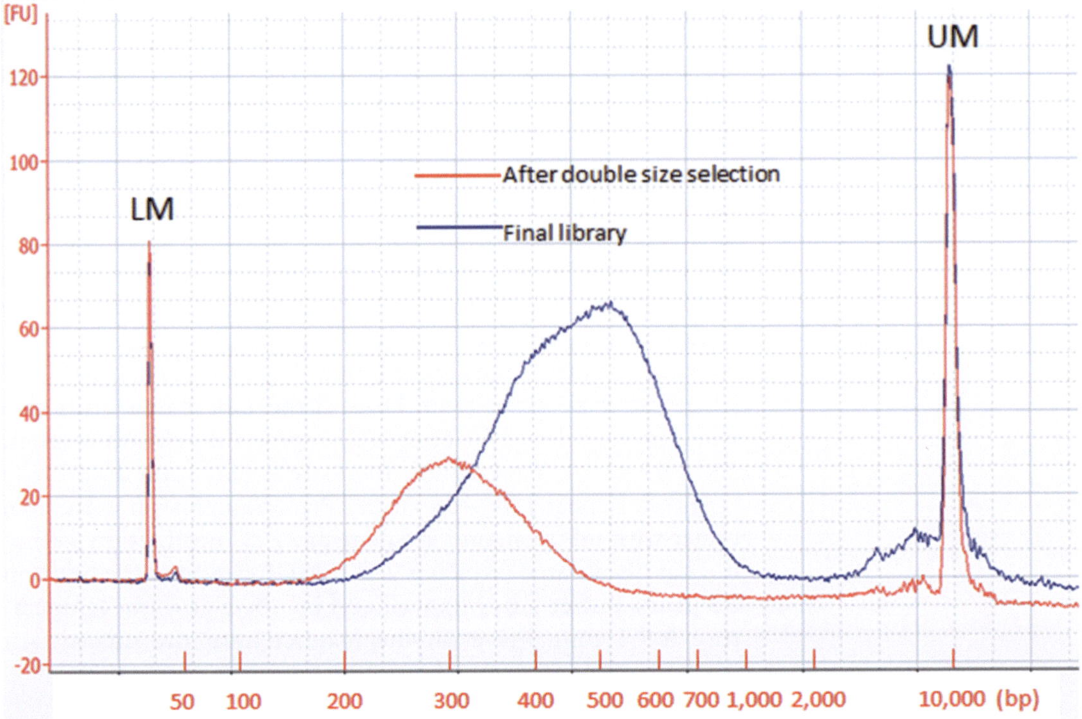

Fig. 2 An example of successful Illumina 300-bp fragment PCR-free library after double size selection (blue) and final library (red) run on Bioanalyzer with High sensitivity kit (Agilent). Note that the peak size shifted up at around 150-bp after adaptor ligation. If the size did not shift, indicating a failure of adapter ligation and the library is not successful

the sample has cooled to 4 °C. Add 1 μL of 18 μM indexed adapter into each tube and mix well. Add 9 μL of 5× Ligation Buffer and 5 μL of Ligase to make 45 μL reaction. Mix well and incubate on a thermal cycler at 20 °C for 15 min.

6. Final Cleanup of Unligated Adapters

   After the ligation is complete, add 5 μL of Buffer EB and 45 μL of the AMPure XP beads (0.9× beads). Mix well and incubate at room temperature for 5 min. Place on the magnet until the supernatant becomes clear. Discard the supernatant. Keeping the tube on the magnet, add 200 μL of 75% ethanol. Incubate for 30 s and discard the ethanol. Repeat this step twice. Place samples on a thermal cycler with lid open, and incubate at 37 °C until residual ethanol has evaporated (for 2–3 min). Resuspend the beads in 52 μL of Buffer EB. Mix well and incubate at room temperature for 1 min. Place on the magnet until the supernatant becomes clear. Transfer 50 μL of the supernatant into a new tube. Add 45 μL of AMPure XP beads (0.9× beads). Mix well and incubate at room temperature for 5 min. Place on magnet until the supernatant becomes clear. Discard the supernatant. Keeping the tube on magnet, add 200 μL of 75% ethanol. Incubate for 30 s and discard the ethanol.

Repeat this step twice. Place samples on a thermal cycler with lid open, and incubate at 37 °C until residual ethanol has evaporated (for 2–3 min). Resuspend the beads in 25 μL of Buffer EB. Mix well and incubate at room temperature for 1 min. Place on the magnet until the supernatant becomes clear. Transfer the supernatant which is the final library.

7. Quality Assessment of Libraries

   Take 1 μL of sample load on Bioanalyzer using High Sensitivity Kit (Fig. 2) or Fragment Analyzer with NGS kit to confirm that the distribution of the DNA is in the targeted size range. After ligation of adapters, the size distribution should shift up at around 150 bp from the size at **step 4**. If unligated adapters are observed at Bioanalyzer trace next to the lower marker, it is required to repeat the final AMPure XP beads cleanup one more time. Record library average size, concentration, and volume to pass the information to sequencing.

3.3 PacBio SMRTbell Library

The standard PacBio SMRTbell library construction starts with 5 μg input. A low input DNA option is also available starting with 1 μg. For <10 kb libraries, size selection can be performed using the AMPure PB beads to remove fragments less the 1.5 kb in length. However, to generate 10 or 20 kb PacBio library a more stringent size selection should be performed using electrophoresis-based approaches such as the BluePippin (Sage Science). The primary disadvantage of this more stringent size selection is yield-loss requiring higher starting input amount (typically 20 μg). For >20 kb sizes, pulsed-field gel electrophoresis such as the Pippin Pulse (Sage Science) or Fragment Analyzer (AATI) should be used. In Fig. 5 we compare the Fragment Analyzer to the Pippin Pulse. Both platforms perform similarly well but the Fragment Analyzer offers significantly shorter operation time, 2 h compared to overnight runs on the Pippin Pulse.

1. DNA Shearing

   Transfer 5 μg DNA (1 μg DNA for low input or 20 μg DNA for above 20 kb insert) with above 100 ng/μL concentration into g-TUBE (Covaris) and spin at 5500 rpm for 60 s in Eppendorf MiniSpin Pulse centrifuge (*see* **Note 3**). Confirm all liquid is passed through to the bottom chamber after spin (*see* **Note 4**). Invert g-TUBE and spin at 5500 rpm (2030×*g*) for 60 s. Confirm all sample is collected in the cap of g-TUBE. Transfer the sheared sample into an 8-strip tube.

2. Concentrate DNA and Remove Short Fragments

   Add 0.45× volume of AMPure PB magnetic beads and tap the tubes to mix. Incubate the tube in Eppendorf ThermoMixer at 500 rpm at 25 °C for 15 min. Place on the magnet until the supernatant becomes clear. Discard the supernatant. Keeping the tube on magnet, add 200 μL of 70% ethanol. Incubate for 30 s and discard the ethanol. Repeat this step twice. Air-dry the beads pellet until residual ethanol has evaporated but do not overdry. Resuspend the beads in 31 μL of PacBio elution buffer. The elution buffer volume is for one time reaction with up to 5 μg input DNA. If the input DNA is more than 5 μg, multiple the elution buffer volume and following reactions according to the number of reactions (such as for the above 20 kb insert with 20 μg input, require four reactions). Mix well and incubate at room temperature for 15 min (15 min for low input). Invert for 2–3 times during the incubation. Place on the magnet until the supernatant becomes clear. Transfer the supernatant into a new tube.

3. Confirmation of Size after Shearing

   Dilute the sheared DNA and load 1 ng on Fragment Analyzer with High Sensitivity Large Fragment 50 Kb Analysis Kit (DNF-464).

4. ExoVII Digestion

   Skip this step for 1 μg input. Add 8 μL of water, 5 μL of DNA Damage Repair Buffer, 0.5 μL of NAD+, 5 μL of ATP High, 0.5 μL of dNTP, and 1 μL of ExoVII. Tap the tube to mix and incubate at 37 °C for 15 min and 4 °C on a thermal cycler. The reaction is to remove the single-stranded DNA.

5. DNA Damage Repair

   Add 2 μL of DNA Damage Repair Mix into ExoVII treated DNA. For 1 μg input, add **step 3** components except for ExoVII and add 2 μL of DNA Damage Repair Mix instead. Tap the tube to mix and incubate at 37 °C for 60 min and 4 °C on a thermal cycler.

6. Repair Ends and Purification

   Add 2.5 μL of End Repair Mix, tap to mix and incubate at 25 °C for 5 min and 4 °C for 1 min on a thermal cycler. Add 24 μL of AMPure PB beads (0.45× beads) and tap to mix. Follow AMPure PB beads purification in **step 2** and elute the DNA in 31 μL of PacBio elution buffer for above 10 kb or 25 μL for above 20 kb target sizes.

7. Blunt Adapter Ligation

   Add 1 μL of Blunt Adapter (20 μM) for above 10 kb insert. If the targeting size is above 20 kb to be size-selected later, use 10 μL of Blunt Adapter per reaction. Tap to mix. Add 4 μL of Template Prep Buffer, 2 μL of ATP Lo, 1 μL of Ligase to top off to 40 μL by water. Tap to mix and incubate at 25 °C for overnight (or 20 min for 1 μg input). Incubate at 65 °C for 10 min to inactivate the ligase followed by 4 °C.

8. Exonuclease to Remove Unligated Adapters and Fragments Lacking Adapters

   Add ExoIII 1 μL and ExoVII 1 μL. Tap to mix and incubate at 37 °C for 1 h and 4 °C. Follow AMPure PB beads purification in **step 2** and elute the DNA in 31 μL of PacBio elution buffer.

9. Size Selection

   This step is to target larger than 20 kb insert. Skip this step to target 10 kb and 1 μg input. Quantify the sample using Qubit DNA High Sensitivity kit and confirm the size by Fragment Analyzer with 1 ng of DNA. According to the results, decide the cutoff size. Each lane of BluePippin cassette is up to 5 μg of DNA in 30 μL. If more than 5 μg of DNA is in the sample, top off the sample to the volume required for the numbers of lanes. Follow the manufacturer's manual to select the cutoff size (Table 1).

   Collect the size-selected library from elution wells after run in one tube. Purify with 1× AMPure PB beads, elute in 30 μL of elution buffer. Confirm the size selection of Fragment Analyzer by loading 1 ng of library (Fig. 3). Treat elution by additional Damage Repair as in **step 5**.

10. Final Purification

    Use AMPure PB beads cleanup as in **step 2** with two 0.45× beads and one 0.4× beads to remove un ligated adapters for 10 kb, two 0.45× beads purification steps for 1 μg input, and one 1× beads purification for size-selected libraries. After the final purification, elute DNA in 15 μL and transfer to a new tube.

11. Quality Assessment of Library

    Take 1 μL of the library and dilute three times in water. Use 1 μL of diluted library to run Fragment Analyzer and another 1 μL to run Qubit High sensitivity kit DNA to qualify and quantify the library (Fig. 4). Typical yield of library is between 10 and 30% of input DNA for non-size-selected libraries and less than 10% for BluePippin size-selected libraries.

3.4 Illumina Sequencing Using HiSeq 2500

The HiSeq 2500 has two different run modes: High Output and Rapid Run, this method description will focus on the High Output run mode utilizing HiSeq v4 reagents.

1. Quantification and multiplexing of Illumina libraries

   The first step of sequencing workflow is to quantify Illumina sequencing library using qPCR and multiplex with other quantified libraries for sequencing. Quantitative real-time polymerase chain reaction (qPCR) is a useful molecular biology technique to quantitatively measure the molecules of a targeted DNA sequence. qPCR is utilized to determine the number of adapter ligated molecules in an Illumina sequencing

Table 1
Recommended BluePippin program for the chosen cutoff size

Cutoff size	Marker	Programs
15 or 20 kb	S1	0.75% DF Marker S1 high-pass 15–20 kb
30 kb	U1	0.75% DF Marker U1 high-pass 30–40 kb

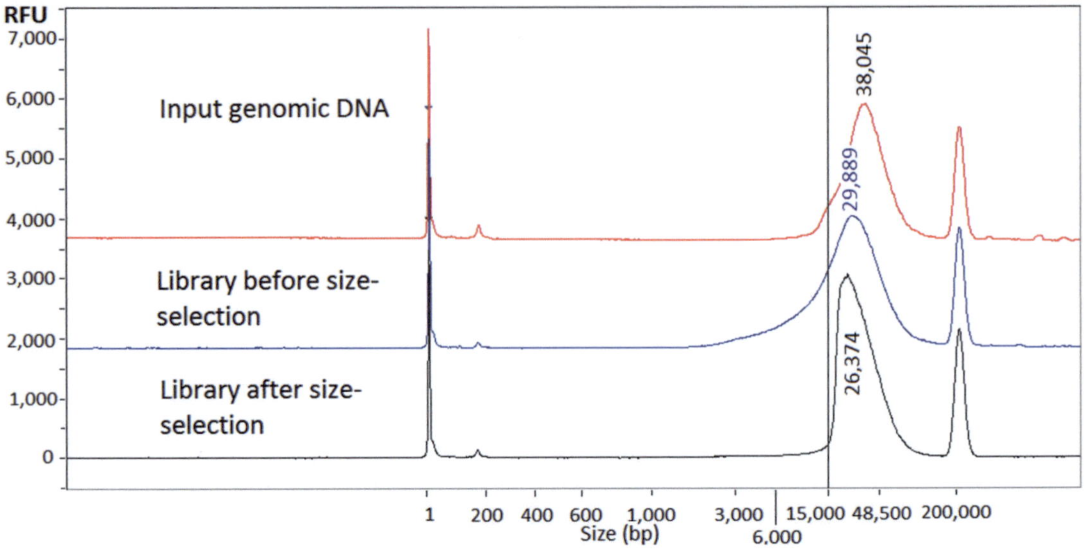

Fig. 3 Successful above 20-kb size-selected PacBio library with 15-kb cutoff by BluePippin run on Fragment Analyzer with High Sensitivity Large Fragment 50 Kb Analysis Kit. Input genomic DNA (top), library before size selection (middle) and library after size selection (bottom) profiles are indicated to show changes in the DNA smear patterns

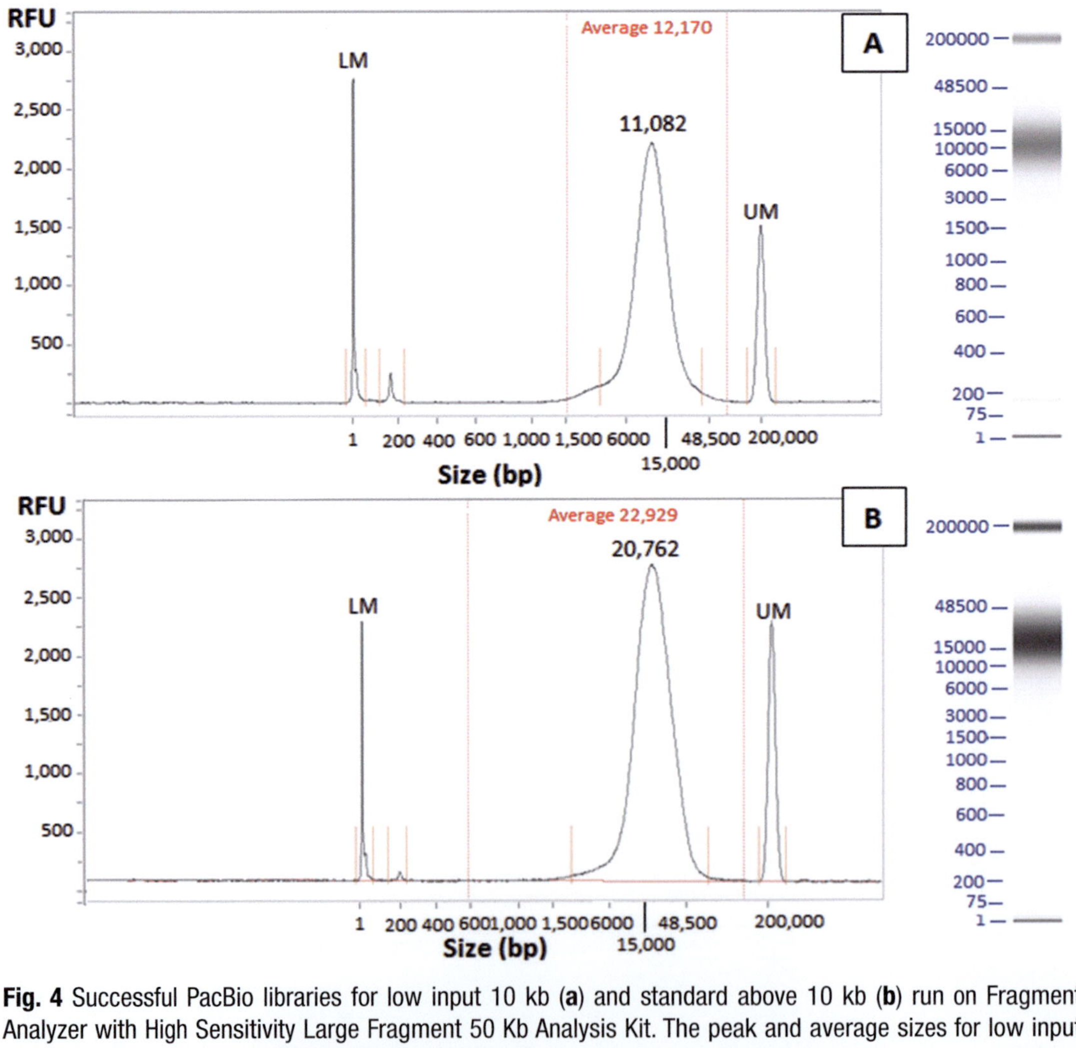

Fig. 4 Successful PacBio libraries for low input 10 kb (**a**) and standard above 10 kb (**b**) run on Fragment Analyzer with High Sensitivity Large Fragment 50 Kb Analysis Kit. The peak and average sizes for low input (1 μg) are generally shorter than standard input (5 μg)

library, which is then used to accurately multiplex libraries into equimolar pools to ensure equal representation during sequencing and to optimally load the libraries onto the Illumina flow cells for clustering [6].

After Illumina library creation the number of template molecules is measured with qPCR using the Kapa SYBR Fast Illumina Library Quantification Kit (KK4854, Kapa Biosystems) optimized for the Roche LightCycler 480 and following the Kapa technical guide [7]. After determining the library template concentration by qPCR, libraries may be pooled together in equimolar ratios. The number of libraries to be pooled together is determined by the amount of sequence data expected to be generated by the Illumina sequencer divided by the amount of sequence data that is needed per library. Illumina system specifications for expected sequencer data output: HiSeq 2500 is shown at Illumina website [8].

2. Clustering Libraries on an Illumina Flow Cell

   The first step of sequencing workflow is to cluster the multiplexed sequencing libraries on an Illumina flow cell using the cBot instrument. Illumina cluster generation is the process by which library templates are hybridized to the surface of the sequencing flow cell and amplified to form clonal clusters of the templates for sequencing. Each resulting clonal cluster on the flow cell will generate a sequencing read. The clustering process for HiSeq 2500 High Output flow cells is performed on the Illumina cBot instrument.

   The Illumina cBot instrument automates the process of clustering the Illumina library templates on the surfaces of the flow cell. The cBot instrument dispenses reagents and performs the amplification reaction. The reagents are provided in the Illumina kit in a ready to use plate that is loaded on the cBot after thawing. Follow the latest cBot System User Guide from Illumina [9] to (a) Prepare the cBot reagent plate, (b) Prepare libraries for clustering, and (c) Prepare cBot and start clustering run.

3. Starting HiSeq 2500 Sequencing Run

   Loading the clustered flow cell on the HiSeq 2500 and starting the sequencing run Illumina's next-generation sequencing (NGS) platforms utilize their sequencing by synthesis (SBS) technology [10] to sequence in parallel the clonal clusters of templates that were formed on the inner-surfaces of the flow cell. Illumina's SBS chemistry uses fluorescently labeled reversible terminator dNTPs [11] that are detected as they are incorporated into growing complementary sequence strand of the cluster templates.

   Illumina's HiSeq 2500 sequencer platform is a high-throughput sequencing system that is capable of generating up to 1 Terabase (Tb) and eight billion paired-end reads of sequence data per 6-day run [8]. The HiSeq 2500 has fully integrated workflows to support the automated SBS sequencing and data analysis of clustered flow cells.

The HiSeq 2500's automated sequencing workflow uses ready to load SBS reagent kits from Illumina. The SBS kits contain the necessary reagents for sequencing a clustered flow cell.

Following the latest HiSeq 2500 System User Guide from Illumina [12], prepare HiSeq SBS reagents by setting up sequencing run following the HiSeq 2500's integrated software prompts: enter run parameters, load and prime reagents, and load the clustered flow cell.

3.5 Pacific Biosciences Sequencing Using RS II

The Pacific Biosciences (PacBio) RS II platform is an NGS system capable of generating very long sequencing reads using PacBio's single molecule, real-time (SMRT) technology [13]. The PacBio RS II features automated liquid handling robotics and high-performance optics for the sequencing of prepared SMRTbell template libraries. The RS II has a run time flexibility from 30 min to 6 h per SMRT Cell, and a single SMRT Cell sequenced with the their P6 enzyme and C4 is capable of generating 1 Gigabase (Gb) and 55,000 reads of sequence data that have average read lengths in excess of 12 kb and some reads in excess of 60 kb [14].

Following the latest RS II Template Prep and Software User Guides from PacBio [15] carry out the steps below:

1. Anneal SMRTbell templates with v2 sequencing primer.
2. Bind P6 sequencing polymerase to SMRTbell templates.
3. Set up sequencing run following the RS II's integrated software prompts: enter run parameters, load SMRT sequencing reagents and consumables, load annealed and bound SMRTbell libraries, and load SMRT Cells.
4. Start the sequencing run.

4 Notes

1. Fragment Analyzer (Advanced Analytical Technologies, Inc.) with High Sensitivity Large Fragment 50 Kb Analysis Kit with running time for about one and a half hours provides an equivalent resolution of fragment sizes as Pippin Pulse System (Sage Science, PP10200) with programmable pulsed-field power for overnight run. Fragment Analyzer's smear analysis is recommended for assessment of sizes in PacBio libraries.
2. Illumina 300-bp fragment libraries can be made in parallel for 8–16 libraries comfortably by using 8-strip tubes and 8-channel pipette. It is recommended to make a master mix of each reaction with buffers and enzymes before adding to DNA to be consistent between samples.

3. According to the DNA quality and concentration at Subheading 3.1, the DNA shearing step should be thoroughly considered before starting. If the DNA quality falls between marginal to pass with above 100 ng/μL concentration, g-TUBE shearing is suitable. If the target size is above 20 kb, marginal quality sample may not be required to be sheared, or use Megaruptor (Diagenode) to shear with 30-kb setting. g-TUBE shearing size is concentration dependent (Fig. 5), while Megaruptor is not which is a better option for samples with lower than

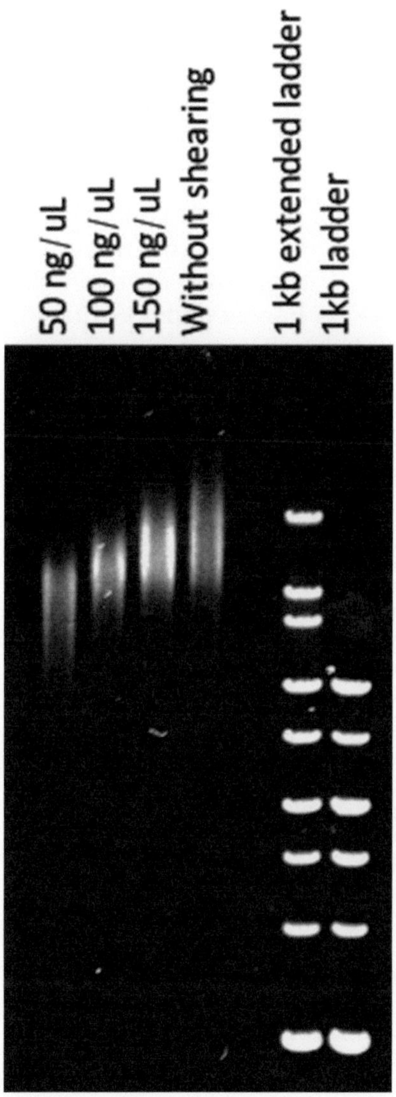

Fig. 5 Different concentrations of DNA were sheared by g-TUBE. Lower concentration than 150 ng/μL sheared too short to target 20-kb insert size. Load ladders (Quick-Load 1 kb DNA Ladder: NEB, N0468S and Quick-Load 1 kb Extend DNA Ladder: NEB, N3239S) and 50 ng of samples on 1% SeaKem GOLD agarose (Lonza, 50150) in 0.5× KBB running buffer (Fisher, 12-100-577) on Pippin Pulse system (Sage Science, PP10200). Select 5–80 kb program and run for 16 h

100 ng/μL. Generally, failed quality samples are not recommended to use for PacBio library constructions. However, if the degraded DNA retains the sizes above 6 kb in majority, nonsheared 6-kb size-selected library could be an option.

4. g-TUBE may have a clogging issue depending on the samples. If not all of the liquid has passed through to the bottom chamber, repeat spin for an additional 60 s at 5500 rpm and visually check. Repeat spin until the entire sample has passed through. If the orifice becomes clogged, reverse the g-TUBE and spin at 5500 rpm for 30 s (or until all passes) to collect samples in the cap. Transfer the sample into new g-TUBE and repeat shearing process.

Acknowledgments

The work conducted by the US Department of Energy Joint Genome Institute, a DOE Office of Science User Facility, was supported by the Office of Science of the US Department of Energy under Contract No. DE-AC02-05CH11231.

References

1. Bentley DR, Balasubramanian S, Swerdlow HP, Smith GP, Milton J, Brown CG, Hall KP, Evers DJ, Barnes CL, Bignell HR et al (2008) Accurate whole human genome sequencing using reversible terminator chemistry. Nature 456(7218):53–59. https://doi.org/10.1038/nature07517
2. Eid J, Fehr A, Gray J, Luong K et al (2009) Real-time DNA sequencing from single polymerase molecules. Science 323(5910):133–138. https://doi.org/10.1126/science.1162986
3. Borgström E, Lundin S, Lundeberg J (2011) Large scale library generation for high throughput sequencing. PLoS One 6(4):e19119. https://doi.org/10.1371/journal.pone.0019119
4. Peng Z, Zhao Z, Nath N et al (2012) Generation of long insert pairs using a Cre-LoxP Inverse PCR approach. PLoS One 7(1):e29437. https://doi.org/10.1371/journal.pone.0029437
5. Hoess RH, Abremski K (1984) Interaction of the bacteriophage P1 recombinase Cre with the recombining site loxP. Proc Natl Acad Sci U S A 81:1026–1029
6. Meyer M, Briggs AW, Maricic T (2008) From micrograms to picograms: quantitative PCR reduces the material demands of high-throughput sequencing. Nucleic Acids Res 36:e5. https://doi.org/10.1093/nar/gkm1095
7. Kapa Biosystems (2016) https://www.kapabiosystems.com/product-applications/products/next-generation-sequencing-2/library-quantification/. Accessed 3 May 2017
8. Illumina, Inc. (2016) http://www.illumina.com/systems/hiseq_2500_1500/performance_specifications.html. Accessed 3 May 2017
9. Illumina, Inc. (2016) http://support.illumina.com/sequencing/sequencing_instruments/cbot.html Accessed 3 May 2017
10. Illumina, Inc. (2016) http://www.illumina.com/technology/next-generation-sequencing/sequencing-technology.html. Accessed 3 May 2017
11. Canard B, Sarfati RS (1994) DNA polymerase fluorescent substrates with reversible 3′-tags. Gene 148(1):1–6. https://doi.org/10.1016/0378-1119(94)90226-7
12. Illumina, Inc. (2016) http://support.illumina.com/sequencing/sequencing_instruments/hiseq_2500.html. Accessed 3 May 2017
13. Pacific Biosciences of California, Inc. (2015–2016) http://www.pacb.com/smrt-science/smrt-sequencing/. Accessed 3 May 2017
14. Pacific Biosciences of California, Inc. (2015–2016) http://www.pacb.com/smrt-science/smrt-sequencing/read-lengths/. Accessed 3 May 2017
15. Pacific Biosciences of California, Inc. (2015–2016) http://www.pacb.com/support/documentation/. Accessed 3 May 2017

Chapter 5

Methods for Genomic Characterization and Maintenance of Anaerobic Fungi

Xuefeng Peng, Candice L. Swift, Michael K. Theodorou, and Michelle A. O'Malley

Abstract

The rapid development of molecular biology and bioinformatics has fueled renewed interests in anaerobic fungi from the phylum *Neocallimastigomycota*. This chapter presents well-established methods for isolation, routine cultivation, and cryopreservation of anaerobic fungi. Moreover, detailed nucleic acid extraction protocols are provided, which should enable readers to isolate high-quality DNA and RNA from a variety of anaerobic fungal culture media for downstream applications such as next-generation sequencing.

Key words Anaerobic fungi, Genomes, Transcriptomes, DNA extraction, RNA extraction, Cultivation, Isolation, Neocallimastigomycota, Next-generation sequencing, Consecutive batch culture

1 Introduction

Since the discovery of chitin in the cell wall of Neocallimastigomycota [1], which led to their reclassification from protozoa to fungi, there has been a large body of published literature that describes their morphology, physiology, ecology, and biochemistry [2]. The rapid development of molecular biology and bioinformatics in the past decade has provided powerful tools for scientists and engineers to gain deeper insight into the functional role of these unusual anaerobic fungi.

In particular, next-generation sequencing approaches have opened the way for comparative genomics, transcriptomics, and metagenomic modeling in these systems for the first time [3]. For example, a recent study identified novel enzyme candidates in anaerobic fungi for lignocellulose breakdown by combining transcriptomics, proteomics, and biochemical characterization [4], which are a powerful combination of tools that also promises to unravel the native syntrophy of anaerobic fungi and methanogenic archaea [5]. All of these methods depend upon successful

Ronald P. de Vries et al. (eds.), *Fungal Genomics: Methods and Protocols*, Methods in Molecular Biology, vol. 1775, https://doi.org/10.1007/978-1-4939-7804-5_5, © Springer Science+Business Media, LLC, part of Springer Nature 2018

cultivation of the anaerobic fungi and effective extraction of high-quality nucleic acid samples. This chapter builds upon established methods for basic maintenance of anaerobic fungi, and highlights new protocols we have developed in the laboratory to extract high molecular weight genomic DNA, as well as intact RNA for next-generation sequencing applications. We also include helpful tips for troubleshooting these experiments that are not easily found in the literature.

The first part of the methods discussed pertains to routine isolation and maintenance of anaerobic fungi based on the consecutive batch culture technique [6]. We then provide detailed protocols for extraction of RNA from batch cultures of anaerobic fungi or mixed microbial consortia that also include prokaryotic organisms (bacteria and archaea). Finally, we focus on two effective cell disruption methods to rapidly isolate high-quality genomic DNA from anaerobic fungi, which relies on simple-to-implement commercial kits.

2 Materials

2.1 For Routine Cultivation

1. Clarified rumen fluid: fresh rumen fluid centrifuged at 3220 × *g* (or sufficient speed to sediment all particles, leaving a tan or greenish colored clarified liquid) for 1 h at 4 °C, separated from the resulting cell pellet. Carefully transfer 75 mL of the supernatant into 120-mL serum bottles and store at −20 °C. Volumes of 1–3 L are routinely clarified at one time.
2. Double clarified rumen fluid: autoclave the clarified rumen fluid at 121 °C for 40 min and store at 4 °C. Before use, the autoclaved rumen fluid is centrifuged at 3220 × *g* (or sufficient speed to sediment all particles, leaving a tan or greenish colored clarified liquid) for 30 min at 4 °C. Only use the resulting supernatant for subsequent media making.
3. Mineral Solution I: Dissolve 3.0 g of dibasic potassium phosphate (K_2HPO_4) in 1 L of water. Filter-sterilize and store at 4 °C.
4. Mineral Solution II: Dissolve 3.0 g of monobasic potassium phosphate (KH_2PO_4), 6.0 g of ammonium sulfate ($(NH_4)_2SO_4$), 6.0 g of sodium chloride (NaCl), and 0.6 g of magnesium sulfate heptahydrate ($MgSO_4{\cdot}7H_2O$) in 800 mL of water. Dissolve 0.6 g of calcium chloride dihydrate ($CaCl_2{\cdot}2H_2O$) in 100 mL of water separately. These two solutions are combined, filter-sterilized, and made up to a final volume of 1 L with water. Store at 4 °C.
5. Resazurin stock solution (1 mg/mL): Dissolve 0.100 g of resazurin sodium salt (redox indicator) in 100 mL of water and filter-sterilize. Store at 4 °C.

6. Chloramphenicol stock solution (10 mg/mL): First completely dissolve 0.50 g of chloramphenicol (C1863 Sigma) in minimal amount of molecular biology grade ethanol (20 mL or less). Add water to a final volume of 50 mL. Filter-sterilize the solution and store at 4 °C.
7. Sodium bicarbonate, yeast extract, and Bacto™ Casitone (or tripticase peptone).
8. Plant material should be air-dried and milled to provide millimeter-sized pieces (generally 2–4 mm). Some examples include reed canary grass, switchgrass, alfalfa stems, and corn stover.
9. A static incubator set at 39 °C, weighing boats, spatulas, beakers, graduated cylinders, 2-L microwave flask (must fit in a microwave), carbon dioxide (CO_2), gas manifolds to distribute CO_2, pipets, 5-mL syringes connected to blunt-end needles, and autoclave are also required. Hungate tubes with butyl rubber stoppers and/or serum bottles outfitted with butyl rubber stoppers and crimp seals are adequate vessels for culturing the anaerobic fungi. It is also advantageous to have access to a pressure transducer manifold for quantifying fungal growth [7].

2.2 For Nucleic Acid (DNA/RNA) Extraction

1. Biospec Mini-Beadbeater-16.
2. Gel-loading pipet tips.
3. 2-mL screw-cap tubes and caps with O-ring.
4. 0.5 mm zirconia/silica beads (Biospec).
5. Microcentrifuge.
6. Vortexer.
7. RNAlater®.
8. Centrifuge.

3 Methods

3.1 Media Preparation

Anaerobic media are required for isolation and maintenance of anaerobic fungi as detailed in this section. Many of the recipes and culture techniques used in rumen microbiology were first described by Hungate [8], followed by a number of modifications [9–11]. Liquid media are typical for routine maintenance; media supplemented with agar (1% w/v) for solidification in roll tubes are often used for fungal isolation procedures.

The following is a recipe for preparing 1 L of Medium C (use amounts shown in Table 1) or "Medium C Minus" ("MC−", use amounts shown in Table 2; *see* **Note 1**) dispensed in 9-mL aliquots. Alternatively, media can also be dispensed in larger volumes

Table 1
Ingredients in Medium C

Ingredients	Per 1000 mL final volume
Yeast extract	2.5 g
Bacto™ Casitone	10.0 g
Sodium bicarbonate	6.0 g
Mineral Solution I	150 mL
Mineral Solution II	150 mL
(Double[a])-Clarified Rumen Fluid	150 mL
L-Cysteine hydrochloride	1.0 g
Resazurin Stock Solution (1 g/L)	1.0 mL
Carbon substrate (e.g., plant, cellubiose)	1% w/v
Optional: Agar (for roll tube preparation)	*10.0 g*
Optional: Glycerol (for cryopreservation)	*150 mL*

[a]Clarified rumen fluid is sufficient for routine maintenance of fungal cultures, whereas double-clarified rumen fluid is recommended for cultures from which DNA will be extracted

Table 2
Ingredients in "Medium C Minus" ("MC−", *see* Note 1)

Ingredients	Per 1000 mL final volume
Yeast extract	0.25 g
Bacto™ Casitone	0.5 g
Sodium bicarbonate	6.0 g
Mineral Solution I	150 mL
Mineral Solution II	150 mL
(Double[a])-Clarified Rumen Fluid	75 mL
L-Cysteine hydrochloride	1.0 g
Resazurin Stock Solution (1 g/L)	1.0 mL
Carbon substrate (e.g., plant, cellubiose)	1% w/v
Optional: Agar (for roll tube preparation)	*10.0 g*
Optional: Glycerol (for cryopreservation)	*150 mL*

[a]Clarified rumen fluid is sufficient for routine maintenance of fungal cultures, whereas double-clarified rumen fluid is recommended for cultures from which DNA will be extracted

into serum bottles with crimp seals. All media should be prepared and aliquoted under a stream of CO_2 to minimize the introduction of oxygen.

1. Weigh out yeast extract, Bacto™ Casitone, and sodium bicarbonate into a 2-L flask.
2. Add 150 mL of Mineral Solutions I and II, and clarified rumen fluid.
3. Microwave for 20 min (*see* **Note 2**).
4. Purge with CO_2 for 10 min (*see* **Notes 3** and **4**).
5. Transfer into a 1-L bottle with 1 g of cysteine in it.
6. Close the lid and let it cool to below 39 °C. (Optional: To speed up the cooling, place the media bottle into an ice bath.)
7. Dispense 9 mL of media into 16-mL Hungate tubes with substrates preweighed and aliquoted in them. For roll tubes, dispense 5 mL of media into 20-mL Balch tubes with 1% (w/v) agar preweighed in them. (a) Use a three-way gas manifold for CO_2 supply during media dispensing (Fig. 1). One of them is placed in the media bottle, and the other two are placed in Hungate tubes. Use blunt-end needles (14 gauge, 6 in. long, Cadence Inc.) at the end of the manifold. For the two needles purging Hungate tubes, bend the ends at about 1 in. length, so that they are not directly blowing at the carbon substrate at the bottom of the Hungate tubes. (a) Use a 10-mL serological

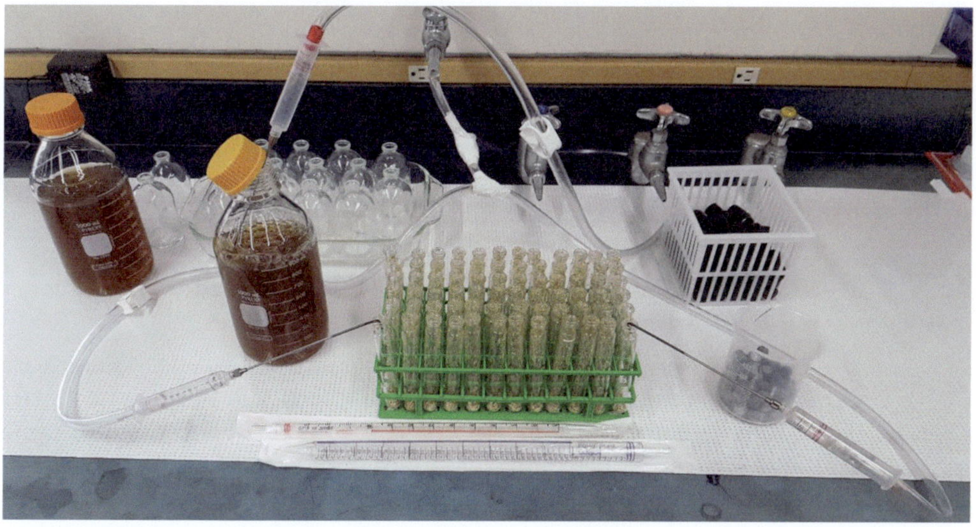

Fig. 1 Media dispensing setup with a three-way gas manifold (clear-colored) to supply carbon dioxide simultaneously to the media bottle and two Hungate tubes to be filled. Also shown: a rack of Hungate tubes with prealiquoted plant material (center); two 1-L bottles of Medium C (left); serological pipets used for dispensing medium (center); black caps and grey butyl rubber septa for sealing Hungate tubes after medium is dispensed (right)

pipet to dispense 9 mL of media into a Hungate while it is purged with CO_2. (b) Place a septum to cover the top of the Hungate tube, and cover it completely as the blunt-end needle is pulled out of this Hungate tube and placed into another one. Perform this step carefully to minimize introducing any air into the headspace of the Hungate tube. (c) Seal the Hungate tube with a plastic screw cap (*see* **Note 5**).

8. Sterilize by autoclaving.
9. Liquid media are ready for use after they are prewarmed to 39 °C. Roll tubes are prepared by melting the solid media with agar (1% w/v) in boiling water and allowing tubes to cool in a water bath at 55 °C. Roll the tubes under a cold water stream to evenly distribute a thin layer of solid media on the inner wall of Balch tubes (*see* **Note 6**).

3.2 Isolation of Anaerobic Fungi

Multiple methods have been used to isolate anaerobic gut fungi from rumen digesta and fecal materials, such as those published by Orpin [12], Bauchop and Mountfort [13], Lowe et al. [14], and Joblin [15]. Here, we describe a straightforward method for isolating anaerobic fungi starting from the fresh fecal materials of large mammalian herbivores, but these methods can be readily adapted to isolate fungi from other sources.

1. Prepare Medium C without plant substrates and Medium C with reed canary grass (or another lignocellulosic substrate). Include chloramphenicol in these media with a final concentration of 0.1 mg/mL in the media.
2. Collect fresh fecal material and transport to laboratory facilities (keep as anaerobic as possible).
3. Prepare the initial inoculum by physically breaking down fecal material and transferring them into Medium C without plant substrates under a stream of CO_2. The final concentration of fecal material in Medium C should be approximately 10% w/v.
4. Using a wide-bore needle (0.2 mm or larger) and syringe, prepare 1:10, 1:100, and 1:1000 dilution of the initial inoculum with Medium C without plant substrates.
5. Inoculate Medium C containing reed canary grass with the initial inoculum and the three serial dilutions, with a final inoculum concentration of 10% v/v. Inoculate three to five replicate tubes for each dilution.
6. Examine growth daily and select enrichment cultures from tubes which show fungal growth. Fungal growth is easily observed by "bubbling" of the grass substrate and/or floating of the grass substrate within the culture tube.
7. Inoculate and evenly distribute 0.1 mL of selected liquid cultures into each roll tube.

8. Examine the growth of fungal colonies over time and select at least three colonies.
9. Under a CO_2 stream, pick the selected fungal colonies and inoculate them into liquid media supplemented with chloramphenicol.
10. To ensure axenic cultures are obtained, repeat **steps 6–9** at least twice more.
11. Putative axenic cultures should be examined using microscopy, and their phylogeny can be determined by sequencing their nuclear ribosomal internal transcribed spacer (ITS) region [16].

3.3 Maintenance of Anaerobic Fungi

Once axenic cultures of anaerobic fungi have been established, they are easily maintained in small batch cultures prepared in gas-tight glass vessels (as detailed in Subheading 3.1). These fungal cultures typically reach mid-exponential phase of growth 3–4 days after inoculation, and are ready to be transferred into fresh medium. The following procedure adapted from Theodorou and colleagues [17] is used for transferring growing fungal cultures into fresh medium.

1. Flame the rubber stoppers of both the inoculum culture and the fresh tube to be inoculated with 100% ethanol.
2. Shake the inoculum culture vigorously to disperse the fungal material.
3. Invert the tube and insert a needle with syringe into the tube and withdraw 1 mL. If the needle is clogged by particles, try clearing the clog by pushing the plunger gently up and down.
4. Inject the 1 mL inoculum into the recipient tube with 9 mL of fresh medium. Invert the recipient tube several times.
5. Incubate cultures at 39 °C.

3.4 Cryopreservation of Anaerobic Fungi

For long-term maintenance of anaerobic fungi, cultures are flash-frozen in liquid nitrogen and stored at −80 °C using glycerol as a cryoprotectant. The following procedure is based on the method published by Solomon et al. [18].

1. Prepare Medium C containing 15% glycerol (Table 1).
2. Grow fungal culture in Medium C for 3–4 days with excess plant substrate (3% w/v) in Hungate tubes.
3. Using a syringe and needle, remove all liquid (~10 mL) in the culture.
4. Inject 10 mL of Medium C containing 15% glycerol into the Hungate tube with the residual plant substrate. Shake gently to mix well.

5. Open the Hungate tube under a stream of CO_2. Transfer 1.8 mL of culture resuspended in 15% glycerol medium into 2 mL screw-top cryovials using pipet tips with tips cut off.
6. Flash freeze in liquid nitrogen and store at −80 °C.

3.5 Reviving Cryopreserved Fungal Stocks

1. Thaw cryopreserved fungal stocks at 39 °C.
2. Under CO_2 streams, or in an anaerobic chamber, remove the liquid media containing 15% glycerol. Leave the plant material behind.
3. Transfer 1 mL of fresh medium into the cryovial.
4. Use cutoff pipet tips to transfer the resuspended culture into a prewarmed culture tube.
5. Add chloramphenicol to prevent bacterial contamination (final concentration 0.1 mg/mL).
6. Incubate at 39 °C and check the growth of the culture daily.

3.6 Determining Growth Curves Using a Pressure Transducer

It is often necessary to determine the relative stage of growth of anaerobic fungal cultures to assist in experimental design and analysis. However, due to the heterogeneity, filamentous nature, and intimate association with plant biomass particles, it is not possible to determine their growth by monitoring the optical density of fungal cultures. Alternatively, measuring the pressure in the headspace of the culture tubes/serum bottles provides a straightforward approximation of the growth of anaerobic fungi, because fermentation gases (predominantly CO_2 and H_2) accumulate as a consequence of growth. This inexpensive and nondestructive method [19] requires a simple pressure transducer (Fig. 2). All pressure measurements should be performed at 39 °C due to pressure sensitivity to temperature fluctuations. Before introducing a needle into a sample tube/bottle, the rubber stopper is sterilized by flaming.

1. Immediately after inoculating a fresh medium tube, release excess pressure in the headspace so that the headspace pressure equals atmospheric pressure on the pressure gauge.
2. Every 6–8 h, measure, record, and release the headspace pressure.
3. Plot accumulated pressure against time. It generally takes at least 6 days to reach stationary phase (Fig. 3).

3.7 Preparation of RNA

Perform protocol using standard best practices for an RNase-free environment (*see* **Note 7**).

1. Invert the culturing vessel (Hungate tubes or serum bottles) several times to break apart the plant substrate with fungal mat.

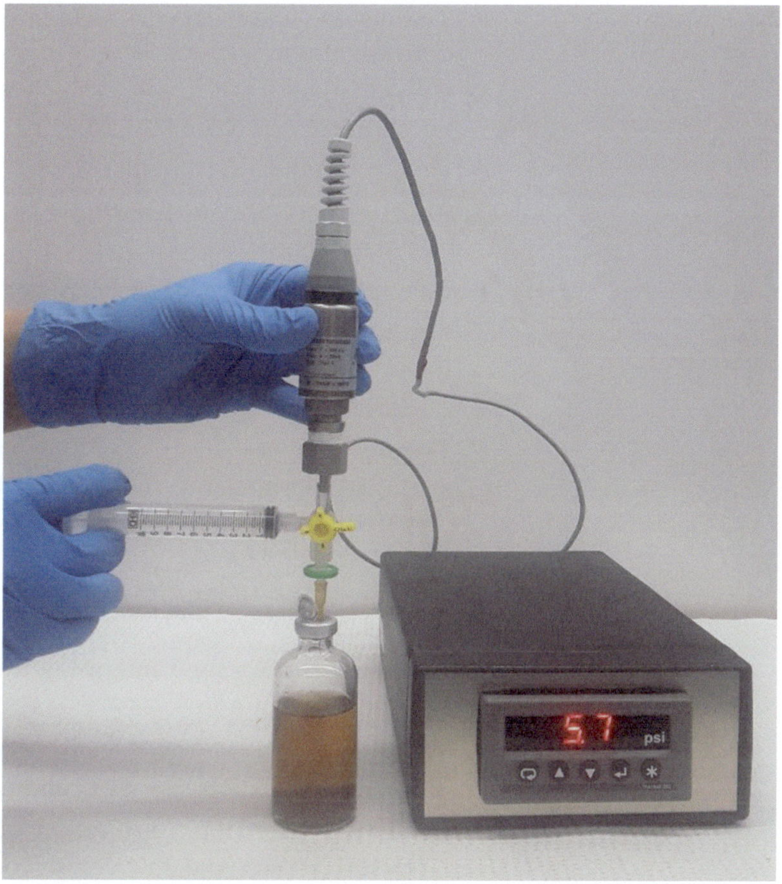

Fig. 2 A pressure transducer assembly with digital display is used to measure the accumulation of gas pressure in the headspace of a fungal culture in a 60-mL serum bottle

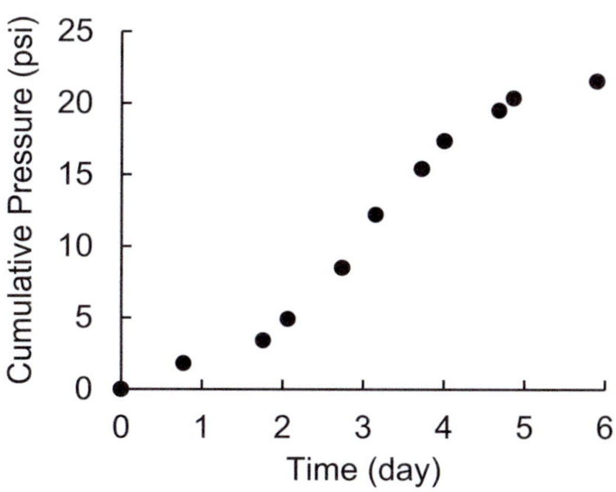

Fig. 3 A typical growth curve of anaerobic fungal culture ***Piromyces*** sp. (maintained on Medium C supplemented with reed canary grass) determined by monitoring the headspace pressure using a pressure transducer

2. Transfer the culture media including the plant substrate into centrifuge tubes (15- or 50-mL, *see* **Note 8**).
3. Centrifuge at 3220 × *g* for 7 min at 4 °C with a swinging bucket rotor (*see* **Note 9**).
4. Decant and discard the supernatant.
5. Add 1 mL of RNAlater® and store at −80 °C if not proceeding to extraction immediately.
6. Thaw samples preserved in RNAlater®, or use fresh samples.
7. Centrifuge at 3220 × *g* for 7 min at 4 °C with a swinging bucket rotor.
8. Decant and discard the supernatant.
9. Transfer all of the substrate and fungal mat into an autoclaved 2-mL screw-cap tube filled with 450 μL of buffer RLT (QIAGEN) and 1.0 mL of 0.5 mm zirconia/silica beads (*see* **Notes 10** and **11**).
10. Briefly vortex to mix the beads, buffer, and sample (*see* **Note 12**).
11. Bead beat samples for 1 min using Mini-Beadbeater-16.
12. Centrifuge at 13,000 × *g* for 3 min.
13. Transfer up to 650 μL of lysate using gel loading pipet tips onto a QIAGEN RNeasy spin column.
14. Follow the protocol "Purification of Total RNA from Plant Cells and Tissues and Filamentous Fungi" from the RNeasy Mini Handbook (QIAGEN, *see* **Note 13**).
15. RNA yields can be measured using Qubit fluorometric quantitation, and RNA quality can be assessed using a TapeStation or Bioanalyzer (Agilent). For next-generation sequencing, we recommend using RNA with a RNA Integrity Number (RIN) > 9.0 (Fig. 4).

3.8 Genomic DNA Extraction from Fungal Cultures Grown on Soluble Substrates

1. The preparation of genomic DNA from fungal cultures depends on the main carbon substrate used in the culture media. If media contain soluble substrates (e.g., cellobiose, glucose), then a gentle bead beating step is used to lyse the fungal cells [18]. Invert the culturing vessel (Hungate tubes or serum bottles) several times to break apart the fungal mat.
2. Transfer the culture media including the plant substrate into centrifuge tubes (15- or 50-mL).
3. Centrifuge at 3220 × *g* for 7 min at 4 °C with a swinging bucket rotor (*see* **Note 9**).
4. Decant and discard the supernatant.
5. Follow the protocol included in the DNeasy PowerPlant Pro DNA Isolation Kit (*see* **Note 14**).

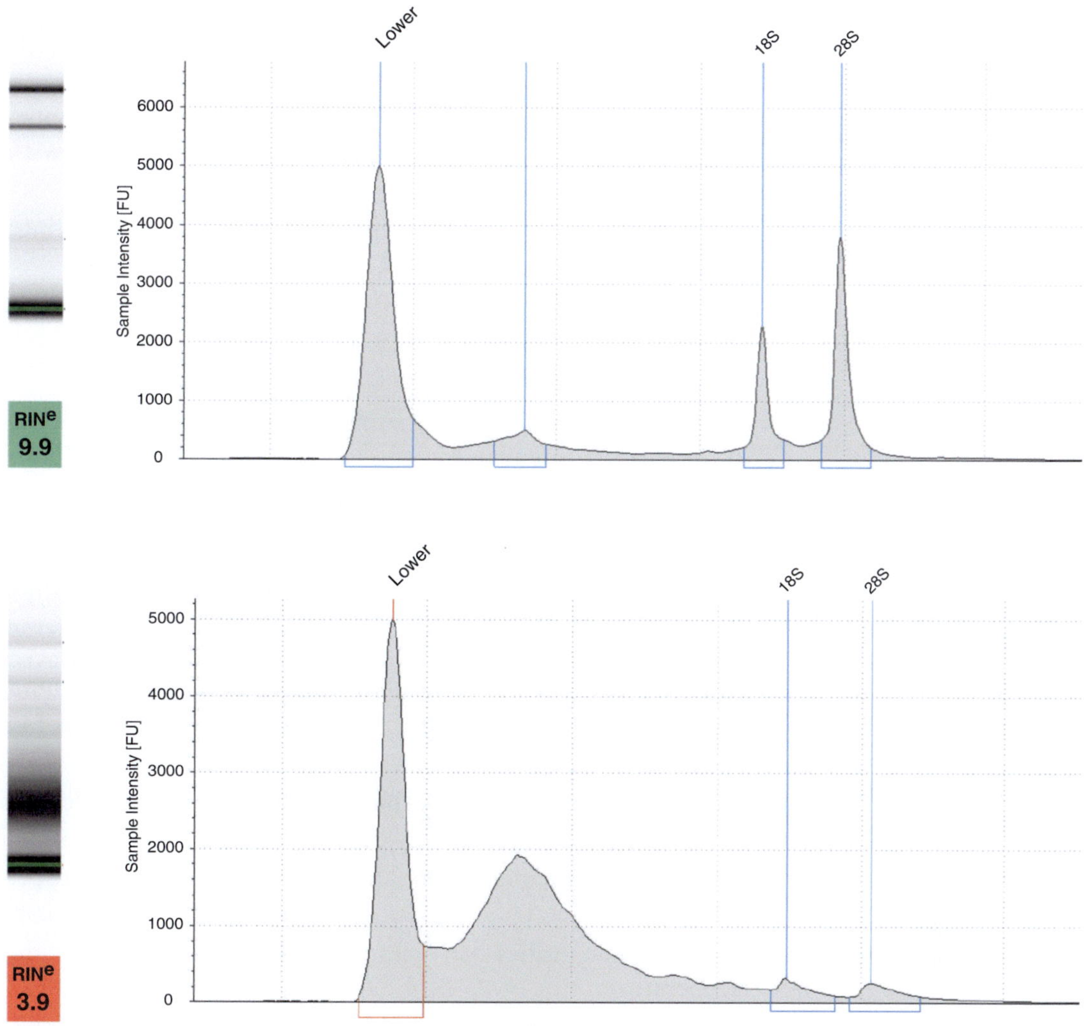

Fig. 4 An example of high RNA quality isolated (RNA Integrity Number, RIN > 9.0) from anaerobic fungal cultures (upper panel), and an example of poor RNA quality (RIN < 6.0) isolated from anaerobic fungal cultures (lower panel). Samples were measured on the Agilent 2200 TapeStation system (Agilent Technologies)

3.9 Genomic DNA Extraction from Fungal Cultures Grown on Plant Substrates

For media containing plant substrates (e.g., reed canary grass, corn stover), a harsher bead beating step with the Mini-Beadbeater-16 is used for cell lysis, because plant substrates absorb a significant part of the bead beating force. In order to reduce background DNA introduced from rumen fluid, double-clarified rumen fluid is recommended (Tables 1 and 2).

We have developed a method to extract RNA and DNA from the same sample based on the QIAGEN AllPrep® DNA/RNA/miRNA Universal Kit. This may prove to be advantageous when the quantity of samples is limiting. For genomic DNA (only) extraction from fungal cultures grown on plant substrates, follow the protocol below for simultaneous preparation of RNA and DNA, and simply disregard the RNA extraction part of the AllPrep® protocol.

1. Invert the culturing vessel (Hungate tubes or serum bottles) several times to break apart the plant substrate with fungal mat.
2. Transfer the culture media including the plant substrate into centrifuge tubes (15- or 50-mL, *see* **Note 8**).
3. Centrifuge at 3220 × *g* for 20 min at 4 °C with a swinging bucket rotor (*see* **Note 9**).
4. Decant and discard the supernatant.
5. Add 1 mL of RNAlater® and store at −80 °C if not proceeding to extraction immediately.
6. Thaw samples preserved in RNAlater®, or use fresh samples.
7. Centrifuge at 3220 × *g* for 20 min at 4 °C with a swinging bucket rotor.
8. Decant and discard the supernatant.
9. Transfer all of the substrate and fungal mat into an autoclaved 2-mL screw-cap tube filled with 500 μL of buffer RLT Plus (QIAGEN) and 1.0 mL of 0.5 mm zirconia/silica beads (*see* **Notes 10** and **11**).
10. Briefly vortex to mix the beads, buffer, and sample (*see* **Note 12**).
11. Bead beat samples for 1.5 min using a Biospec Mini-Beadbeater-16.
12. Place sample tubes on ice for 1.5 min to lower the temperature.
13. Bead beat samples for another 1.5 min using a Biospec Mini-Beadbeater-16.
14. Centrifuge at 13,000 × *g* for 3 min.
15. Transfer up to 650 μL of lysate using gel loading pipet tips onto a QIAGEN AllPrep® DNA Mini spin column (*see* **Notes 15** and **16**).
16. Follow the protocol "Simultaneous Purification of Genomic DNA and Total RNA, including miRNA, from Cells" from the AllPrep® DNA/RNA/miRNA Universal handbook (QIAGEN).
17. DNA yields can be measured using Qubit fluorometric quantitation, and DNA quality can be assessed using a TapeStation or Bioanalyzer (Agilent). For next-generation sequencing, we recommend using DNA with a minimal degree of shearing (Fig. 5).

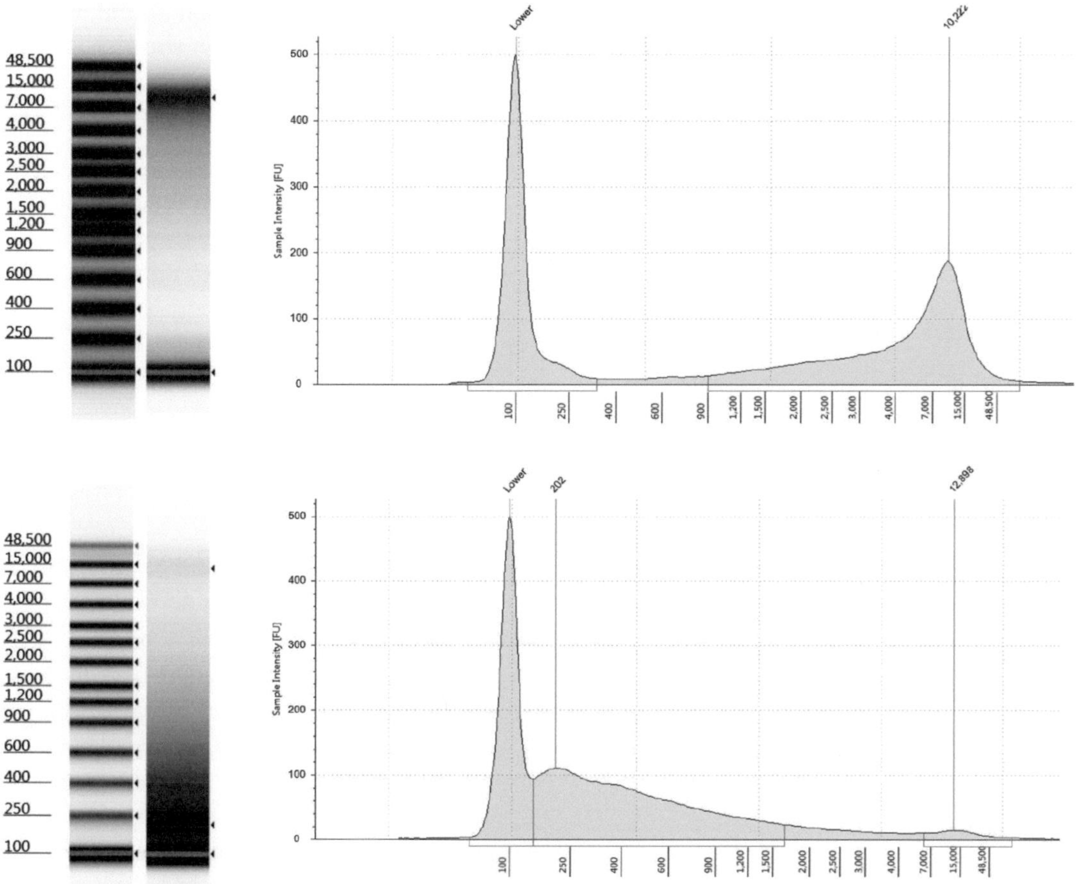

Fig. 5 An example of high DNA quality isolated from fungal cultures (upper panel), and an example of poor, overly sheared DNA isolated from fungal cultures (lower panel). Samples were measured on the Agilent 2200 TapeStation system (Agilent Technologies). The genomic DNA ladders are marked by their sizes on the left, and fungal DNA samples are shown on the right

4 Notes

1. The "Medium C Minus" ("MC−") was developed for experiments that require measurements of media composition using high performance liquid chromatography (HPLC). With reduced concentrations of yeast extract (1/10), casitone (1/20), and rumen fluids (1/2) compared to Medium C, MC− contains lower concentrations of key primary metabolites, such as formate, acetate, and hence lower background signal on the HPLC. Both Medium C and MC− include rumen fluid and are undefined. A defined medium, Medium 2, is described by Lowe and colleagues [14].
2. Boiling for 20 min will remove approximately 200 mL of water, so before heating the total volume of the medium

solution should be 1200 mL in order to reach a final volume of 1000 mL. After microwaving, the solution should be boiling and pink.

3. It is not necessary to purge with CO_2 for a long time. Slow CO_2 flow rate works better than high flow rate.
4. Alternatively, add a few chunks of dry ice into the solution.
5. Media are usually pink immediately after dispensing into Hungate tubes, but the pink color should disappear after autoclaving.
6. Before rolling the tubes under a cold water stream, avoiding small bubbles in the media will facilitate fungal colony identification.
7. This protocol is equally effective for fungal growth on soluble substrates (e.g., 5 g/L glucose) and insoluble substrates (e.g., 0.1 g reed canary grass).
8. Depending on the purpose of the RNA analysis experiment, it may be necessary to perform this step in an anaerobic environment, such as an anaerobic chamber or a glove bag.
9. Alternatively, the sample can be centrifuged for 1 h at 20,000 × *g* using a fixed angle rotor. This results in the separation of the less dense fungal mat on top of the plant substrate. This can be advantageous to reduce the amount of sample to process.
10. We find that a metal spatula with a flat end works best to transfer samples into 2-mL screw-cap tubes for bead beating.
11. We found comparable results in RNA yield by liquid nitrogen grinding compared to bead beating for 1 min using Mini-Beadbeater-16.
12. Vortex both orientations of the tube (cap down and cap up) in order to fully mix.
13. Perform "Optional On-Column DNA Digestion with the RNase-Free DNase Set" if performing RT-qPCR.
14. Use the Phenolic Separation Solution at Subheading 3.8, **step 1**.
15. If the total volume of lysate from a sample was greater than 650 μL, repeat this step until all lysate has passed through the AllPrep® DNA Mini spin column in order to maximize DNA yield. Generally the amount of DNA from fungal cultures <50 mL in volume is not sufficient to overload the AllPrep® DNA Mini spin column.
16. Generally RNA yields are high enough from just 650 μL of lysate that it is not necessary to save the flow-through from all of the lysate for RNA purification.

References

1. Orpin CG (1977) The occurrence of chitin in the cell walls of the rumen organisms Neocallimastix frontalis, Piromonas communis and Sphaeromonas communis. Microbiology 99:215–218. https://doi.org/10.1099/00221287-99-1-215
2. Mountfort DO, Orpin CG (1994) Anaerobic fungi: biology, ecology, and function. Marcel Dekker Inc, New York
3. Solomon KV, Haitjema CH, Thompson DA, O'Malley MA (2014) Extracting data from the muck: deriving biological insight from complex microbial communities and non-model organisms with next generation sequencing. Curr Opin Biotechnol 28:103–110. https://doi.org/10.1016/j.copbio.2014.01.007
4. Solomon KV, Haitjema CH, Henske JK, et al (2016) Early-branching gut fungi possess a large, comprehensive array of biomass-degrading enzymes. Science aad1431. https://doi.org/10.1126/science.aad1431
5. Peng X, Gilmore SP, O'Malley MA (2016) Microbial communities for bioprocessing: lessons learned from nature. Curr Opin Chem Eng 14:103–109. https://doi.org/10.1016/j.coche.2016.09.003
6. Gascoyne DJ, Theodorou MK (1988) Consecutive batch culture—a novel technique for the in vitro study of mixed microbial populations from the rumen. Anim Feed Sci Technol 21:183–189. https://doi.org/10.1016/0377-8401(88)90099-5
7. Haitjema CH, Solomon KV, Henske JK et al (2014) Anaerobic gut fungi: advances in isolation, culture, and cellulolytic enzyme discovery for biofuel production. Biotechnol Bioeng 111:1471–1482. https://doi.org/10.1002/bit.25264
8. Hungate RE (1969) Chapter IV A roll tube method for cultivation of strict anaerobes. In: Norris JR, Ribbons DW (eds) Methods in microbiology. Academic, New York, pp 117–132
9. Bryant MP (1972) Commentary on the Hungate technique for culture of anaerobic bacteria. Am J Clin Nutr 25:1324–1328
10. Miller TL, Wolin MJ (1974) A serum bottle modification of the Hungate technique for cultivating obligate anaerobes. Appl Microbiol 27:985–987
11. Balch WE, Wolfe RS (1976) New approach to the cultivation of methanogenic bacteria: 2-mercaptoethanesulfonic acid (HS-CoM)-dependent growth of Methanobacterium ruminantium in a pressureized atmosphere. Appl Environ Microbiol 32:781–791
12. Orpin CG (1975) Studies on the rumen flagellate Neocallimastix frontalis. Microbiology 91:249–262. https://doi.org/10.1099/00221287-91-2-249
13. Bauchop T, Mountfort DO (1981) Cellulose fermentation by a rumen anaerobic fungus in both the absence and the presence of rumen methanogens. Appl Environ Microbiol 42:1103–1110
14. Lowe SE, Theodorou MK, Trinci APJ, Hespell RB (1985) Growth of anaerobic rumen fungi on defined and semi-defined media lacking rumen fluid. Microbiology 131:2225–2229. https://doi.org/10.1099/00221287-131-9-2225
15. Joblin KN (1981) Isolation, enumeration, and maintenance of rumen anaerobic fungi in roll tubes. Appl Environ Microbiol 42:1119–1122
16. Schoch CL, Seifert KA, Huhndorf S et al (2012) Nuclear ribosomal internal transcribed spacer (ITS) region as a universal DNA barcode marker for Fungi. Proc Natl Acad Sci U S A 109:6241–6246. https://doi.org/10.1073/pnas.1117018109
17. Theodorou MK, Brookman J, Trinci APJ (2005) Anaerobic fungi. In: Makkar HPS, McSweeney CS (eds) Methods in gut microbial ecology for ruminants. Springer, Netherlands, pp 55–66
18. Solomon KV, Henske JK, Theodorou MK, O'Malley MA (2016) Robust and effective methodologies for cryopreservation and DNA extraction from anaerobic gut fungi. Anaerobe 38:39–46. https://doi.org/10.1016/j.anaerobe.2015.11.008
19. Theodorou MK, Davies DR, Nielsen BB et al (1995) Determination of growth of anaerobic fungi on soluble and cellulosic substrates using a pressure transducer. Microbiology 141:671–678. https://doi.org/10.1099/13500872-141-3-671

Chapter 6

Efficient Extraction Method for High Quality Fungal RNA from Complex Lignocellulosic Substrates

Miia R. Mäkelä and Kristiina Hildén

Abstract

Here we describe an efficient and reproducible method for the extraction of fungal RNA from complex lignocellulose containing materials. The fungal cells are snap-frozen and disrupted in chaotropic guanidinium thiocyanate buffer, after which the extracted RNA is isolated by using CsCl gradient ultracentrifugation. By lowering the pH of the extraction buffer, the procedure is also suitable for sample materials rich in humic acids. The method results in high quantity and quality RNA that is separated from endogenous contaminants (e.g., RNases) and substances derived from plant biomass (e.g., colored aromatic compounds). In addition, no further steps such as DNase treatment are needed. The extracted RNA is highly suitable for downstream gene expression analyses such as RNA sequencing.

Key words Lignocellulose, Plant biomass, Fungi, Basidiomycetes, Cesium chloride, RNA extraction, Ultracentrifugation

1 Introduction

Postgenomic transcriptome analyses focusing to understand how fungi thrive on their natural carbon and energy sources have highlighted the need for the extraction of high-quality RNA with sufficient quantity from diverse materials including complex plant biomass substrates. These substrates, including wood, plant and soil litter, and compost, are rich in polysaccharides and organic compounds such as phenolics and humic acids that can interfere with RNA isolation and hamper downstream analyses [1–4]. In addition, protocols that are applicable on a wide range of substrates often have low reproducibility [5–7].

CsCl density gradient centrifugation for RNA extraction has been introduced several decades ago [8]. During the ultracentrifugation, molecules present in the extracted sample migrate to the CsCl layer with the same density. Due to the high density of RNA molecules, there is no CsCl concentration to which RNA would migrate, but instead it will pellet. It is a procedure that efficiently

Ronald P. de Vries et al. (eds.), *Fungal Genomics: Methods and Protocols*, Methods in Molecular Biology, vol. 1775, https://doi.org/10.1007/978-1-4939-7804-5_6,

separates RNA from RNases and other proteins, as well as DNA, and therefore, no additional DNase treatment is needed [9, 10]. Here we describe a method in which CsCl ultracentrifugation is used, in addition to the separation of RNA from molecules that originate from fungal cells, to efficiently remove substances that are derived from plant biomass containing substrates. In addition, the method is highly reproducible on different lignocellulosic substrates.

Fungal mycelium containing sample is snap-frozen in liquid nitrogen to suppress endogenous RNase activity. The frozen sample is disrupted by grinding it with mortar and pestle and extracted with a chaotropic guanidinium thiocyanate buffer that further inhibits the RNase activity. Guanidinium thiocyanate is a strong denaturant since both guanidinium anion and thiocyanate cation are strong chaotropic agents. In addition, reducing agent β-mercaptoethanol is used in the extraction buffer to break disulfide bonds to eliminate RNases released during cell lysis [9]. For humic acid-rich substrates, such as soil litter and compost, the guanidinium thiocyanate buffer with acidic pH (pH 5.0) should be used [6], and additional chloroform–isoamyl alcohol extractions are usually needed before CsCl ultracentrifugation [11]. The extract is pipetted onto CsCl cushion and ultracentrifuged for 21 h using a swinging bucket rotor, after which a transparent RNA pellet will be separated in the bottom of the centrifuge tube. The isolated RNA is of high quality and quantity as well as good integrity.

2 Materials

Prepare all solutions using RNase-free water and analytical grade RNase-free reagents. Use only RNase-free laboratory plasticware and pipette tips. Cover glassware as well as mortar and pestle with aluminum foil and dry heat-sterilize them at 180 °C for at least 3 h. Prepare all reagents at room temperature and store them at 4 °C, unless otherwise indicated. When disposing of wastes, carefully follow all waste disposal regulations.

1. RNase-free water: double distilled water treated with 0.1% (v/v) diethylpyrocarbonate (DEPC). Prepare 1% DEPC solution by mixing 1 mL of DEPC stock solution with 99 mL ethanol. Pipet 1 mL of 1% DEPC to a 1 L graduated glass. Fill up to 1 L with double distilled water. Mix with magnetic stirrer for 1 h in a fume hood. Keep the solution at 37 °C overnight. Autoclave at 121 °C, 1 bar, 15 min. Store at room temperature.
2. Guanidinium thiocyanate buffer: 4 M guanidinium thiocyanate in 25 mM Na-citrate buffer, pH 5.0 or 7.0 (*see* **Note 1**). Weigh

500 g guanidinium thiocyanate and 5 g *N*-lauroylsarcosine sodium salt to a 1L graduated glass. Add 25 mL 25 mM Na-citrate buffer, pH 5.0 or 7.0 and 14 mL β-mercaptoethanol. Fill up to 900 mL with RNase-free water. Mix with magnetic stirrer at 37 °C overnight. Check pH and adjust it with 1 M NaOH or 1 M HCl if necessary. Make up to 1 L with RNase-free water. Filter the guanidinium thiocyanate buffer with bottle-top sterile filter unit with the pore size 0.2 μm. Store at 4 °C.

3. Chloroform–isoamyl alcohol: mixture of chloroform and isoamyl alcohol at 24:1 ratio. Pipet 10 mL isoamyl alcohol to a 250 mL graduated glass. Fill up to 250 mL with chloroform. Store at room temperature.
4. Cesium chloride (CsCl) buffer: 5.7 M CsCl, pH 5.0. Weigh 958 g CsCl to a 1 L graduated glass. Add 25 mL 25 mM Na-citrate buffer, pH 5.0 and 2 mL 10% DEPC in EtOH. Fill up to 1 L with distilled water. Mix 1 h with magnetic stirrer. Keep at 37 °C overnight. Autoclave at 121 °C, 1 bar, 15 min. Store at room temperature.

3 Methods

Conduct out all steps at room temperature unless otherwise stated. Use RNase-free solutions, laboratory ware, and pipette tips.

3.1 RNA Extraction

1. Grind the fungal mycelium containing sample to fine powder in mortar and pestle (*see* **Note 2**) under liquid nitrogen (*see* **Note 3**).
2. Transfer the powder to a 50 mL conical centrifuge tube that contains 10 mL guanidinium thiocyanate buffer (pH 5.0 or 7.0).
3. Ensure that all clumps are dispersed by mixing or vortexing.
4. Incubate at room temperature for 10 min.
5. Centrifuge at 11,000 × *g* at 4 °C for 10 min.
6. Pipet the supernatant to an RNase-free tube.
7. In case the extracted sample is colorful or expected to contain coextracted humic acids (*see* **Note 4**), add 1 volume chloroform–isoamyl alcohol to the supernatant.
8. Mix the tube by inversion and incubate at room temperature for 2 min.
9. Centrifuge at 11,000 × *g* at 4 °C for 15 min.
10. Pipet the upper phase to an RNase-free tube.
11. Repeat the **steps 7–10**.
12. Proceed immediately with CsCl ultracentrifugation (*see* **Note 5**) or store the extract at −80 °C.

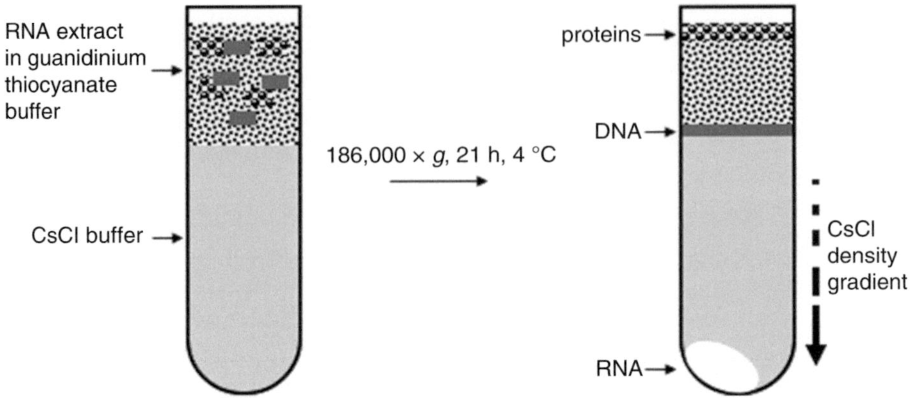

Fig. 1 Schematic presentation of RNA extraction using CsCl ultracentrifugation

3.2 CsCl Ultracentrifugation

1. Carefully pipet 2 mL CsCl buffer (pH 5.0) to the bottom of polyallomer ultracentrifuge tubes. Make sure that no liquid is dispersed on the upper part of the tube walls.
2. Pipet the RNA extract onto the CsCl cushion (Fig. 1).
3. Fill the ultracentrifuge tube with the guanidinium thiocyanate buffer.
4. Balance the tubes with the guanidinium thiocyanate buffer so that there is at maximum 0.1 g difference between the opposite tubes.
5. Centrifuge at 186,000 × *g*, for 21 h, at 4 °C using a swinging bucket rotor.
6. After the centrifugation, remove the guanidinium thiocyanate buffer by pipetting.
7. Remove CsCl buffer by pipetting or quickly inverting the centrifuge tube.
8. Cut all but the bottom 1 cm off from the centrifuge tube.
9. Rinse the transparent RNA pellet with 100 μL DEPC-treated water.
10. Dissolve the RNA pellet in 50 μL DEPC-treated water and transfer it to an RNase-free tube.
11. Rinse the bottom of the centrifuge tube with 50 μL DEPC-treated water and combine it with the RNA sample.
12. Store the RNA sample at −80 °C prior to downstream analyses.

4 Notes

1. 25 mM Na-citrate buffer of pH 7.0 is suitable for most of the lignocellulose containing samples. However, if humic acids containing material, such as plant and soil litter or compost, is

used for RNA extraction, 25 mM Na-citrate buffer of pH 5.0 should be used.

2. Laboratory homogenizers that allow efficient cooling of the sample can also be used for the disruption.
3. With some fungal species with high endogenous RNase activity, it is important to freeze the mycelium containing sample immediately after harvesting and proceed with the RNA extraction.
4. When RNA is isolated from for example fungal mycelium grown on soil litter or compost, usually colored compounds such as humic acids that originate from the plant biomass substrate are coextracted with the nucleic acids. These compounds also interfere with CsCl ultracentrifugation and therefore additional cleanup step with chloroform–isoamyl alcohol is needed.
5. Samples extracted from substrates that contain high amounts of humic acids, such as compost, should not be frozen, but processed immediately with CsCl ultracentrifugation.

References

1. England LS, Trevors JT (2003) The microbial DNA cycle in soil. Riv Biol 96:317–326
2. England LS, Trevors JT, Holmes SB (2001) Extraction and detection of baculoviral DNA from lake water, detritus and forest litter. J Appl Microbiol 90:630–636
3. Sayler GS, Fleming JT, Nivens DE (2001) Gene expression monitoring in soils by mRNA analysis and gene lux fusions. Curr Opin Biotechnol 12:455–460
4. Trevors JT (1996) Nucleic acids in the environment. Curr Opin Biotechnol 7:331–336
5. Leite GM, Magan N, Medina Á (2012) Comparison of different bead-beating RNA extraction strategies: an optimized method for filamentous fungi. J Microbiol Methods 88:413–418
6. Mettel C, Kim Y, Shrestha PM et al (2010) Extraction of mRNA from soil. Appl Environ Microbiol 76:5995–6000
7. Wang Y, Hayatsu M, Fujii T (2012) Extraction of bacterial RNA from soil: challenges and solutions. Microbes Environ 27:111–121
8. Kurland CG (1960) Molecular characterization of ribonucleic acid from *Escherichia coli* ribosomes: I. Isolation and molecular weights. J Mol Biol 2:83–91
9. Chirgwin JM, Przybyla AE, MacDonald RJ et al (1979) Isolation of biologically active ribonucleic acid from sources enriched in ribonuclease. Biochemistry 18:5294–5299
10. Glišin V, Crkvenjakov R, Byus C (1974) Ribonucleic acid isolated by cesium chloride centrifugation. Biochemistry 13:2633–2637
11. Patyshakuliyeva A, Mäkelä MR, Sietiö O-M et al (2014) An improved and reproducible protocol for the extraction of high quality fungal RNA from plant biomass substrates. Fungal Genet Biol 72:201–206

Chapter 7

Isolation of High Quality RNA for Metatranscriptomic Analysis of Lignocellulose Digestion in the Rumen

R. J. Gruninger, R. J. Forster, and T. A. McAllister

Abstract

Metatranscriptomics can be used to examine both the composition of a microbial community as well as its metabolic activity under a particular set of conditions and complement metagenomic studies. The availability of low-cost, high-throughput next-generation sequencing has led to a rapid increase in the number of metatranscriptomic studies being undertaken. One of the primary difficulties when conducting transcriptomics is the ability to isolate high-quality RNA from samples of interest. The application of metatranscriptomics in rumen microbiology is still relatively novel but there is a significant push toward applying this technology in this field. In this protocol, we outline the method that is used routinely in our laboratory to purify high quality RNA from rumen contents that are suitable for metatranscriptomic sequencing using RNA-seq.

Key words Metatranscriptomics, RNA-Seq, Carbohydrate active enzyme, Rumen

1 Introduction

Lignocellulosic biomass is the most abundant renewable carbon source and can be used to produce biofuels, low-cost livestock feed, and high-value chemicals [1]. The potential value of this resource has led to intensive research efforts to develop cost effective methods to break down lignocellulose. One of the best characterized microbial communities that can effectively degrade biomass is harbored within the rumen [2]. The rumen microbiome consists of a complex anaerobic microbial community of bacteria, archaea, protozoa, and fungi. The metabolic activity of these microbial symbionts converts complex fibrous substrates into volatile fatty acids and microbial protein that are used by the ruminant host for maintenance, growth, and lactation [3]. Even for the relatively intensively studied rumen microbial community it is estimated that more than 85% of its members have still not been cultivated [2]. This unexplored microbial diversity represents an untapped source of potentially novel and unique enzymatic activities and metabolic pathways that can be applied to industrial

Ronald P. de Vries et al. (eds.), *Fungal Genomics: Methods and Protocols*, Methods in Molecular Biology, vol. 1775, https://doi.org/10.1007/978-1-4939-7804-5_7,

biomass conversion. The fastidious nature of rumen microbes has led to many researchers now using culture-independent next-generation sequencing approaches to better understand the function of the rumen microbiome under a number of conditions. Both metagenomics and metatranscriptomic studies of the rumen have been reported in the literature [4, 5]. Metatranscriptomics in particular has been a powerful tool for uncovering the mechanisms that are employed by lignocellulose degrading microbes to break down the plant cell wall, and has been used to identify essential carbohydrate active enzymes (CAZymes) involved in this process [5, 6].

One of the major challenges of conducting metatranscriptomic studies is the ability to obtain sufficient quantities of high quality, intact RNA. In this chapter we outline the methodology that is used in our lab to isolate high quality RNA from the complex microbial community in the rumen. This protocol assumes that the user either has direct access to ruminant animals to obtain a fresh rumen sample or is working with a collaborator that has access to animals. The purified total RNA is suitable for use in metatranscriptomic studies to examine the role the microbes, and the enzymes that they express, that are involved in the degradation of the plant cell wall. This method will focus on sample collection, isolation of crude RNA, column cleanup of isolated RNA, and sample quantification and quality control.

2 Materials

2.1 Rumen Sampling

1. Liquid nitrogen (*see* **Note 1**).
2. 6″ × 6″ labeled aluminum foil squares.
3. 250 mL centrifuge bottle.
4. Cannulated ruminant.
5. Top loading balance.
6. Long tweezers.
7. Mortar (400 mL capacity) and pestle.
8. 50 mL Falcon tubes.

2.2 RNA Isolation and Purification

1. Chloroform.
2. Isopropanol.
3. 2 mL Eppendorf tubes.
4. RNase *AWAY* (Ambion Cat #: 10328011).
5. Clean PCR hood with dedicated pipettes.
6. Nuclease-free, filter pipette tips.
7. Microfuge.
8. Absolute (95% or 99%) ethanol.
9. 75% (v/v) RNase-free ethanol (to prepare 10 mL add 2.5 mL of RNAse-free H_2O to 7.5 mL of absolute ethanol).

10. MEGAclear Kit (Ambion Cat #: AM1908).
11. TRIzol Reagent (Life Technologies Cat #: 15596026).

2.3 RNA Quality Control

1. Agilent 2100 Bioanalyzer.
2. RNA 6000 Nano kit (Agilent Cat #: 5067-1511).

3 Methods

3.1 Sample Collection from Rumen

1. Using a 250 mL centrifuge bottle take a sample of ruminal contents from the reticulum, ventral, caudal, and dorsal–ventral sac of the reticulorumen (Fig. 1; *see* **Note 2**).
2. Pour sample into 500 mL glass beaker and mix thoroughly to ensure that the sample is homogeneous (*see* **Note 3**).
3. Remove rumen contents from beaker and weigh out ~5 g of contents onto labeled tin foil boat and float boat on liquid nitrogen to freeze sample (Fig. 2; *see* **Note 4**).
4. When Sample is frozen the labeled, tin-foil boat can be folded up and stored at −80 °C until processed.
5. When ready to process rumen samples, transfer the frozen sample into a prechilled mortar and cover with liquid nitrogen. Grind the sample to a fine powder for 5 min adding more liquid nitrogen as needed to ensure that the sample does not thaw during grinding.

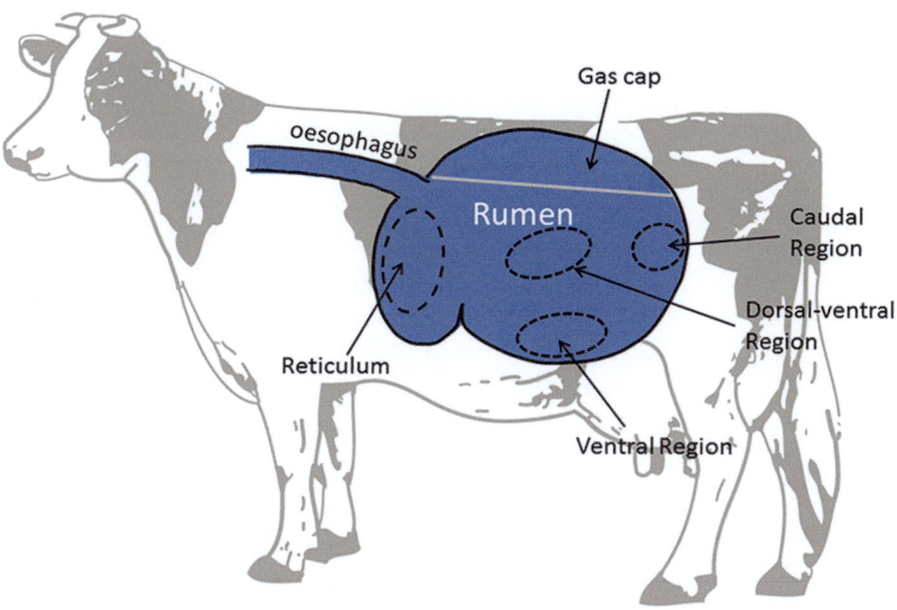

Fig. 1 Sampling sites within the rumen to obtain a representative sample. The region were gases accumulate at the top of the rumen is referred to as the gas cap. Depending on the composition of the diet there is often a layer of solid digesta at the interface (indicated by the grey line). The solids are mixed with rumen fluid and circulated by peristaltic muscle contractions of the rumen wall

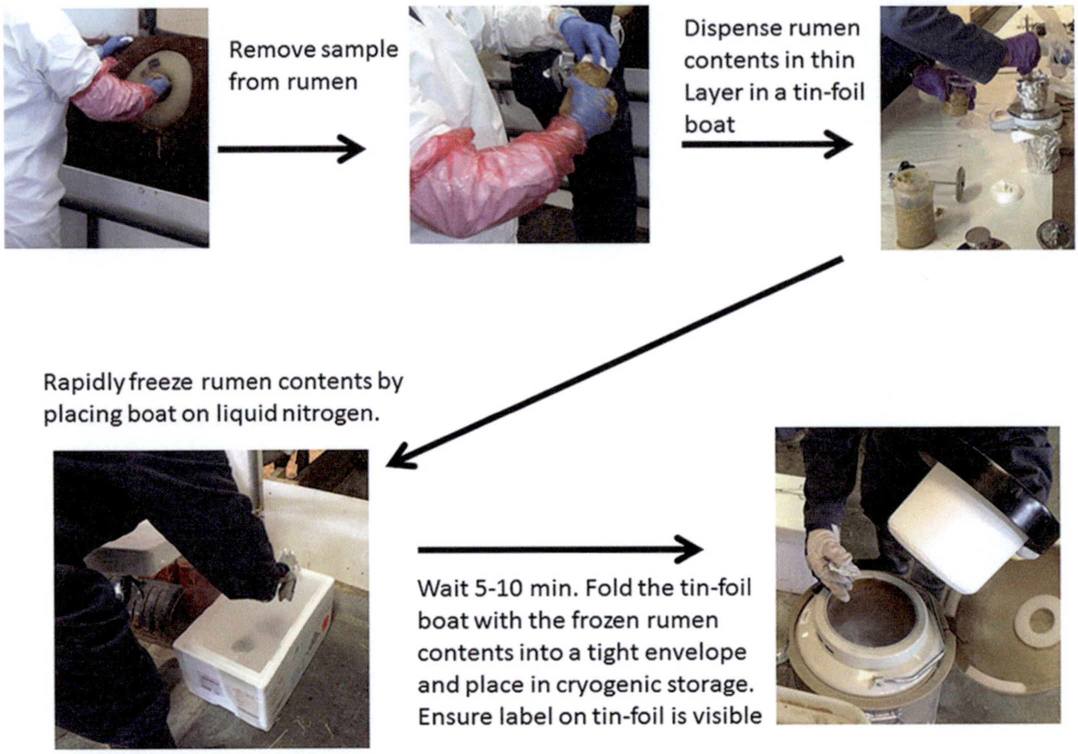

Fig. 2 Rumen sampling workflow and technique to rapidly freeze rumen samples. Tin foil boats are formed by folding an approximately 6″ × 6″ sheet of aluminum foil around the bottom of a 500 mL beaker. Be sure to clearly label the foil with permanent marker on both sides. These can then be separated into labeled Ziploc bags for long-term storage

6. Suspend the ground powder in liquid nitrogen and carefully transfer equal amount of powder (approximately) into four labeled 50 mL preweighed Falcon tubes (*see* **Note 5**).
7. Allow liquid nitrogen to boil off and store tubes at −80 °C.
8. After all of the liquid nitrogen has boiled away, tubes containing the ground rumen contents are weighed. The amount of rumen contents in each tube is calculated by subtracting the mass of the empty tube from the mass of the tube containing the ground sample.

3.2 RNA Isolation and Cleaning

1. Add TRIzol reagent to the ground sample at a ratio of 1 mL of TRIzol for every 100 mg of sample and mixture to homogenize at RT for 5–10 min to allow RNA to dissociate from proteins (*see* **Note 6**).
2. Using a wide bore tip, make 1.2 mL aliquots of the TRIzol–Rumen contents suspension in labeled 2 mL Eppendorf tubes. Aliquots can be stored at −80 °C until ready to process (*see* **Note 7**).

3. Centrifuge 1.5 mL TRIzol–Rumen Content suspension in microfuge at max speed to pellet undissolved rumen solids (*see* **Note 8**).
4. Carefully decant the supernatant into a new 2 mL tube being careful not to disturb the pellet.
5. Add 0.35 mL of room temperature 100% chloroform to decanted supernatant.
6. After ensuring tubes are securely closed, shake vigorously for 15 s, and incubate at room temperature for 2–3 min.
7. Centrifuge mixture in a microfuge at 4 °C at max speed for 15 min to separate aqueous and organic phases.
8. Carefully remove the aqueous top layer (~800–900 μL) and transfer to a new 2 mL Eppendorf tube. Be very careful not to transfer any of organic layer, or the white precipitate located at the interface of the two layers. You will likely not be able to transfer all of the aqueous phase.
9. Precipitate RNA from the aqueous phase by adding 0.75 mL of 100% isopropanol at room temperature.
10. Invert tubes 5× and incubate mixture at room temperature for 10 min.
11. Centrifuge mixture at 4 °C in a microfuge at max speed for 10 min to pellet precipitated RNA.
12. Carefully decant supernatant being careful not to disturb any pelleted RNA (*see* **Note 9**).
13. Add 1 mL of nuclease-free 75% ethanol and gently suspend and wash RNA pellet.
14. Centrifuge at max speed for 5 min to pellet washed RNA and carefully decant supernatant.
15. Allow residual ethanol to evaporate from RNA pellet but do not completely dry as this can negatively impact the solubility of the RNA.

3.3 Column Cleanup of Total RNA

1. Resuspend RNA pellet in 100 μL of RNase-free water and proceed to column purification of total RNA using MEGAclear kit (*see* **Note 10**).
2. Add 350 μL of binding buffer (provided in kit) and 250 μL of 100% ethanol (supplied by user) to 100 μL RNA from previous step.
3. Pipette 700 μL RNA solution from **step 2** into MEGAclear spin column and centrifuge at 10,000 × *g* for 30 s to bind RNA to column. Discard flow through.
4. Wash bound RNA 2× with 500 μL of wash buffer. For each wash, centrifuge sample for 30 s at 10,000 × *g* and discard flow-through.

5. After discarding the second wash, centrifuge the empty column for an additional 60 s to remove all residual ethanol.
6. Elute RNA with 50 μL of elution buffer. Be sure to repeat the elution a second time so that a total of 100 μL of column cleaned RNA has been collected in the same collection tube. The final volume of purified RNA will be 100 μL (*see* **Note 11**).

3.4 Quality Control of Purified RNA

1. Quality check and determine RNA concentration using an Agilent Bioanalyzer 2100 and 6000 RNA nano kit (Agilent) according to the manufacturer's instructions.
2. RNA samples that have an RNA integrity number (RIN, Fig. 3) of less than 7 are not suitable for analysis by RNA-seq and should be reisolated (*see* **Note 12**).

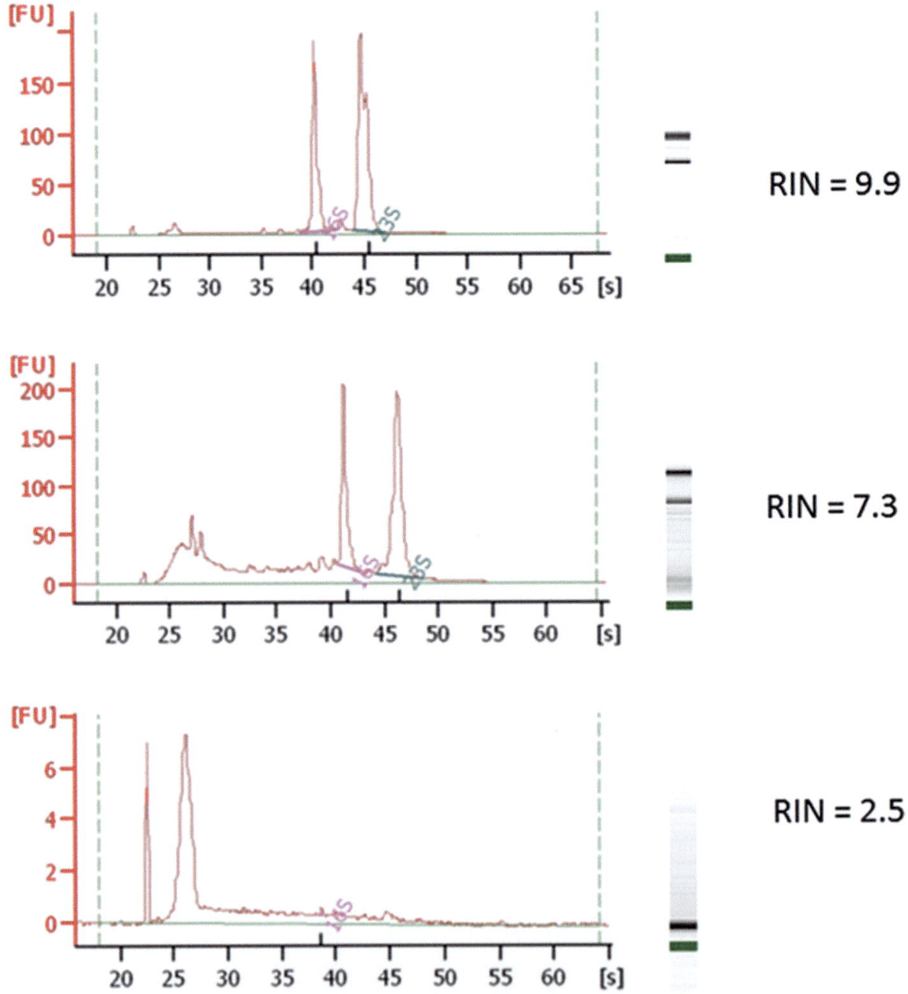

Fig. 3 Representative traces of metagenomic RNA isolated from rumen samples. The electrophoresis trace of fluorescence intensity (FU) versus retention time (s), and corresponding gel banding patterns are shown for samples with an RNA integrity number (RIN) of 9.9 (near perfect with no degradation), 7.3 (some degradation but sample still acceptable for RNA-seq analysis), and 2.5 (almost completely degraded. Not suitable for use)

4 Notes

1. Ensure that caution and proper personal protective equipment is used by personal who are working with liquid nitrogen. It is extremely cold and can cause severe burns.
2. Researchers can obtain samples of rumen contents using a stomach tub. However, this method is less effective at sampling both liquid and solid phases and may not provide a representative sample of the entire rumen.
3. If researchers are only interested in the fiber-adherent microbes, rumen samples can be separated into liquid and solid phases using a Bodum coffee plunger or by filtering the liquid through several layers of nylon cloth and rumen liquid can be discarded prior to sampling from the solid digesta.
4. Samples are frozen in liquid nitrogen immediately after removal from the animal. Freezing samples in tinfoil boats allows for rapid cooling and easy transfer to the mortar and pestle. It is essential that once frozen, the sample remains at cryogenic temperatures until TRIzol is added. This will ensure that nucleases in the sample do not degrade the RNA.
5. To reduce the amount of TRIzol needed, ground samples can be aliquoted into more tubes. This will reduce the amount of sample per tube. Unused material can be stored at −80 °C but there is risk that the sample quality can deteriorate over time.
6. TRIzol contains phenol and guanidine isothiocyanate and can cause chemical burns. Extreme care should be taken when working with this reagent. Always wear gloves, proper personal protection equipment, and work in a fume hood.
7. RNase is ubiquitous in the environment and contamination of samples can easily occur. To minimize the likelihood of contaminating your sample always wear gloves, a lab coat and work with RNase-free H_2O, buffers, tubes and pipette tips. Solutions such as RNase *AWAY* can inactivate enzymes that might be present on work spaces and equipment and should be applied prior to working with RNA samples.
8. We have found that if this solid material is not removed from the TRIzol prior to adding chloroform the quality of the isolated RNA is compromised.
9. The RNA pellet may be difficult to see or not visible at all at this point.
10. Placing tubes in a heating block at 50–60 °C will help dissolve pellet.
11. Carrying out both elution steps significantly improves yield. In our experience the concentration of the first and second elutions are similar, so skipping this step will result in a significant decrease in overall yield.

12. The RNA integrity number, or RIN, provides an objective standardized, repeatable numerical indicator of RNA quality. Calculation of the RIN value compares the shape of the electrophoresis trace to a standard database and assigns a number from 1 (lowest quality) to 10 (highest quality) to the RNA sample [7].

References

1. Chundawat SP, Beckham GT, Himmel ME, Dale BE (2011) Deconstruction of lignocellulosic biomass to fuels and chemicals. Annu Rev Chem Biomol Eng 2:121–145
2. Flint HJ, Bayer EA, Rincon MT, Lamed R, White BA (2008) Polysaccharide utilization by gut bacteria: potential for new insights from genomic analysis. Nat Rev Microbiol 6:121–131
3. Morgavi DP, Kelly WJ, Janssen PH, Attwood GT (2013) Rumen microbial (meta) genomics and its application to ruminant production. Animal 7(s1):184–201
4. Hess M, Sczyrba A, Egan R, Kim TW, Chokhawala H, Schroth G, Luo S, Clark DS, Chen F, Zhang T, Mackie RI, Pennacchio LA, Tringe SG, Visel A, Woyke T, Wang Z, Rubin EM (2011) Metagenomic discovery of biomass-degrading genes and genomes from cow rumen. Science 331(6016):463–467
5. Dai X, Tian Y, Li J, Su X, Wang X, Zhao S, Liu L, Luo Y, Liu D, Zheng H, Wang J (2015) Metatranscriptomic analyses of plant cell wall polysaccharide degradation by microorganisms in the cow rumen. Appl Environ Microbiol 81(4):1375–1386
6. Qi M, Wang P, O'Toole N, Barboza PS, Ungerfeld E, Leigh MB, Selinger LB, Butler G, Tsang A, McAllister TA, Forster RJ (2011) Snapshot of the eukaryotic gene expression in muskoxen rumen—a metatranscriptomic approach. PLoS One 6(5):e20521
7. Schroeder A, Mueller O, Stocker S, Salowsky R, Leiber M, Gassmann M, Lightfoot S, Menzel W, Granzow M, Ragg T (2006) The RIN: an RNA integrity number for assigning integrity values to RNA measurements. BMC Mol Biol 7:3

Chapter 8

Fungal Transcriptomics

Vasanth R. Singan, Rita C. Kuo, and Cindy Chen

Abstract

Transcriptomics, the study of RNA molecules, provides in-depth understanding of cellular functions and the genomic landscape of transcription. Transcriptomics refers to the study of all classes on RNA molecules including mRNA, tRNA, and siRNA. In this chapter, we specifically focus on mRNA, which encodes the protein-coding portion of the genomic DNA. We discuss the use of mRNA in annotation of genomes and in studying differential regulation of genes under experimental conditions.

Key words Transcriptomics, RNA-Seq, Transcriptome assembly, Gene expression analysis

1 Introduction

The advent of sequencing technologies including RNA-Seq has brought about a paradigm shift in the way RNA molecules have been analyzed [1]. Earlier technologies like microarrays are limited in scope and the advancement in sequencing technologies has enabled rapid discovery and accurate quantification of RNA in various studies [2]. mRNA sequencing can be used to aid in gene discovery [3]. Contig assembly and subsequent gene annotation helps in revealing the location of transcription, identifying isoforms [2, 3] and further aids in building gene models and developing tools for users to analyze and explore genomics in a large-scale data set (e.g., MycoCosm portal [4]). Additionally, abundance of reads from specific transcripts can be used in studying differential gene regulation between experimental conditions [5].

There are methods for sequencing full-length cDNA (e.g., PacBio Iso-Seq and direct RNA sequencing using Oxford NanoPore) or fragmented cDNA (e.g., Illumina RNA-Seq). In this chapter, we will focus on Illumina RNA-Seq as it has the most well developed pipeline for fungal data analysis at JGI. Library kits from different venders (e.g., KAPA and New England Biolabs) are available to make Illumina RNA-Seq libraries. We chose to demonstrate the protocol using the Illumina kit, which should always

Ronald P. de Vries et al. (eds.), *Fungal Genomics: Methods and Protocols*, Methods in Molecular Biology, vol. 1775,
https://doi.org/10.1007/978-1-4939-7804-5_8, © Springer Science+Business Media, LLC, part of Springer Nature 2018

produce reasonable sequencing results. The protocol is optimized from manufacturer's manual [6] and the library creation methods and sequencing is identical for both assembly and differential expression analysis (counting). In general, mRNA is selected or rRNA is depleted from total RNA, then the remaining mRNA is fragmented and converted to cDNA fragments with Illumina adpaters attached to both ends for sequencing [6]. Computational tools are available for assembly and exploration of the sequence data. In this chapter, we discuss RNA-Seq library creation and computational approaches used for transcriptome assembly and transcript counting using a few example commands.

2 Materials

2.1 Library Preparation and RNA-Seq

1. 10 ng–1 μg of total RNA is required. Good quality, undegraded RNA is strongly suggested.
2. Reagents and Kits: TrueSeq Stranded RNA Kit (Illumina, Inc.) or TrueSeq Stranded RNA Kit with Yeast Ribo Zero (Illumina, Inc.), SuperScript II Reverse Transcriptase (ThermoFisher Scientific), and AMPure XP beads (Agencourt). For the reagents of library qPCR and Illumina sequencing, please refer to Subheadings 2.4 and 2.5 in Chapter 4.

2.2 Transcriptome Assembly

1. bbtools (https://sourceforge.net/projects/bbmap/) for preprocessing of the input reads (*see* **Note 1**).
2. Trinity [7] for de novo assembly.

2.3 Transcript Counting

1. bbtools (https://sourceforge.net/projects/bbmap/) for preprocessing of the input reads.
2. HISAT [8] for mapping reads to reference genome.
3. samtools [9] to create sorted BAM alignment files.
4. featureCounts [10] to extract read counts.
5. DESeq2 [11] for differential expression analysis.

3 Methods

3.1 Library Preparation for RNA-Seq

This protocol is designed to produce RNA-Seq with average insert size around 300 bp for sequencing on the Illumina platform using a 2×150bp recipe. Please refer to Subheading 3.4 in Chapter 4 for the procedures of library qPCR and Illumina sequencing.

3.1.1 mRNA Selection vs. rRNA Depletion

In order to reduce rRNA contamination in your sequencing data, mRNA (PolyA) selection or rRNA depletion is required before cDNA synthesis. Some species work better using mRNA selection protocol (start with Subheading "mRNA Selection and Fragmentation"), whereas others work better using rRNA depletion protocol (start with Subheading "RiboZero Depletion and RNA Fragmentation"). Normally the mRNA selection protocol works for fungal RNA-Seq. Among the 3437 fungal samples that we have sequenced at JGI, the rRNA contamination is about 9.4% on average and about 80% of the 3437 samples had low rRNA contamination (<10%).

mRNA Selection and Fragmentation

1. Bring the volume of total RNA to 50 μL with nuclease-free water.
2. Resuspend the RNA Purification Beads (Oligo-dT beads).
3. Add 50 μL of the resuspended RNA Purification Beads to the sample and mix gently by pipetting (*see* **Note 2**).
4. Incubate at 65 °C for 5 min and then remove the sample from the thermocycler when it reaches 4 °C.
5. Place the sample at room temperature (RT) for 5 min.
6. Place the sample on a magnetic stand until liquid is clear, remove and discard the supernatant.
7. Add 200 μL of Beas Washing Buffer to the beads and mix gently by pipetting (*see* **Note 2**).
8. Remove and discard supernatant (as **step 6**).
9. Add 50 μL of Elution Buffer. Mix gently by pipetting (*see* **Note 2**).
10. Incubate the sample at 80 °C for 2 min and hold at 25 °C to elute mRNA from the beads.
11. When the thermocycler reaches 25 °C add 50 μL of *Bead Binding Buffer* to reduce rRNA nonspecific binding on beads. Mix gently by pipetting (*see* **Note 2**).
12. Incubate the sample at RT for 5 min and remove the supernatant (as **step 6**).
13. Add 200 μL of Bead Washing Buffer, mix gently (*see* **Note 2**) and remove the supernatant (as **step 6**).
14. Add 18.5 μL of Fragment, Prime, Finish Mix. Vortex and quickly spin the sample.
15. Incubate the sample at 94 °C for 2 min and put the sample on ice immediately after the incubation (*see* **Note 3**).
16. Put the sample on a magnetic stand and transfer 17 μL of the supernatant to a new PCR tube.

RiboZero Depletion and RNA Fragmentation

Preparation of RiboZero Beads

1. Pipette 225 μL of the RiboZero beads to a 1.5 mL tube.
2. Use a magnetic stand to remove the supernatant.

3. Add 225 μL of RNAse-free water to the beads and vortex to wash the beads.
4. Place the tube on a magnetic stand. Discard the supernatant and resuspend the beads with 60 μL of the Resuspension Solution.
5. Transfer 65 μL of the washed beads to a new PCR tube and add 1 μL of RNase Inhibitor.

Hybridization of rRNA and RiboZero rRNA Removal Solution

1. Add *x* μL of RNase-free water to your total RNA in a PCR tube to make the volume at 28 μL.
2. Add 4 μL of RiboZero Reaction Buffer and 8 μL of RiboZero rRNA Removal Solution to the sample and mix gently (*see* **Note 2**).
3. Incubate the sample at 68 °C for 10 min and transfer the sample to RT immediately.

rRNA Removal with Beads

1. Transfer the treated RNA (40 μL) to the RiboZero beads. Mix gently.
2. Incubate the treated RNA in the beads at RT for 5 min.
3. Do a quick vortex and incubate the beads at 50 °C for 5 min.
4. Place the tube on a magnetic stand and *transfer the supernatant* to a new *1.5 mL* tube.
5. Purify the sample with 160 μL of AMPure XP beads.
6. Add 11 μL of Elution Buffer to elute mRNA, transfer 8.5 μL to a new PCR tube, and keep 1 μL for QC (*see* **Note 4**).

RNA Fragmentation

1. Add 8.5 μL of Elute–Prime–Fragment Mix to the sample.
2. Incubate the sample at 94 °C for 2 min and put the sample on ice immediately after the incubation (*see* **Note 3**).

3.1.2 cDNA Synthesis

1. Add 1 μL of SuperScript II to 9 μL of First Strand Master Mix. Pulse vortex and quickly spin.
2. Take 8 μL of the mix and add to the sample. Mix well.
3. Incubate the sample at 25 °C for 10 min, 42 °C for 15 min and 70 °C for 10 min and hold at 4 °C.
4. Add 5 μL of Resuspension Buffer and 20 μL of SMM Master Mix to the sample and mix thoroughly.
5. Incubate at 16 °C for 1 h.
6. Purify the sample with 90 μL of AMPure XP beads and elute cDNA with 16.5 μL of Resuspension Buffer (*see* **Note 5**).
7. Transfer 15 μL of the supernatant to a new tube.

3.1.3 A-Tailing and Adaptor Ligation

1. Add 2.5 μL of Resuspension Buffer and 12.5 μL of A-Tailing Mix to the sample.
2. Incubate at 37 °C for 30 min, 70 °C for 5 min and place the sample on ice.
3. Add 2.5 μL and Resuspension Buffer and 2.5 μL of Illumina Index Adapter to the sample. Mix well.
4. Add 2.5 μL of Ligation Mix and incubate the sample at 30 °C for 10 min.
5. Remove the sample from the thermocycler and add 5 μL of Stop Ligation Buffer.
6. Purify the adapter-ligated cDNA with 42 μL of AMPure XP Beads and elute the sample with 51.5 μL of Resuspension Buffer (*see* **Note 5**).
7. Transfer 50 μL of the supernatant to a new tube.
8. Purify the sample again with 50 μL of AMPure XP Beads (*see* **Note 5**) and elute the sample with 21.5 μL of Resuspension Buffer.
9. Transfer 20 μL of the supernatant to a new PCR tube.

3.1.4 Library Enrichment and Purification

1. Add 5 μL of PCR Primer Cocktail and 25 μL of PCR Master Mix to sample. Mix well.
2. Place the sample on a thermocycler and run following program: 98 °C for 30 s; 8 cycles of 98 °C for 10 s, 60 °C for 30 s, and 72 °C for 30 s (*see* **Note 6**); 72 °C for 5 min; 4 °C on hold.
3. Purify the amplified library with 45 μL of AMPure XP beads and then elute DNA with 25 μL of Resuspension Buffer.
4. Transfer 24 μL of the supernatant to a new tube. Run QC to verify the average size and the concentration of the library.

3.2 Transcriptome Assembly

RNA-Seq reads can be assembled into longer contigs representing complete transcripts, their fragments or isoforms. Here we describe ab initio protocols for transcriptome assembly using Trinity [6]. Trinity partitions RNA-seq data into individual de Bruijn graphs and attempts to reconstruct full-length splicing isoforms. The Trinity fasta file can be used for gene prediction using PASA as described in Chapter 15.

Preprocessing of reads including trimming of adapter sequences and low quality sequences is critical for obtaining optimal assemblies. Additionally spike-in sequences and contaminants should be removed.

1. Trim adapters from raw reads using bbduk. An example command and parameters used are described as follows:

```
bbduk.sh in=raw.fastq.gz out=trimmed.fastq.gz
ref=adapters.fa k=23 ktrim=r minlen=51 mink=11 hdist=1
```

Parameters: k=23, Kmer length used for finding adapters/contaminants. Contaminants/adapters shorter than k will not be found. ktrim=r, trim reads to the right, to remove bases matching reference kmers. minlen=51, reads shorter than 51 bp after trimming will be discarded. mink=11, look for shorter kmers at the reads tops down to 11 bp. hdist=1, maximum hamming distance for ref kmers (substitutions only).

2. Trim for low quality sequences using BBDuk. An example command and parameters used are as follows:

```
bbduk.sh in=trimmed.fastq.gz out=filtered.fastq.gz qtrim=r
trimq=6 minlength=51 hdist=1
```

Parameters: qtrim=r, trim reads to the right, to remove bases below quality score. trimq=6, regions with average quality score below 6 will be trimmed.

3. Spike-in sequences and contaminants can also be filtered out using BBDuk (*see* **Note 7**). An example command and parameters used are as follows:

```
bbduk.sh in=trimmed.fastq.gz out=filtered.fastq.gz
ref=contaminants.fa k=31 hdist=1
```

Parameters: k, Kmer length used for finding adapters/contaminants. Contaminants/adapters shorter than k will not be found. hdist, maximum hamming distance for ref kmers (substitutions only).

4. Trinity assembly: The filtered fastq file is used as input for transcriptome assembly. This protocol uses trinity RNA-Seq assembler (version 2.1.1) for transcriptome assembly (*see* **Note 8**). The parameters provided here are suggestive and must be customized based on each organism for optimal assembly. A folder called trinity_out_dir is created and the assembled transcripts are written to a file called Trinity.fasta (*see* **Note 9**). An example Trinity command and parameters used are as follows:

```
Trinity --max_memory 36G --jaccard_clip --seqType fq--
normalize_reads --run_as_paired --CPU 8 --bflyCalculateCPU
--min_per_id_same_path 95 --full_cleanup --single
filtered.fastq.gz
```

Parameters: --max_memory, maximum memory to use. --jaccard_clip, reads are paired and expect high gene density with UTR overlap. --seqType, sequence type. --normalize_reads, run in silico normalization of reads. --run_as_paired, input sequence is paired. --CPU, number of CPUs to use. --bflyCalculateCPU, calculate CPUs for Butterfly based on 80% of max_memory. --min_per_id_same_path, minimum percent identity for two paths to be merged into single paths. --full_cleanup, only retain Trinity fasta file.

3.3 Analysis of Differential Gene Expression (Counting)

Abundance of reads generated from RNA samples can provide a direct readout of the levels of expression of transcripts. This can be leveraged to identify differential expression of genes under experimental conditions. Below is a protocol based on mapping Illumina reads using HISAT and determining differentially expressed genes using DESeq2.

1. Filter read following **steps 1–3** listed in Subheading 3.2.
2. Reformat the filtered fastq file to split the interleaved files into read1 and read2. The reformat.sh script is part of the bbtools package. An example command and parameters used are described as follows:

   ```
   reformat.sh in=filtered.fastq.gz out1=read1.fq.gz
   out2=read2.fq.gz
   ```

   Parameters: `in`, input fastq file. `out1`, output read1 of the interleaved fastq file. `out2`, output read2 of the interleaved fastq file.
3. Build a hisat index of the genome reference used for mapping. An example index command and parameters used are as follows:

   ```
   hisat-build genome_reference.fasta bt2_base
   ```

   Parameters: *bt2_base*, the basename of the index files to write.
4. Map preprocessed reads to the genome reference using HISAT version 0.1.4-beta [8]. Subsequently pipe the result to samtools [9] to create sorted BAM files. An example mapping command and parameters used are as follows:

   ```
   hisat -p 8 -k 1 -x bt2_base -1 read1.fq.gz -2 read2.
   fq.gz | samtools view -bS - | samtools sort -
   mapped_hits
   ```

   Parameters: `-p`, number of alignment threads. `-k`, number of alignments per reads. `-x`, index file basename. `-1`, file with #1 mates of a paired input. `-2`, file with #2 mates of a paired input. `samtools view`, SAM<->BAM conversion. `-bS`, output BAM, input is SAM. `Samtools sort`, sort alignment file. `mapped_hits`, name of output bam.
5. Extract counts from the alignment bam file using featureCounts [10]. An annotation file in the gff3 format is needed. An example command and parameters used are described as follows:

   ```
   featureCounts -a annotation.gff3 -s 2 -p --primary
   -T 8 -t gene -g ID -o counts.txt mapped_hits.bam
   ```

   Parameters: `-a`, name of the annotation file. `-s`, perform strand-specific counting (2, reverse strand). `-p`, counts as fragments. `--primary`, count only primary alignments. `-T`, number of threads. `-t`, specify feature type in the gff3 file. `-g`, specify attribute type in the gff3 file. `-o`, name of the output file.

Table 1
Example count matrix file

Gene ID	Condition 1 Replicate 1	Condition 1 Replicate 2	Condition 1 Replicate 3	Condition 2 Replicate 1	Condition 2 Replicate 2	Condition 2 Replicate 3
G001	200	195	199	186	174	179
G002	1201	1230	1194	6545	6602	6599
G003	107	112	114	118	109	111
G004	0	0	0	245	226	233

Read counts from each of the replicates are listed as individual columns in the matrix file

6. Follow above **steps 1–5** to create counts file for each of the replicates across all condition samples to be analyzed (*see* **Note 10**). Create a count matrix file as show in Table 1 with rows representing genes and each column representing a sample replicate with its raw read counts.
7. Provide the count matrix file as input to DESeq2 version 1.10.0 [11] and generate a results table (DEG_results.txt) with fold-change and *p*-value for each gene between the conditions tested (*see* **Note 11**). All the following commands should be executed within the R environment or an R script should be created for execution.

```
countData <- read.table('count_matrix.txt',
sep="\t", comment.char="", header=TRUE, row.names=1,
stringsAsFactors=FALSE)
groups <- as.numeric(unlist(strsplit
("1,1,1,2,2,2", ",")))
colData <- data.frame(row.names=colnames(countData),
condition=as.factor(groups))
dds <- DESeqDataSetFromMatrix(countData=
countData, colData=colData, design=~condition)
dds <- DESeq(dds, betaPrior=FALSE, quiet=TRUE,
parallel=TRUE)
res <- as.data.frame(results(dds, contrast=c("condi
tion","1","2")))
write.table(res, file='DEG_results.txt',
sep="\t", row.names=TRUE, quote=FALSE)
```

4 Notes

1. Many of the computational tools are in constant development and newer versions with bug fixes maybe available. You should always check for the latest versions and use appropriate parameters.

2. It is important to mix the sample gently at these steps to keep the mRNA intact.
3. The incubation time and temperature are critical for RNA fragmentation. If you want to produce shorter libraries for 1×75 run, you can increase the incubation time to 10 min.
4. Use Bioanalyzer (Agilent), Fragment Analyze (Advanced Analytical), or TapeStation (Agilent) to verify rRNA contamination. If you still see high peaks of rRNA, you may want to repeat the procedure of RiboZero depletion or try mRNA isolation instead.
5. Bead purification steps:
   (a) Add *x* μL of well-mixed AMPure XP beads to the sample. Mix well.
   (b) Incubate at RT for 5 min.
   (c) Place the sample on a magnetic stand until liquid is clear.
   (d) Remove the supernatant.
   (e) Keep the sample on the magnetic stand. Add 200 μL of fresh 75% EtoH and pipette gently up and down 5–6 times.
   (f) Remove 75% EtOH and repeat the wash one more time.
   (g) Add *x* μL of Resuspension Buffer to the sample and mix well.
   (h) Incubate at RT for 5 min. Quick vortex and spin.
   (i) Place the sample on a magnetic stand.
   (j) Transfer *X* μL of the supernatant to a new tube.
6. PCR cycles can be increased if your total RNA is less than 1 μg. You want to keep the PCR cycles as low as possible to maintain minimum PCR bias and high data complexity. Some people increase the PCR cycles to 15 when the input RNA is very little, but data complexity will decrease and PCR bias will increase drastically.
7. BBtools is a collection of software that can do a variety of sequence processing including filtering/trimming. Detailed help/usage guides for tools within bbtools including bbduk can be obtained by using the –help parameter. For example, you can obtain the parameters for bbduk using the following command: `bbduk.sh -help`
8. Trinity is designed with many components that can be run in parallel on large distributed computing networks. Parallel jobs can be run on computing grids. Detailed help/usage guides for trinity can be obtained by executing the following command:

```
Trinity --show_full_usage_info
```

9. Quality of transcriptome assemblies can be assessed using various tools. Assembled contigs can be mapped to respective genomes/known protein databases to assess full-length reconstruction of transcripts. Tools like BUSCO [12] can also be used to assess the quality of transcript assembly.
10. A minimum of three replicates per condition is recommended for obtaining statistically significant differential expression analysis results.
11. The results file DEG_results.txt has the base mean, $\log_2$ fold change, *p*-value and adjusted *p*-value between conditions 1 and 2 for each of the genes. Thresholds for fold-change and adjusted *p*-value can be established to determine significance of differential expression. Refer to the DESeq2 manual on various visualization options including heat map representations of expression profiles.

Acknowledgments

The work conducted by the US Department of Energy Joint Genome Institute, a DOE Office of Science User Facility, is supported under Contract No. DE-AC02-05CH11231.

References

1. Wang Z, Gerstein M, Snyder M (2009) RNA-Seq: a revolutionary tool for transcriptomics. Nat Rev Genet 10:57–63. https://doi.org/10.1038/nrg2484
2. Wilhelm BT, Marguerat S, Watt S et al (2008) Dynamic repertoire of a eukaryotic transcriptome surveyed at single-nucleotide resolution. Nature 453:1239–1243
3. Nagalakshmi U, Wang Z, Waern K et al (2008) The transcriptional landscape of the yeast genome defined by RNA sequencing. Science 320:1344–1349
4. Grigoriev IV, Nikitin R, Haridas S et al (2013) MycoCosm portal: gearing up for 1000 fungal genomes. Nucleic Acids Res 42(D1): D699–D704
5. Mortazavi A, Williams BA, McCue K et al (2008) Mapping and quantifying mammalian transcriptomes by RNA-Seq. Nat Methods 5:621–628
6. Illumina, Inc. https://support.illumina.com/sequencing/sequencing_kits/truseq_stranded_mrna_ht_sample_prep_kit/documentation.html. Accessed 22 Aug 2017
7. Grabherr MG, Haas BJ, Yassour M et al (2011) Full-length transcriptome assembly from RNA-seq data without a reference genome. Nat Biotechnol 29(7):644–652. https://doi.org/10.1038/nbt.1883
8. Kim D, Langmead B, Salzberg SL (2015) HISAT: a fast spliced aligner with low memory requirements. Nat Methods 12:357–360. https://doi.org/10.1038/nmeth.3317
9. Li H, Handsaker B, Wysoker A et al (2009) The Sequence alignment/map (SAM) format and SAMtools. Bioinformatics 25:2078–2079. https://doi.org/10.1093/bioinformatics/btp352
10. Liao Y, Smyth GK, Shi W (2014) featureCounts: an efficient general purpose program for assigning sequence reads to genomic features. Bioinformatics 30(7):923–930
11. Love MI, Huber W, Anders S (2014) Moderated estimation of fold change and dispersion for RNA-seq data with DESeq2. Genome Biol 15:550
12. Simão FA, Waterhouse RM, Ioannidis P et al (2015) BUSCO: assessing genome assembly and annotation completeness with single-copy orthologs. Bioinformatics 31(19):3210–3212. https://doi.org/10.1093/bioinformatics/btv351

Chapter 9

Mass Spectrometry-Based Proteomics

Marcos Rafael Di Falco

Abstract

Proteomics is the large-scale analysis of proteins rendered possible by modern mass spectrometry analysis methods capable of identifying thousands of peptides/proteins in a fast high-throughput manner. Here I describe protocols for the preparation of fungal culture protein samples for mass spectrometry-based proteomics analysis including protein sample cleanup, proteolytic digestion, LC-MS/MS separation, and database search protein identification.

Key words In-gel digestion, In-solution digestion, LC-MS/MS, Database search, Fungal proteome, Microcapillary column packing

1 Introduction

Technological and computational advances in the era of whole-genome sequencing are continuously contributing to increase the number of assembled and structurally annotated fungal genomes. As an example, while less than 50 fungal genomes were sequenced a decade ago, a Uniprot query of publicly available fungal proteomes in 2016 returned 686 entries [1]. Such unprecedented amount of protein sequence database information has bolstered the use of mass spectrometry-based proteomics as a tool to analyze biological systems exhaustively.

Thanks to refinements in protein preparation and separation methodologies as well as improvements in mass spectrometry instruments that can acquire tens of high-resolution spectra per second with sub-ppm mass accuracy, mass spectrometry-based proteomics analyses are consistently providing results with extensive proteome coverages. These technical advances are playing a prominent role in fungal biology studies. Results from proteomic studies used in characterizing changes in secreted fungal proteins in cultures supplemented with various lignocellulosic substrates are paramount to the identification of enzymes involved in biomass degradation [2, 3]. The combination of targeted peptide/protein

Ronald P. de Vries et al. (eds.), *Fungal Genomics: Methods and Protocols*, Methods in Molecular Biology, vol. 1775,
https://doi.org/10.1007/978-1-4939-7804-5_9,

enrichment techniques and mass spectrometry analysis can be used to characterize specific subsets of fungal proteins such as glycoproteins or phosphoproteins [4, 5]. Quantitative proteomics analysis of production patterns in protein–protein interaction networks of fungi in response to culture challenges can reveal changes in proteins responsible for secondary metabolite production [6]. Proteomics experiments that yield extensive coverage of fungal proteomes are being used as a "proteogenomic" approach for the validation and refinement of computationally predicted gene models [7]. The success of such proteomics experiments rely on the use of liquid-chromatography tandem mass spectrometry (LC-MS/MS) as a powerful technique that allows the analysis and identification of thousands of proteins in complex mixtures and is the cornerstone of modern-day proteomics experiments. During a typical proteomics analysis workflow, proteins are digested with an endoprotease (usually trypsin), the resulting peptides are analyzed by LC-MS/MS and peptide sequences are assigned to MS/MS fragmentation spectra using a protein database computational search program. Finally, the proteins that make up the sample are inferred from the identified peptides (*see* Fig. 1). This approach to identification of proteins in a mixture is known as bottom-up shotgun proteomics [8].

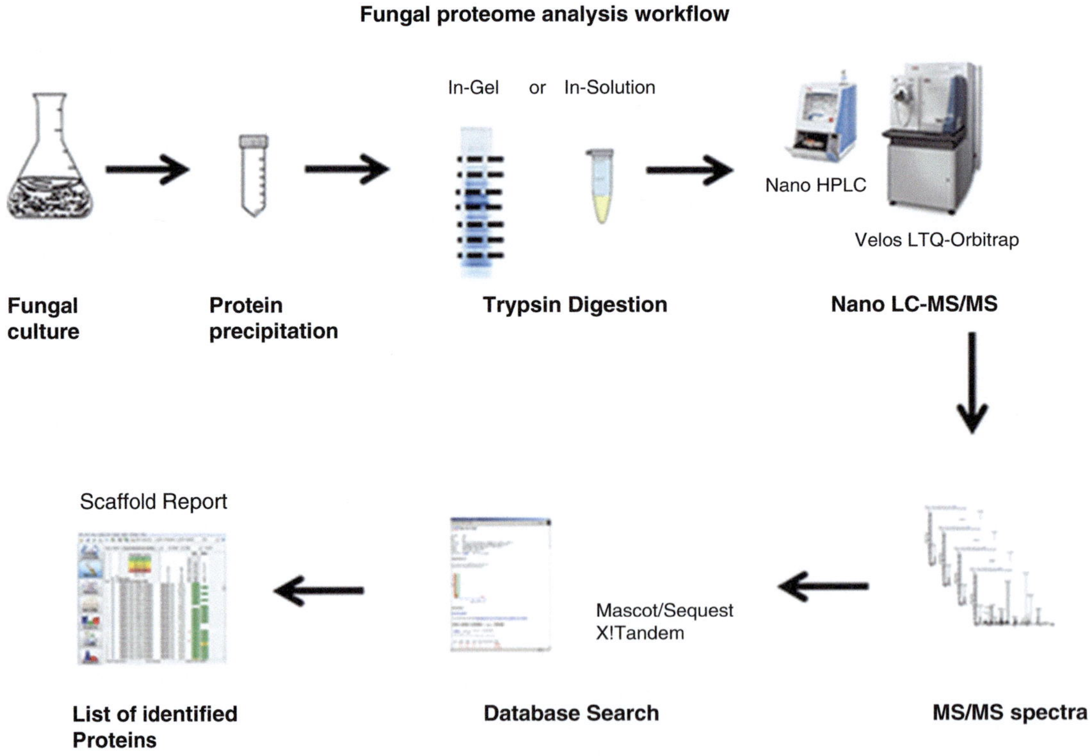

Fig. 1 Typical proteomics workflow used in the large scale analysis of fungal proteins

An in-gel digestion workflow provides an efficient way to fractionate and reduce protein sample complexity prior to LC-MS/MS analysis and can be used to identify thousands of proteins. Samples are separated by SDS-PAGE and whole gel lanes cut into slices that are individually processed for protein identification. This approach is time-consuming since it requires digestion and analysis of dozens of gel samples but it is capable of providing extensive identification of proteins even if present in low abundance (attomoles of material loaded on gel). Alternatively, one may opt for an in-solution proteomic workflow which involves the digestion of proteins in a liquid sample followed by the direct characterization of the resulting peptides by LC-MS/MS. This is a more direct and faster process best used with samples of lower complexity (up to a few hundred proteins). Factors such as the level of sample complexity, amount of protein material, and available resources in terms of personnel and instrumentation have to be considered when choosing a proteomics workflow.

This chapter provides protocols and experimental suggestions for carrying proteomics experiments using LC-MS/MS protein identification and includes methods for protein sample cleanup and concentration, protein determination and proteolytic trypsin digestion as well as LC-MS/MS setup and database searching protein identification guidelines.

2 Materials

All solutions should be prepared using LC-MS grade solvents including: water, acetone, methanol, and acetonitrile. Great care should be taken to avoid the contamination of samples with keratins. These protein contaminants can greatly mask the signal of sample protein components of interest. Ideally, all chemicals used should be of the highest purity available and preferably stored in a dust-free lab section reserved for the preparation of proteomics samples. Powder-free gloves, arm sleeves, hair net, goggles, mask and clean lab coat should be worn when handling chemicals and samples and preferably carry out all digestion steps under a laminar flow hood.

2.1 TCA-Protein Precipitation

1. 100% (w/v) trichloroacetic acid (TCA). Add 227 mL of water to 500 g of TCA.
2. Acetone.
3. 200 mM ammonium bicarbonate (NH_4HCO_3) pH 8.2 in water. Dissolve 0.395 g of NH_4HCO_3 in 20 mL of water, adjust the pH to 8.2 and add water to a final volume of 25 mL.
4. 0.1% (w/v) anionic acid labile surfactant (AALS) dissolved in 200 mM NH_4HCO_3, pH 8.2.

2.2 Endoprotease (Trypsin) Protein Digestion

1. Acetonitrile.
2. 100 mM dithiothreitol (DTT) in 100 mM NH_4HCO_3, pH 8.2.
3. 500 mM iodoacetamide in 100 mM NH_4HCO_3, pH 8.2.
4. 0.1% (w/v) AALS dissolved in 100 mM NH_4HCO_3, pH 8.2.
5. Mass spectrometry grade trypsin aliquot solubilized at 100 ng/μL in 50 mM acetic acid.
6. 20% formic acid solution in water.
7. Heating block or water bath set at 55 and 37 °C.
8. Siliconized low-retention 0.5 mL microcentrifuge tubes.
9. C18 solid phase extraction (SPE) pipette tip cartridges.

2.3 Microcapillary Column Packing

1. Bulk C18 reversed-phase 5 μm resin.
2. 90% methanol in water.
3. 360 μm outer diameter/75 μm internal diameter fritted tapered fused-silica microcapillary tubing.
4. Pneumatic microcapillary column packing apparatus (Pressure bomb).

3 Methods

Mass spectrometry analysis of peptides is acutely sensitive to interferences from salts and polymeric detergents. These contaminants can reduce ionization efficiency, overwhelm a spectrum and mask ions of interest, and, importantly, negatively affect the integrity and efficiency of the chromatographic columns used for peptide separation. The TCA-protein precipitation method is an effective way to concentrate protein from culture media supernatants while ridding the samples of interfering substances and is amenable to subsequent in-solution digestion. However, when processing samples that contain even trace amounts of ionic detergents (i.e., sodium dodecyl sulfate, deoxycholate, sarkosyl) or other commonly used surfactants (Triton, Tween, NP-40) present at concentrations higher than 0.1%, an in-gel digestion approach should be considered after TCA-precipitation for the subsequent preparation of samples for LC-MS/MS analysis. It is necessary to accurately determine the concentration of total protein in all samples so as to adjust the ratio of trypsin to sample protein during digestion but more importantly to determine proper and meaningful differential expression values when carrying out comparative proteomics analyses. Protease digestion of proteins generates a mixture containing hundreds to many thousands of peptides. In order to maximize the number of peptide identifications it is necessary to separate or fractionate this complex peptide mixture using either single or multiple dimension high pressure liquid chromatography (HPLC) separation. Ideally, since mass spectrometry detection is dependent on analyte concentration, HPLC systems that deliver very small volumes (e.g., nano-flow) of mobile phase (50–300 nL/min) provide the highest analytical sensitivity and are thus optimally suited for LC-MS/MS proteomics experiments.

3.1 TCA-Protein Precipitation

1. Perform a sample centrifugation at 3200 × *g* for 10 min at 4 °C in order to clear the sample of any particulate material (*see* **Note 1**).
2. Mix one volume of ice cold 100% (w/v) TCA to four volumes of sample supernatant, mix and incubate for 10 min at 4 °C.
3. Pellet precipitated protein by centrifugation at 16,000 × *g* for 10 min at 4 °C. Remove supernatant by decanting as soon as the centrifugation is complete.
4. Break up the protein pellet by adding 1 mL chilled 100% acetone (−20 °C) and pipetting up and down (*see* **Note 2**).
5. Repeat **steps 3** and **4** at least twice.
6. Allow the pellet to air-dry for 5 min in a fume hood (*see* **Note 3**).
7. Resolubilize the protein material in 0.1% AALS II, 200 mM NH_4HCO_3 pH 8.5 (50–200 μL) (*see* **Note 4**).

3.2 Excision of Gel Bands or Spots (2D-Gels) for In-Gel Digestion

1. Wash gel in 1% acetic acid solution.
2. Excise bands using a clean scalpel. Gel bands should not have an area larger than 50 mm^2 per digestion reaction.
3. Cut gel bands into cubes no larger than 1 mm^3.
4. Transfer gel pieces into a 0.5 mL microfuge tube containing enough 1% acetic acid to cover the gel pieces. Alternatively, a polypropylene conical-bottom 96-well plate can be used for processing large numbers of gel bands or spots (*see* **Note 5**).

3.3 Gel Washing (Coomassie Destaining), and Cysteine Reduction and Alkylation

1. Remove the 1% acetic acid solution from tubes containing the gel pieces.
2. Dispense enough 100 mM NH_4HCO_3 solution to cover the gel pieces and incubate for 10 min (all incubations can be performed at room temperature except for **steps** 7 and **8**).
3. Dispense a volume of acetonitrile equivalent to the volume of 100 mM NH_4HCO_3 added in **step 2** and incubate for 5 min.
4. Remove all liquid.
5. Repeat **steps 2–4**.
6. After removing all liquid from the tubes, add a volume of acetonitrile that is at least twice the volume of the gel pieces and incubate for 5 min. The gel pieces will shrink, become white and clump together.

7. Remove all liquid and add 10 mM DTT solution in 100 mM NH_4HCO_3 (enough to cover the gel pieces) and incubate for 30 min at 37 °C.
8. Add a volume of 50 mM iodoacetamide in 100 mM NH_4HCO_3 that is equivalent to the volume used in **step 7** and incubate for 20 min at 37 °C.
9. Dispense a volume of acetonitrile equal to double of the volume added in **step 8** and incubate for 5 min.
10. Remove all liquid and repeat **steps 2–4**.
11. Shrink gel pieces as in **step 6**.
12. Remove all liquid and let gel pieces air-dry for 5 min before proceeding with addition of trypsin digestion solution.

3.4 In-Gel Trypsin Digestion

1. Rehydrate gel pieces in trypsin digestion solution containing 50 mM NH_4HCO_3 and 6 ng/μL of trypsin (keep the trypsin digestion solution at 4 °C until it is added to the gel pieces). The volume of solution to be added should be equivalent to the volume of the hydrated gel pieces.
2. Let the gel pieces swell up for 15 min and add enough liquid of 50 mM NH_4HCO_3 solution to cover the gel pieces.
3. Incubate the gel pieces at 37 °C for 4 h. (It may be more practical, timewise, to carry out this incubation step overnight.) (*see* **Notes 6** and **7**).

3.5 Peptide Extraction

1. Add 15–25 μL of extraction buffer solution consisting of 2% acetonitrile and 1% formic acid in water to each tube and incubate for 15 min. (All incubations are done at room temperature.)
2. Collect the liquid and place it in a siliconized 0.5 mL microfuge tube. (These low retention tubes help minimize loss of peptide material.)
3. Add 1 volume of extraction buffer followed by 1 volume of acetonitrile and incubate for 20 min. The total amount of liquid should be enough to completely cover the gel pieces.
4. Collect the supernatant and combine it to the supernatant collected in **step 2**.
5. Add 1 volume of extraction buffer then 3 volumes of acetonitrile such that the combined liquid completely immerses the gel pieces. Incubate for 20 min.
6. Collect the supernatant and combine it with the liquid collected in **steps 2** and **4**.
7. Dry down the pooled extracts completely using a vacuum centrifuge. This step ensures the removal of the volatile NH_4HCO_3

salt and reduces the organic solvent content that would otherwise reduce peptide binding efficiency during the HPLC separation.

8. Resolubilize the dried peptide extracts with a solution of 5% acetonitrile and 0.1% formic acid in water by vortexing vigorously for at least 5 min.

3.6 In-Solution Digestion

1. In a 0.5 mL siliconized microfuge tube, add 25 μL 100 mM NH_4HCO_3, 5 μL 0.1% AALS II solution in 100 mM NH_4HCO_3, 5 μL of 50 mM DTT, and 10 μL of protein sample solution (~5 μg of total protein is acceptable) and incubate at 55 °C for 20 min.
2. Add 5 μL of 250 mM iodoacetamide in 100 mM NH_4HCO_3 solution and incubate at 55 °C for 20 min in the dark.
3. Add 5 μL trypsin digestion solution consisting of 100 mM NH_4HCO_3, and 40 ng/μL of trypsin (keep the trypsin digestion solution at 4 °C until it is added to the gel pieces). Incubate for a minimum of 4 h at 37 °C.
4. Add a required volume of 20% formic acid to reach a final concentration of 1% in the final digestion solution volume and incubate for at least 30 min at room temperature before proceeding to the sample desalting step. This acidification step stops the digestion reaction and breaks down the detergent used as a denaturant (*see* **Notes 8** and **9**).

3.7 In-Solution Digest Desalting

1. Equilibrate the C18 SPE pipette tip cartridge by first aspirating and discarding a volume of acetonitrile (typically 10 μL) followed by drawing and discarding a volume of 0.1% formic acid.
2. Bind the digest peptides to the C18 resin by pipetting 10 μL of digestion solution up and down into the same sample tube at least ten times.
3. Wash the resin bound peptides by first drawing and then discarding 10 μL of 0.1% formic acid. Repeat this at least five times.
4. Elute the peptides by drawing 5–10 μL of 80% acetonitrile, 0.1% formic acid then discarding into a new siliconized 0.5 mL microfuge tube.
5. Since the binding capacity of these C18 tips is between 5 and 10 μg of peptide material, **steps 1–4** may be repeated to ensure complete recovery of digest material.
6. Dilute the eluate solution with 0.1% formic acid solution to ensure that the final acetonitrile concentration is at 5% v/v. Alternatively, samples can be vacuum dried and resolubilized using a solution of 5% acetonitrile, 0.1% formic acid.

3.8 Microcapillary Column Packing

1. Prepare a slurry solution of C18 resin by placing about 20 μL of dry weight resin in 1 mL of 90% methanol inside a 2 mL glass vial along with a micro stirring bar.
2. Load an appropriate amount of reversed-phase C18 resin onto a fritted, pulled tip, fused silica microcapillary using a pressure bomb apparatus (*see* **Note 10**). The sample loading capacity of 10 cm × 75 μm internal diameter capillary column packed with C18 resin is about 500 ng of total digested protein and should be sufficient to detect low femtomole amount of peptide.

3.9 (Nano) LC-MS/MS Analysis

1. Install the packed microcapillary column connected to a nano-HPLC system onto the mass spectrometer nano-electrospray source.
2. Load an appropriate volume of sample onto the column using the autosampler module of the nano-HPLC system. For injection volumes between 1 and 5 μL, the sample can be delivered directly to the microcapillary column (direct injection). For larger injection volumes (5–20 μL), the sample can be injected onto a larger bore trap column (i.e., 5 × 0.3 mm) at higher flow rates (10–40 μL/min) by means of a multiport valve connection to allow for faster loading times and online sample desalting/washing.
3. Start the peptide separation and LC-MS/MS acquisition method using an HPLC method of gradient composition and length appropriate for the level of sample complexity (*see* **Note 11**).

3.10 LC-MS/MS Data Acquisition and Analysis

As peptides are chromatographically separated and eluted from the microcapillary column they become protonated during the electrospray ionization process and are introduced into the mass spectrometer. A typical acquisition method for LC-MS/MS based bottom-up proteomics mass spectrometry experiment includes a set of criteria that automatically drives the selection of ions for subsequent tandem mass spectrometry analysis. In the first instance, the mass spectrometer performs an MS scan across a set m/z range (typically 350–1800) then a number of ions (between 6 and 10) that surpass a certain intensity threshold and have a specified charge state (between 2 and 4) are selected individually for successive MS/MS analysis. In order to maximize ion sampling efficiency, the previously selected m/z values are excluded from reselection until a specified amount of time has elapsed. This selection process is repeated after every full range MS scan thus resulting in the acquisition of thousands of MS/MS spectra.

3.11 Peptide and Protein Identification

Spectra acquired following collisional induced dissociation (fragmentation) of tryptic peptides are predominantly populated by y" and to a lesser extent b" type ions. It is possible to visually determine the amino acid sequence of peptides by measuring the mass differences between fragment ions. This process is time consuming and complicated by the presence of multiple fragment ion types including those generated by side chain fragmentation, losses of water and/or ammonia molecules and ions with multiple charge states. Fortunately, various bioinformatics tools are available to facilitate the task of matching an MS/MS spectrum to an amino acid sequence. These bioinformatics programs consist of database search algorithms that use a probabilistic approach to assign the likeliest match between a fragmentation spectrum and a peptide sequence derived from a database of in silico protease-specific digested proteins.

The analysis of samples by tandem mass spectrometry generates raw spectrum data files that are converted into *m/z* peak lists which are subsequently used for protein database searching. Most mass spectrometer instrument vendors provide proprietary software packages designed to optimize the process of converting raw spectral signals into peak lists of meaningful monoisotopic *m/z* values along with their respective signal intensities and charge states. Some of these raw data treatment processes involve the discrimination of true peaks from background noise and the resolution and disentanglement of isotope clusters. It is crucial that the *m/z* peak list data be of the highest quality as it can greatly impact the level of confidence associated with the peptide identification analysis. The commercial computational tools with the longest history and most popularity in the field of proteomics for both the conversion of MS/MS raw spectral data into peak lists and database searching to identify peptides and subsequently proteins are Mascot [9] and SEQUEST [10]. Other proteomics commercial software packages include Protein Pilot, MassLynx, and SpectrumMill. These software tools require that the protein fragments selected for MS/MS analysis have a representative sequence in the database used for peptide matching. PEAKS [11] is another powerful commercial software package that has the advantage of carrying out de novo sequencing and thus provides the ability to identify peptides (and proteins) whose sequences may not have been contained in the database used for the original search. Alternative, open source tools are available for database searching such as X!Tandem [12], MaxQuant [13], and OMSSA [14].

There are a number of variables that can influence the success or failure of protein identification and determine the final list and/or coverage of proteins reported by a database search engine. These variables include the size of the protein database, the search algorithm used for protein identification, the method and specificity of digestion method (chemical or enzymatic), the completeness of protein digestion, whether reduction and alkylation of sample was used, the presence and number of posttranslational modifications queried, and instrument-specific parameters such as mass accuracy and resolution.

Once database searching is completed and a list of identified proteins is reported, careful downstream analysis of the results must take place. This time-consuming process requires extensive manual interpretation and validation which is usually the bottleneck of a proteomics study. For example, contamination of samples with keratin is a common problem in proteomic analysis. It is advisable to include a list of common protein contaminants to the protein sequence database used for MS/MS characterization so that these ubiquitous proteins can be detected and manually screened out prior to data interpretation. A database of such contaminants, designated as common Repository of Adventitious Proteins (cRAP), has been assembled by the global proteome machine group and is accessible online at ftp://ftp.thegpm.org/fasta/cRAP.

In order to promote the dissemination of high quality proteomics data, reporting guidelines established by the HUPO proteomics standards initiative have been adopted as the minimal requirements for publication by many journals [15]. For example, only proteins that have been identified with a minimum number of two unique peptide sequences should be reported. For those proteins that have been identified using the minimum required number of peptides, the quality of the MS/MS spectral fragment matches should be visually validated.

The ultimate goal of a proteomics experiment goes beyond characterizing the complement of proteins being produced in a cell, tissue, or organism; it aims at comparing changes in protein production across varying experimental conditions. A review by Domon and Aebersold [16] provides an overview of current quantitative proteomics methods along with their advantages and disadvantages. How closely these results approximate the true changes in a biological system is rigorously dependent on a number of factors including: the quality and reproducibility of biological samples, data quality and acquisition reproducibility, inclusion of proper internal quality control and reference standards, and the number of replicate experiments required to obtain statistically significant results [17]. There are a number of commercial software packages, such as Scaffold, Progenesis and Proteome Discoverer, that can be used to facilitate the processing of quantitative proteomics experiments as well as for reporting and visualizing changes in protein production. For an overview of some of the currently available proteomics data visualization tools the reader may consider reading a review article by Oveland [18].

4 Notes

1. When working with cell culture samples, the initial supernatant clearing step is carried out at lower speeds so as to prevent potential cell lysis and release of intracellular proteins into the supernatant.

2. The acetone resuspended pellet can be transferred to a 1.5 mL microcentrifuge tube after the first acetone wash.
3. Acetone is used to wash the protein pellet free of TCA while keeping proteins in insoluble form. It is advisable not to let the acetone washed pellet dry completely as this may prevent some proteins from being resolubilized.
4. The advantage of using an acid labile detergent for the solubilization of proteins is the ability to break down the detergent and rid the sample of a potential polymeric interfering component by simply acidifying the sample after the digestion step. NH_4HCO_3 is added to the solubilization solution to counteract the effects of any trace amount of TCA and keep the sample at a pH that is compatible with subsequent trypsin digestion conditions.
5. The amount of sample available and the composition of the sample preparation should be considered when choosing a digestion method. In-gel digestion usually leads to greater loss of sample material than in-solution digestion. This is because it involves a greater number of sample manipulation steps where sample losses can occur such as the peptide extraction step from the gel pieces after digestion. On the other hand, in-gel digestions generally produce peptide solutions that are considerably less contaminated by LC-MS interfering substances, such as salts and detergents, than in-solution digests.
6. Trypsin is the endoprotease of choice for the preparation of samples to be analyzed using bottom-up proteomic strategy. This enzyme has high cleavage specificity at arginine and lysine amino acids which results in the generation of a protonated C-terminal peptide end that promotes the generation of y" fragment ions during collisionally induced dissociation tandem mass spectrometry. If a particular protein of interest is not amenable to trypsin digestion, either because of the absence or large abundance of target amino acid cleavage sites, endoproteases with different cleavage specificities such as chymotrypsin, AspN or GluC may be considered.
7. Usually, a complete analysis of samples processed using an in-gel digestion involves SDS-PAGE separation followed by whole lane fractionation into multiple gel pieces that require individual LC-MS/MS injections. This process is much more time consuming and costly both in terms of data processing and analytical time in comparison to an in-solution digestion followed by single dimension HPLC separation. It is therefore advisable to consider both the complexity of the sample composition and the expectation in terms the number of protein identifications and coverage that are required for generating meaningful results. For example, when analyzing secreted proteins of fungal cultures grown under different substrates, it is

not expected to see changes in more than 50–150 proteins. An in-solution digest analysis using a 90 min HPLC gradient LC-MS/MS method would be sufficient to detect and provide differential expression of the majority of the relevant secreted proteins. On the other hand, if the goal is to validate a particular gene prediction model in a proteogenomics-type experiment by looking at thousands of predicted proteins, it would be advantageous to analyze both intracellularly and extracellularly produced proteins using a gel fractionation workflow that would yield more extensive proteome coverage than an in-solution approach.

8. Should an in-solution digest approach be chosen, it is highly recommended to run a small aliquot of protein sample preparation (~3 μg of total protein) on an SDS-PAGE gel and visualize the protein bands by silver staining in order to assess the quality of the protein sample preparation, the level of sample complexity and the range of protein expression. These parameters need to be evaluated since they can guide the choice of database search option, the HPLC gradient length and additional sample preparation steps. For example, the presence of smearing could indicate protein degradation or presence of highly glycosylated protein species. If considerable protein degradation is suspected, one may consider using a semitryptic enzyme database search option in order to increase peptide identification and consequently protein coverage. In extreme cases, it may be required to add protease inhibitors during the sample preparation (prior or up to the TCA-protein precipitation step) to help reduce protein degradation. Extensive protein glycosylation results in decreased protein coverage due to limited digestion site accessibility or modification of the expected mass of tryptic peptide sequences. In such cases, treatment with glycosidases that remove the oligosaccharide moieties attached to proteins may be considered.

9. It is particularly important to estimate the relative amounts of protein components when considering an in-solution digest sample preparation workflow. Since all proteins are digested together, those present in molar excess a greater than 20–30-fold than the lowest component can mask proteins of lower concentration. In such cases it is recommended to use an in-gel digestion approach where gel sections of greater intensity can be analyzed separately from the areas of low staining intensity.

10. A pressure bomb is a stainless steel chamber that is pressurized with an inert gas at ~1000 psi (*see* Fig. 2). A vial containing a slurry solution of C18 resin in either methanol or ACETONITRILE is contained within the chamber. A microcapillary is inserted from the top of the pressure cell through a ferruled nut until it nearly reaches the bottom of the

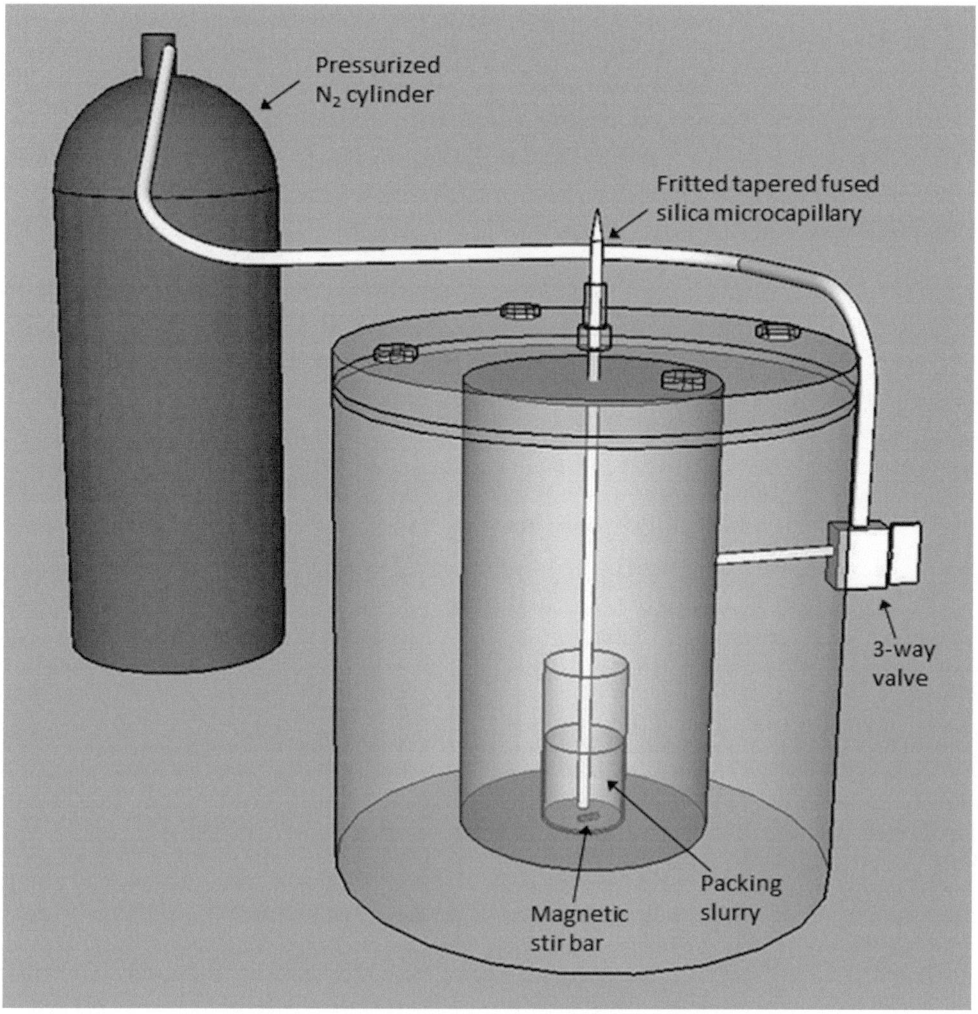

Fig. 2 Example of a pressure bomb setup used for packing microcapillary columns

vial. The nut is tightened to hold the capillary in place and when the cell is pressurized the C18 resin along with solvent is pushed through the capillary. While the solvent exits the microcapillary, the resin becomes packed since it is prevented from being expelled by the tapered or fritted end.

11. In order to maximize both MS/MS acquisition sampling efficiency and sensitivity an appropriate length of HPLC separation gradient should be selected to match the complexity of protein sample components. When choosing a gradient length it is important to keep in mind that longer gradients improve peptide mixture resolution by increasing peptide elution separation. Conversely, it is also also important to consider that peak intensity and therefore sensitivity decreases with increased gradient length. Therefore, in practice, one might choose a

short 10 min separation gradient when trying to identify a mixture of a dozen protein components that are barely detectable on Coomassie-stained gel but a longer 90 min gradient when analyzing a mixture of hundreds of proteins.

Acknowledgment

I am grateful for the critical review of the manuscript by Sylvester Palys and the technical drawing of Fig. 2 by Denis Legault. The author is supported by Genome Canada and Génome Québec.

References

1. The UniProt Consortium (2015) UniProt: a hub for protein information. Nucleic Acids Res 43:D204–D212
2. Kolbusz MA, Di Falco M et al (2014) Transcriptome and exoproteome analysis of utilization of plant-derived biomass by Myceliophthora thermophila. Fungal Genet Biol 29(72):10–20
3. Benoit I, Culleton H et al (2015) Closely related fungi employ diverse enzymatic strategies to degrade plant biomass. Biotechnol Biofuels 8:107. https://doi.org/10.1186/s13068-015-0285-0
4. Wang L, Aryal UK et al (2012) Mapping N-linked glycosylation sites in the secretome and whole cells of *Aspergillus niger* using hydrazide chemistry and mass spectrometry. J Proteome Res 11:143–156
5. Ren S, Yang M et al (2016) Global phosphoproteomic analysis reveals the involvement of phosphorylation in aflatoxins biosynthesis in the pathogenic fungus Aspergillus flavus. Sci Rep 6:34078. https://doi.org/10.1038/srep34078
6. Bai Y, Wang S et al (2015) Integrative analyses reveal transcriptome-proteome correlation in biological pathways and secondary metabolism clusters in *A. flavus* in response to temperature. Sci Rep 5:14582. https://doi.org/10.1038/srep14582
7. Bringans S, Hane JK et al (2009) Deep proteogenomics; high throughput gene validation by multidimensional liquid chromatography and mass spectrometry of proteins from the fungal wheat pathogen *Stagonospora nodorum*. BMC Bioinformatics 10:301. https://doi.org/10.1186/1471-2105-10-301
8. Wu CC, MacCoss MJ (2002) Shotgun proteomics: tools for the analysis of complex biological systems. Curr Opin Mol Ther 4(3):242–250
9. Perkins DN, Pappin DJ et al (1999) Probability-based protein identification by searching sequence databases using mass spectrometry data. Electrophoresis 20:3551–3567
10. Eng JK, McCormack AL, Yates JR (1994) An approach to correlate tandem mass spectral data of peptides with amino acid sequences in a protein database. J Am Soc Mass Spectrom 5:976–989
11. Ma B, Zhang K et al (2003) PEAKS: powerful software for peptide de novo sequencing by tandem mass spectrometry. Rapid Commun Mass Spectrom 17(20):2337–2342
12. Craig R, Bevis RC (2004) TANDEM: matching proteins with tandem mass spectra. Bioinformatics 20:1466–1467
13. Cox J, Mann M (2008) MaxQuant enables high peptide identification rates, individualized ppb-range mass accuracies and proteome-wide protein quantification. Nat Biotechnol 26:1367–1372
14. Geer LY, Markey SP et al (2004) Open mass spectrometry search algorithm. J Proteome Res 3(5):958–964
15. Taylor CF, Paton NW et al (2007) The minimum information about a proteomics experiment (MIAPE). Nat Biotechnol 25(8):887–893
16. Domon B, Aebersold R (2010) Options and considerations when selecting a quantitative proteomics strategy. Nat Biotechnol 28:710–721
17. Nilsson T, Mann M et al (2010) Mass spectrometry in high-throughput proteomics: ready for the big time. Nat Methods 7(9):681–685
18. Overland E, Muth T et al (2015) Viewing the proteome: how to visualize proteomics data? Proteomics 15:1341–1355. https://doi.org/10.1002/pmic.201400412

Chapter 10

Mass Spectrometry-Based Metabolomics

Young-Mo Kim and Heino M. Heyman

Abstract

Metabolomics based on mass spectrometry can provide quantitative and qualitative information of the pool of metabolites (metabolome) present intracellularly or extracellularly in a given biological system. A typical metabolomics workflow requires several key steps such as quick and robust sample preparations with quenching of metabolism, chemical derivatization if needed, instrumental measurement, data-processing with/without database information and further statistical analysis and interpretation. Here, we introduce general metabolomics workflows for global and targeted analyses using gas chromatography or liquid chromatography coupled with mass spectrometers.

Key words Mass spectrometry, Gas chromatography, Liquid chromatography, Data processing, Library construction, Metabolite extraction, Derivatization

1 Introduction

Metabolomics can be challenging with increasing biological complexity, but it is a rapidly developing bioanalytical suite of tools to measure all metabolites (metabolome) in a given biological or environmental system [1]. While typical analytics for systems biology like genomics, transcriptomics, and proteomics measurement rely on the information of sequences based on monomer constituents, the metabolomics requires individual measurement of target metabolite with elevated sensitivity and accuracy, also with better separation of complex mixtures of metabolites. NMR and mass spectrometry (MS) are considered as preferred analytical platforms, especially MS is regarded as a powerful tool when it is coupled with gas chromatography (GC) or liquid chromatography (LC) to separate inherent chemically diverse metabolite mixtures prior to the MS measurement [2]. Metabolomics can utilize many different types of MS analyzer such as single or triple quadrupole, ion trap, time of flight (TOF), Fourier-transformation ion cyclotron resonance (FT-ICR), and Orbitrap. New technologies for separation and measurement are being developed and improved to achieve

Ronald P. de Vries et al. (eds.), *Fungal Genomics: Methods and Protocols*, Methods in Molecular Biology, vol. 1775, https://doi.org/10.1007/978-1-4939-7804-5_10, © Springer Science+Business Media, LLC, part of Springer Nature 2018

better MS resolution, chromatographic separation with reduced time [3]. In addition, ion mobility mass spectrometry (IMS) has more recently been introduced as a technique that provides a new dimension of separation in a drift tube with reactive gas phase can be applied to metabolomics research [4].

The instrumental analysis is the most critical part of the metabolomics workflow and is required to be performed accurately. However, sample preparation is also important to capture the metabolome profiles at a given moment as a representation of the change in the metabolism. It must be emphasized that metabolome changes will inevitability occur even during fast sample processing, thus highlighting the importance of a rapid and consistent sampling procedure for metabolomics studies. The sampling process usually has two parts, metabolism quenching, and extraction of metabolites. A quick filtration method with a membrane filter can generate biomass samples for metabolomics with minimum exposure time from a culture suspension, while traditional biomass harvesting method using centrifugation is still used to get many samples [5]. The addition of aqueous cold solvents can help to minimize metabolome changes during the centrifugation. Spent medium samples can be used to obtain extracellular metabolome information from the fungal culture through filtration or centrifugation. It is recommended to use a chemically defined medium for extracellular metabolome analysis, rather than a complex and nutrient-rich medium such as yeast extract or peptone added. Metabolites are generally extracted using a definite solution which would contain a specific ratio of organic solvents and water (e.g., chloroform–methanol–water, 8:4:3) [6]. The extraction involves denaturing the proteins which is facilitated by the organic solvent and results in the metabolites leaking out, collected and dried down. Occasionally a physical disruption (e.g., ultrasonication, supercritical fluid extraction (SFE), microwave-assisted extraction (MAE), and pressurized solvent extraction (PSE)) of biomass (i.e., polymers, cellulose, hemicellulose, and lignin) might be needed to effectively extract the necessary metabolome from the biological sample. Dried metabolites can be reconstituted and injected to LC-MS directly or they can be analyzed by GC-MS with proper chemical derivatizations. A large number of metabolites are not volatile as the natural form, however, derivatization (chemically changing compounds that volatilize poorly and are thermally unstable) changes the metabolite into a more amenable form of the metabolites which then can be analyzed by GC-MS [7].

Data processing and interpretation are challenging, but a major component in the post-instrumental analyses. Several freely available or commercial software programs were developed and they have pros and cons based on the number of samples, the richness of metabolome, the major chemical category of metabolites in samples such as lipid, carbohydrates or amino acids [8, 9]. It is thus advisable that the available software or platforms are tested at the onset of any project. The quality of metabolomics data is highly dependent on an accurate and reliable database with wide coverage. In the current Omics era, significant emphasis is being placed on proper visualization and metabolic pathway mapping of the metabolomic data after thorough statistical analysis. This process is part of data interpretation and is best done together with other omics information from the same samples. Here, we introduce the general scheme and protocols for global/targeted metabolomics based on GC-MS and LC-MS analytical platform after extracting metabolites using MPLEx extraction [6]. These protocols provide brief guidelines for fungal metabolomics research.

2 Materials

2.1 Metabolism Quenching and Metabolite Extraction

Vacuum filtration apparatus connected with vacuum pump, membrane filter (nylon type with pore size 0.45 μm, centrifuge with refrigerator, solvent-leaching free microcentrifuge tube (0.7, 1.5, or 2 mL), liquid nitrogen, solvent (methanol, chloroform, acetonitrile, nanopure water), phosphate buffer (or bicarbonate buffer) with 100–150 mM, glass beads or tube-type grinder for pulverization.

2.2 Instrumentation

1. Gas chromatography–Mass spectrometry: Any type of benchtop GC-MS equipped with a capillary GC column. Nonpolar or low-polarity columns are required for analyzing chemically derivatized metabolome samples (e.g., 1% or 5% of phenyl polysiloxane with 99% or 95% of methylsiloxane). Ultrapure helium gas is needed to run metabolome samples, and its flow calibration is necessary to match between flow rate and corresponding GC inlet pressure. Low-resolution GC-MS (single quadrupole or ion trap analyzer) has typically scan range of 50–1000 m/z of mass spectrum with speed of 3–5 Hz and shows around 1000 mass resolving power at 200 m/z. And high-resolution GC-MS (TOF or Orbitrap, electric and magnetic sector MS) has 30,000 to 120,000 mass resolving power with faster scan speed, however, these instruments are substantially more expensive than low-resolution GC-MS. Derivatizing reagents and compound mixture for calculating retention index are required (MSTFA, BSTFA, methoxyamine, ethoxyamine, and other for derivatization; alkane mixture or fatty acid methyl ester mixture (FAME) for calculation of retention index). Internal standard (a compound that does not exist naturally or is 13C isotope-labeled) is preferred.
2. Liquid chromatography–Mass Spectrometry: Any type of LC-MS system equipped with reverse phase (C18) and/or

hydrophilic interaction phase (HILIC) column, eluent solvents: pure water, methanol, acetonitrile, formic acid (or acetic acid), electrospray ionization (ESI) source, MS analyzers: triple quadrupole type is preferred for MS/MS analysis for accurate identification and quantification (targeted metabolomics), but higher MS resolving power machines can be used for targeted and discovery metabolomics (Orbitrap, FT-ICR, ion trap, linear ion trap, TOF).

2.3 Data Analysis

1. Data processing software: Compound Discoverer (Thermo), Mass Profiler Professional (Agilent), MetaboScape (Bruker), Progenesis QI (Waters), AMDIS, Metabolite Detector, MZmine, and many others.
2. Commercially available or public database: MassBank, MassBank of North America (MoNA), NIST Mass Spectral Library, Wiley GC-MS library, Golm Database, Fiehn metabolomics database.
3. In-house database: A method for optimized separation and measurement can be used to build an in-house metabolomics database for both GC-MS and LC-MS. To get better metabolome coverage, a large number of chemical standards are needed, however, this type of database can provide accurate identification for library matching.
4. Statistical analysis: For thorough statistical analysis of GC-MS and LC-MS metabolomics data MetaboAnalyst, SIMCA, and R can be used.

3 Methods

3.1 Sample Preparation

3.1.1 Filtration Method

1. Measure the optical density of fungal culture or weight of biomass (*see* **Note 1**).
2. Prepare the cell suspension and filter through the membrane filter using vacuum filtration (Fig. 1).
3. Rinse the cells on the membrane with the same volume of phosphate or bicarbonate buffer prepared (100–150 mM).
4. Bend the membrane using forceps and transfer the filter into a microcentrifuge tube.
5. Put the tube into liquid nitrogen for flash freezing and keep the samples in −70 °C freezer until extraction.

3.1.2 Centrifugation Method

1. Measure the optical density of fungal culture or weight of biomass (*see* **Note 1**).
2. Add cold 40% methanol into the suspension and centrifuge the culture at 15,000 × *g* for 5 min at 4 °C.
3. Discard the supernatant and add phosphate buffer in the tube.

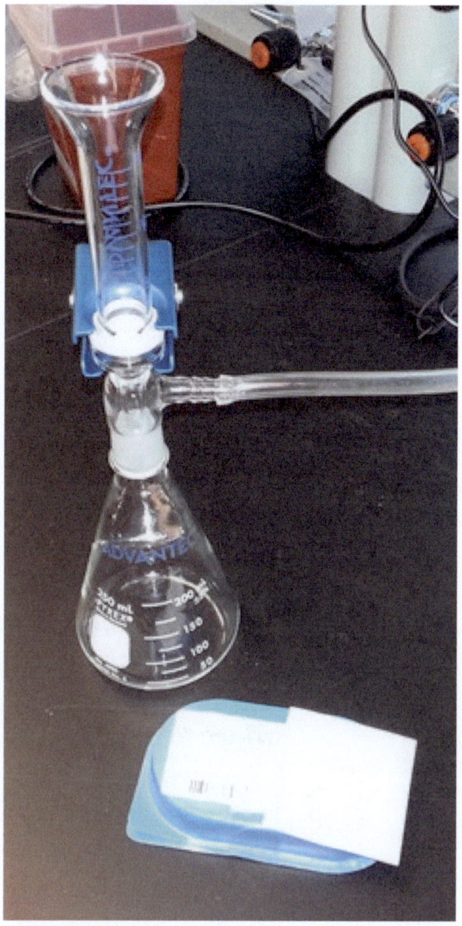

Fig. 1 Vacuum filtration apparatus. A membrane filter equipped with a vacuum filtration apparatus can harvest biomass quickly with minor metabolism changes during the sample handling

4. Vortex briefly and centrifuge again.
5. Discard the supernatant and freeze the samples in liquid nitrogen until metabolite extraction.

3.1.3 Metabolite Extraction

1. Prepare the chloroform–methanol extraction solvent (2:1), store it in a cold freezer (−20 °C) as well as filtered ultrananopure water.
2. (a) Keep the frozen samples on ice to avoid rapid thawing. With a cell scraper peel the biomass layer from the filter and put into a microcentrifuge tube if the samples were harvested by filtration method. If biomass is not thick, then use a syringe to take them down into a tube by spraying solvent mixture (chloroform–methanol). Vortex the tube vigorously for a 1 min. Put the tube on ice for 1 min and vortex again for 1 min.

(b) For the samples collected by centrifugation, directly add 100 μL of nanopure water and 400 μL of cold solvent mixture. Vortex the tube vigorously for a 1 min. Put the tube on ice for a minute and vortex again for 1 min.

3. After homogenization and extraction, centrifuge samples at 15,000 × *g*, for 5 min at 4 °C. Centrifugation will assist and improve the phase separation of the sample.
4. After the phase separation, transfer the top aqueous layer and bottom organic layer separately into clean vials. Evaporate the samples to dryness before continuing with pretreatment of the samples.

3.2 Pretreatment for MS Analysis

3.2.1 Derivatization for GC-MS

1. Weigh 30 mg of methoxyamine hydrochloride and dissolve it into 1 mL of pyridine (30 mg/mL).
2. Vortex the solution to make methoxyamine completely dissolved.
3. Prepare blank control samples and retention time standard (e.g., alkane mixture (C8–C30, every C2), or fatty acid methyl esters mixtures (FAMEs: C8–C28) in hexane.
4. Add 20 μL methoxyamine hydrochloride solution with a glass syringe to the dried extracted metabolites, blank, and retention time standard in glass vials, seal vials with cap (*see* **Note 2**).
5. Vortex briefly and incubate in thermomixer for 90 min at 37 °C with shaking (1200 rpm).
6. After incubation, briefly spin down the vials to collect all liquid part at the bottom.
7. Use a glass syringe to add 80 μL of MSTFA into the methoxyaminated samples, blank, and retention time standard. Close the caps and vortex briefly (*see* **Note 3**).
8. Incubate in the thermomixer for 30 min at 37 °C with shaking (1200 rpm).
9. Briefly spin down vials to move the liquid to the bottom.
10. Transfer the derivatized metabolite solution into a glass insert in a glass sample vial for GC-MS analysis.

3.2.2 Solvent Reconstitution for LC-MS

1. Reconstitute the dried metabolite extracts with a solvent solution that resembles the starting conditions of the LC-MS analysis [e.g., Reverse phase (C18)—80:20 water–acetonitrile; HILIC (ZIC-HILIC)—85:15 acetonitrile–water].
2. Vortex the sample vigorously and centrifuge to spin down any undissolved debris in the samples (samples can also be filtered using 0.45/0.22 μm PTFE syringe filters).
3. Transfer the clear liquid portion into a glass type sample vial (put a glass insert if needed).

3.3 Instrumental Analysis

3.3.1 GC-MS

1. Tune and Calibrate MS before analysis to make sure the machine reads MS data correctly and check helium gas pressure (*see* **Note 4**).
2. Set up the optimized running parameters in GC; Injector: temperature, pressure, flowrate; Oven: oven temperature, ramping rate, holding time, the temperature of MS transfer line, which was used to run standard compounds to build database (if in-house library used).
3. Set up the optimized running parameters in MS; mass scan range (50–600 *m/z*), Ion source temperature: 250°C, Ionization energy: 70 eV, which was used to run standard compounds to build database (if in-house library used).
4. Transfer samples to sample tray and place them in a randomized order to minimize instrumental artifacts.
5. For large samples sets requiring more than 24 h to be analyzed, batching will be required. Running batches will require additional randomization and QC samples and/or internal standards to account for any type of variation that could be introduced. Necessary blanks and retention time standards need to be included in each batch being run on subsequent days.
6. Run the blank and samples.

3.3.2 LC-MS

1. Check the machine status and set up the running parameters (composition and level of eluent solvents, pump condition, stability of flow and ESI spray tip, LC column connection) (*see* **Note 4**).
2. Queue samples on the sample rack of a temperature-controlled autosampler. The samples should be kept a constant temperature throughout the analysis.
3. The run order needs to be randomized to reduce any instrumentational artifacts.
4. Run the blank and samples with the optimized method which was used to build database.
5. LC-MS analysis using both RP and HILIC separation modes as well as both positive and negative ionization is recommended for optimal coverage of the metabolome.

3.4 Data-Processing

3.4.1 GC-MS Data Processing

1. Check all the data files were correctly obtained from the analysis (Fig. 2). If internal standard(s) was spiked, make sure their retention time and peak intensity values are consistent through the sample analysis.
2. Convert the vendor specific MS data format to general MS format if required.
3. Upload all the data files in a chosen software and do an alignment of retention time.

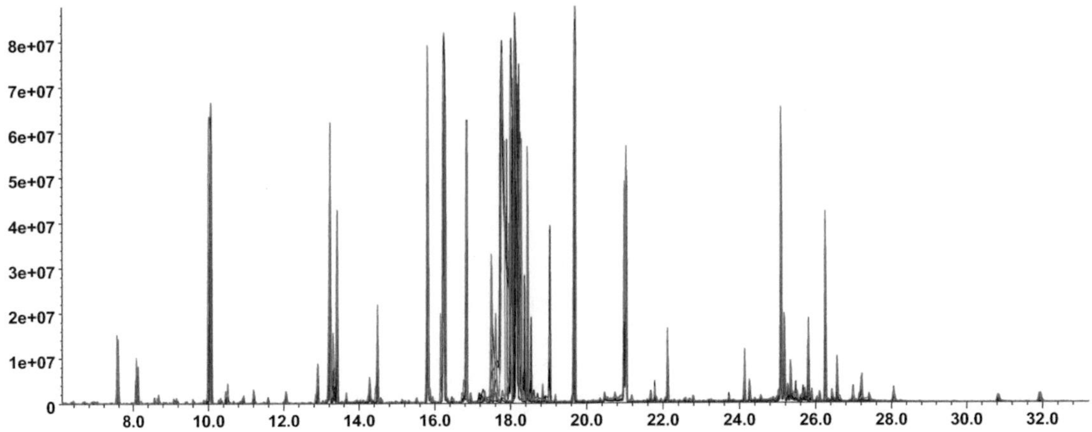

Fig. 2 GC-MS chromatogram. A typical GC-MS chromatogram was obtained from the analysis of intracellular metabolites extracted from glucose grown oleaginous yeast, *Yarrowia lipolytica*

4. Align the retention time information to all the data files, and do a metabolite identification process against an available library (if available) (*see* **Notes 5** and **6**).

3.4.2 LC-MS Data Processing

1. Check all the data files were correctly obtained from the analysis. If internal standard(s) was spiked, make sure their retention time and peak intensity values are consistent throughout the sample analysis.
2. Convert the MS data format of acquired sample runs to the target format which is required for data processing in selected software program (not required with vendor software).
3. Process the data (peak picking, retention time alignment, and deconvolution) to get retention time and intensity of peaks with unique m/z.
4. Transfer the data into spreadsheet to align the peak area information and cross-check.
5. Make identification based on the retention time and correct m/z values of peaks against an available library (*see* **Notes 5** and **6**).

3.5 Data Interpretation and Visualization

1. Choose what information will be included for data interpretation. Global metabolomics will include all the peaks detected from the MS analysis, or only identified metabolite peaks can be used for statistical analysis if necessary.
2. Remove all the information of peak which has similar or higher signals compared with the blank controls. Normalize data using the internal standard (if applicable), or z-score transformation is generally preferred for the comparison of metabolite signals between samples.

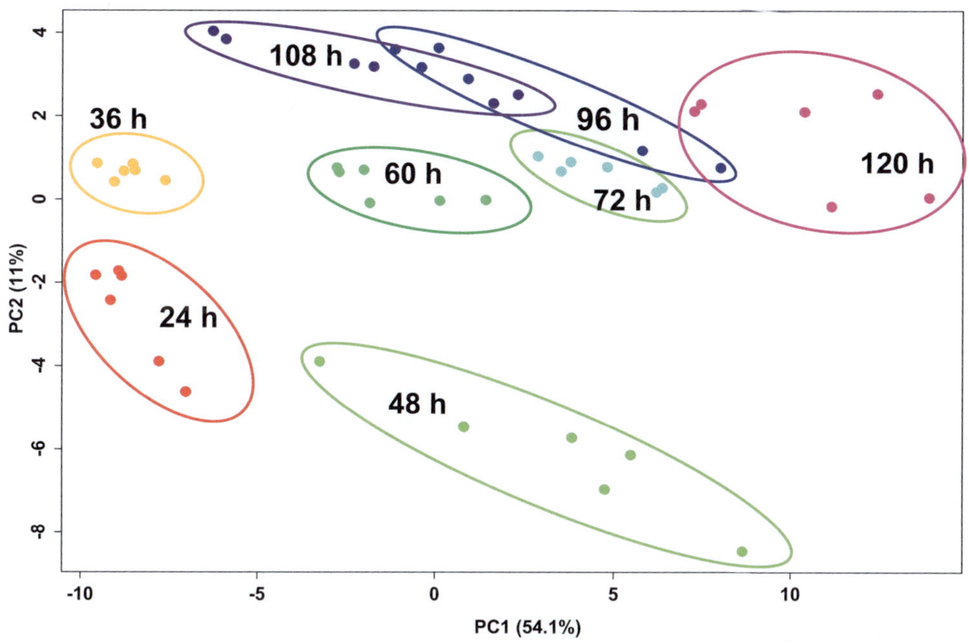

Fig. 3 PCA plot from intracellular metabolome analysis. The information of all identified and unidentified metabolites were used in the principal component analysis to see profile changes. The intracellular metabolites from *Yarrowia lipolytica* clearly show separate clusters over incubation time periods (24–120 h)

3. Process the MS data by normalizing using median and/or log transformation (e.g., $\log_2$) and/or scaling (e.g., pareto).
4. Perform univariate data analysis (UVDA) *t*-test or ANOVA analysis on the data to compare detect (identified or nonidentified) features between the samples. Execute multivariate data analysis (MVDA) analysis using principal component analysis (PCA) (Fig. 3), partial least squares discriminant analysis (PLS-DA), and orthogonal partial least squares (OPLS-DA). Data can be visualized using a heatmap (Fig. 4) or PCA plot (Fig. 3). UVDA and MVDA are used to identify biological significant features/metabolites.
5. Accurate metabolite identification is required for integrated analysis with other omics data such as proteomics, transcriptomics, and genomics to be shown together in pathway mapping and enrichment analysis. Statistically significant metabolites, proteins and transcripts can be cross-checked at the beginning and coverages can be expanded to related metabolites in the enriched pathways.

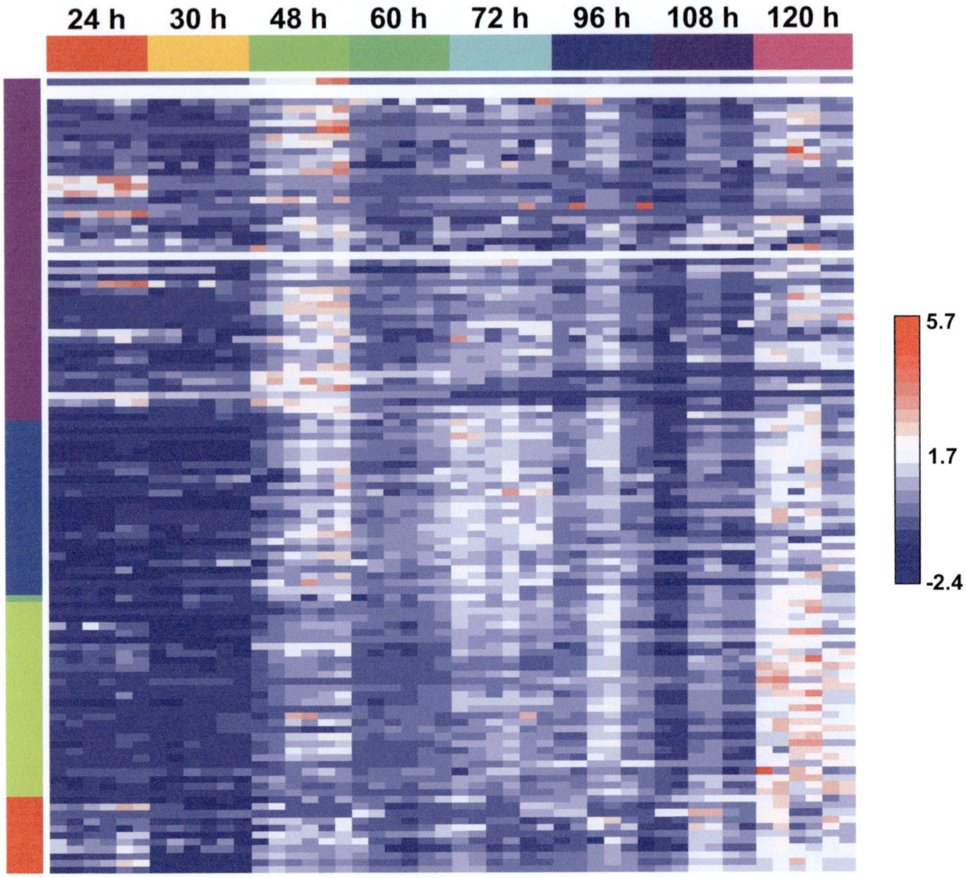

Fig. 4 Heat map analysis from intracellular metabolome changes. All the peak area values of identified and unidentified metabolites were used in the heat map analysis for data visualization. The values were *z*-score transformed and showed in relative abundance color scale

4 Notes

1. A fungal culture produced in the lab mostly has either homogeneous biomass (e.g., yeast) or filamentous and multicellular biomass (e.g., mushroom) which has atypical structures. Attainment same amount of biomass of fungal biomass is a key step to obtaining normalized MS signal of metabolites. A normalization of biomass is a critical step if the fungal samples were collected from the environment.
2. An excessive amount of salts or carbohydrate can make crystal forms of metabolite if they were dried after extraction. Sonicating samples will help to disperse the crystalline mass into the derivatizing reagent (for GC-MS) or reconstitution solvents (for LC-MS).
3. Several different derivatization methods are available for metabolomics for GC-MS system. A combination of methoxy-

amination with trimethylsilylation is one of the most popular methods due to its wide metabolome coverage.

4. Parameters and setting values can be obtained from publications and reports based on manufacturer. The major vendors have various terminologies and controllable information (Major vendors: Thermo, Agilent, Waters, Bruker, Shimadzu, SCIEX, LECO, and JEOL).
5. Building a database is generally considered an expensive and time-consuming work, but an in-house developed database is more accurate and has more reliable information on retention time and fragmentation of each individual metabolite. Establishing an in-house library requires both time and financial commitment. Some metabolite mixture kits are available (Major providers: Sigma-Aldrich, IROA Technologies, Cambridge Isotope Laboratories, Cayman Chemical, and Avanti Polar Lipids).
6. Mass spectrometer manufacturing companies provide metabolomics databases which can be obtained commercially for GC-MS based metabolomics (e.g., Agilent, LECO). Additionally, NIST Mass Spectral Library (various versions are available) and Wiley Registry MS databases for GC-MS can be used to match against MS spectrum. This method can give a level 2 of identification of metabolites (Metabolomics Standards Initiative) [10]. Due to the nature of LC-MS analysis, construction of standardized and generally applicable MS database is less favorably expected, thus highlighting the need for an in-house library using chemical standards to get retention time and m/z values. If unknown peaks are biologically significant and of interest, then MS/MS identification is preferred, and public library can be used to compare with. (Representatives: Golm Database, MassBank, and MoNA—MassBank of North America.)

Acknowledgment

This chapter is based upon work supported by the US Department of Energy, Office of Science, Office of Biological and Environmental Research, Genomic Science program, under Award Number DE-SC0008744. This is also enabled by the Environmental Molecular Sciences Laboratory (EMSL), a national scientific user facility sponsored by the US Department of Energy's (DOE) Office of Biological and Environmental Research (OBER), and located at the Pacific Northwest National Laboratory (PNNL). PNNL is a multiprogram national laboratory operated by Battelle for the DOE under Contract DE-AC05-76RLO 1830.

References

1. Nicholson JK, Lindon JC, Holmes E (1999) "Metabonomics": understanding the metabolic responses of living systems to pathophysiological stimuli via multivariate statistical analysis of biological NMR spectroscopic data. Xenobiotica 29:1181–1189. https://doi.org/10.1080/004982599238047
2. Fiehn O (2002) Metabolomics – the link between genotypes and phenotypes. Plant Mol Biol 48:155–171. https://doi.org/10.1023/A:1013713905833
3. Zhang A, Sun H, Wang P et al (2012) Modern analytical techniques in metabolomics analysis. Analyst 137:293–300. https://doi.org/10.1039/C1AN15605E
4. Kyle JE, Casey CP, Stratton KG et al (2016) Comparing identified and statistically significant lipids and polar metabolites in 15-year old serum and dried blood spot samples for longitudinal studies. Rapid Commun Mass Spectrom. https://doi.org/10.1002/rcm.7808
5. Bennett BD, Yuan J, Kimball EH, Rabinowitz JD (2008) Absolute quantitation of intracellular metabolite concentrations by an isotope ratio-based approach. Nat Protoc 3:1299–1311. https://doi.org/10.1038/nprot.2008.107
6. Nakayasu ES, Nicora CD, Sims AC, et al (2016) MPLEx: a robust and universal protocol for single-sample integrative proteomic, metabolomic, and lipidomic analyses. mSystems 1(3). pii: e00043-16
7. Lai Z, Fiehn O (2016) Mass spectral fragmentation of trimethylsilylated small molecules. Mass Spectrom Rev. https://doi.org/10.1002/mas.21518
8. Coble JB, Fraga CG (2014) Comparative evaluation of preprocessing freeware on chromatography/mass spectrometry data for signature discovery. J Chromatogr A 1358:155–164. https://doi.org/10.1016/j.chroma.2014.06.100
9. Niu W, Knight E, Xia Q, McGarvey BD (2014) Comparative evaluation of eight software programs for alignment of gas chromatography–mass spectrometry chromatograms in metabolomics experiments. J Chromatogr A 1374:199–206. https://doi.org/10.1016/j.chroma.2014.11.005
10. Sumner LW, Amberg A, Barrett D et al (2007) Proposed minimum reporting standards for chemical analysis. Metabolomics 3:211–221. https://doi.org/10.1007/s11306-007-0082-2

Chapter 11

Genome Editing: CRISPR-Cas9

Jakob B. Hoof, Christina S. Nødvig, and Uffe H. Mortensen

Abstract

In the present chapter, we present the protocols and guidelines to facilitate implementation of CRISPR-Cas9 technology in fungi where few or no genetic tools are in place. Hence, we firstly explain how to identify dominant markers for genetic transformation. Secondly, we provide a guide for construction of Cas9/sgRNA episomal expression vectors. Thirdly, we present how to mutagenize reporter genes to explore the efficiency of CRISPR-Cas9 in the relevant fungus and to ease subsequent CRISPR-mediated genetic engineering. Lastly, we describe how to make CRISPR-mediated marker-dependent and marker-free gene targeting.

Key words CRISPR, Gene targeting, Gene editing, Mutagenesis, Nonmodel fungi

1 Introduction

More and more genomes of filamentous fungi are being fully sequenced, and this provides a multitude of exciting research opportunities. For example, it sets the stage for gene targeting (*see* **Note 1**). However, the fact that genetic tools are developed only for a few model species and that most fungi show low gene-targeting frequencies [1] provide significant barriers toward analyzing nonmodel species by reverse genetics. In the last couple of years, CRISPR-Cas9-based technologies have revolutionized the gene-editing field by exploiting that highly specific DNA double strand breaks (DNA DSBs) introduced by the Cas9 endonuclease allows efficient introduction of site specific genetic modifications [2, 3]. Due to the fact that Cas9 can easily be programmed to cleave specific DNA sequences in the genome, CRISPR promises to serve as a versatile genetic-engineering tool that can be implemented in model as well as nonconventional fungal species.

The CRISPR-Cas9 system originates from an archaeal and bacterial adaptive immune system that protects the organism from invading DNA, e.g., phages and plasmids. In the natural systems, the Cas9 nuclease forms a complex with two RNAs, a constant

Ronald P. de Vries et al. (eds.), *Fungal Genomics: Methods and Protocols*, Methods in Molecular Biology, vol. 1775, https://doi.org/10.1007/978-1-4939-7804-5_11,

trans-activating CRISPR RNA (tracrRNA) and a CRISPR RNA (crRNA), which is composed by a constant section and a 20 nucleotides (nt) variable section termed the protospacer. The specificity of Cas9 is determined by the protospacer that guides Cas9 to a target site containing a complementary sequence. Importantly, Cas9 only produces a DNA DSB at the target locus if it is followed by a protospacer adjacent motif (PAM, *see* Fig. 1). Most CRISPR gene editing has been performed by using the Cas9 variant from *Streptococcus pyogenes* [2]; and for this Cas9 species the PAM sequence is NGG (or for less efficient cleavage, NAG). Hence, target restrictions are low and almost any sequence region can be engineered using *S. pyogenes* Cas9 (*see* **Note 2**). Importantly, the complexity of the natural three component system has been simplified into a two component system by linking the crRNA with the tracrRNA via a hairpin loop resulting in a chimeric single-chain guide-RNA (sgRNA, Fig. 1) [4]. The ability of introducing specific DNA DSBs into the genome by Cas9 sets the stage for different types of gene editing driven by one of the two main DNA DSB repairing pathways, i.e., nonhomologous end-joining (NHEJ, *see* **Note 3**) or homologous recombination (HR, *see* **Note 4**).

We have previously developed a versatile CRISPR-Cas9 system for fungi [5], which has successfully been used in a range of fungal species [6–8]. In our system, CRISPR has been adapted to fungi by employing an *A. niger* codon-optimized version of the *S. pyogenes cas9* gene, which has been extended by a sequence encoding the SV40 nuclear localization sequence (NLS) [9, 10]. The second component of the system, the sgRNA, is liberated by ribozymes from a larger polymerase II derived transcript [11]. The *cas9* and sgRNA genes are both controlled by fungal polymerase II promoters and terminators (*see* **Note 5**, Fig. 2) and are harbored on bacterial/fungal

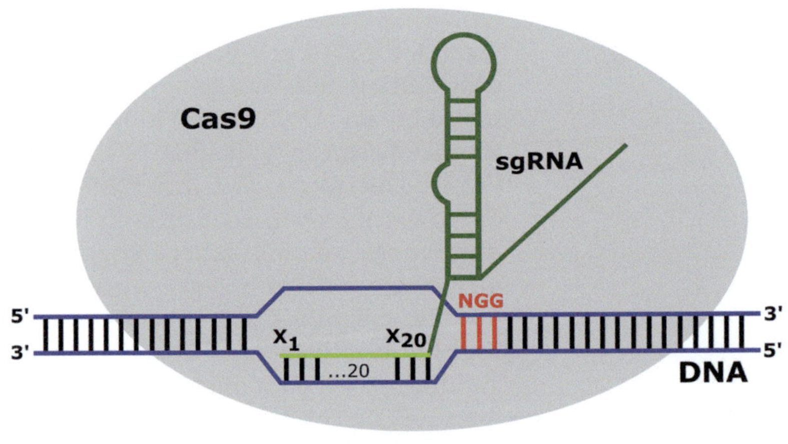

Fig. 1 The Cas9/sgRNA complex binding to a DNA sequence complementary to the protospacer adjacent to the PAM. The X_1 to X_{20} denotes the protospacer start and end, whereas NGG highlighted in red shows the PAM sequence

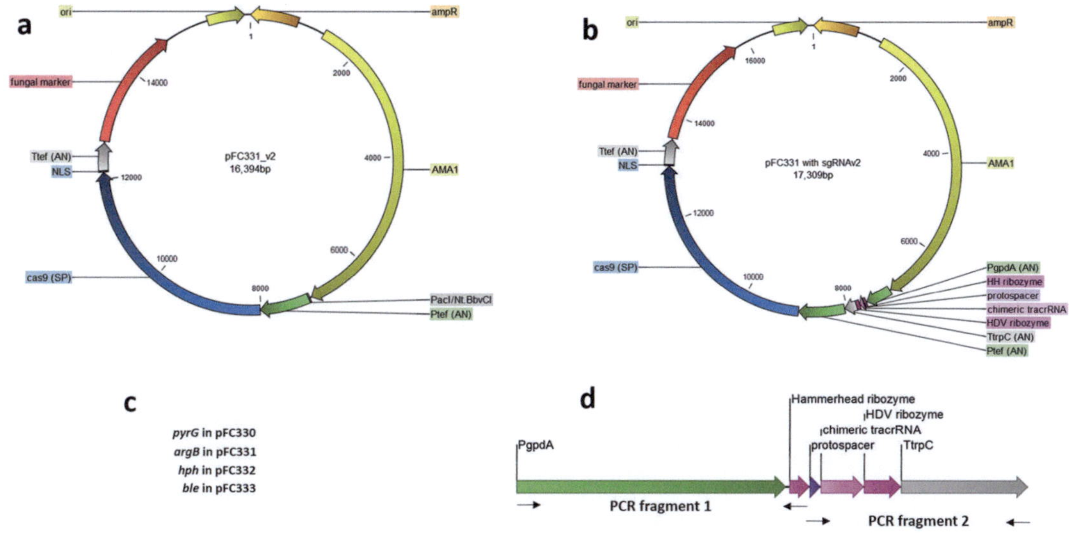

Fig. 2 A CRISPR plasmid without (**a**) and with (**b**) an sgRNA expression cassette. The CRISPR vector set, pFC330-333, covers four different markers. pFC330 contains *A. fumigatus pyrG* including its promoter and terminator, and pFC331 (sequence represented in panel **a**) contains *A. nidulans argB* including its promoter and terminator. pFC332 and pFC333 harbor *hph* (hygromycin resistance) and *ble* (bleomycin resistance), respectively. Both resistance genes are under the control of *A. nidulans* (AN) *PtrpC* and *TtrpC*. The remaining parts of all vectors are identical. An enlargement of an sgRNA expression cassette (**d**) showing the order and nature of all elements

shuttle vectors, which we call CRISPR plasmids. The CRISPR plasmids contain the *A. nidulans* AMA1 element [12] (*see* **Note 6**), which contain one of currently four different fungal selection markers (Fig. 2). Each CRISPR experiment requires construction of a unique CRISPR plasmid in which the desirable sgRNA gene has been inserted into a PacI/Nt.BbvCI cassette by USER cloning [13]. The sgRNA gene is generated via two independent PCR reactions using a total of four primers of which two are constant in all experiments. The materials and methods described below fit with this system, but the methods can easily be adapted to other vectors and cloning systems if desirable (*see* **Note 7**).

2 Materials

2.1 Growth Medium

We use minimal medium (MM) for our cultivation of *Aspergillus* species. If another medium is preferred, then this medium should support spore germination, vegetative growth, asexual sporulation, and the functionality of the antibiotic.

1. Trace metal solution (1000 ml): 0.8 g $FeSO_4 \cdot 7H_2O$, 0.4 g $CuSO_4 \cdot 5H_2O$, 8 g $ZnSO_4 \cdot 7H_2O$, 0.8 g $MnSO_4 \cdot 2H_2O$, 0.04 g $Na_2B_4O_7 \cdot 10H_2O$, 0.8 g $Na_2MoO_4 \cdot 2H_2O$—ultrapure water to 1000 ml volume.

2. 20× Nitrate salts solution (1000 ml): 120 g $NaNO_3$, 10.4 g KCl, 10.4 g $MgSO_4{\cdot}7H_2O$, 30.4 g KH_2PO_4—ultrapure water to 1000 ml volume.
3. Liquid MM (1000 ml): 10 g glucose, 50 ml 20× nitrate salts solution (*see* Subheading 2.1, **item 2**), 1 ml trace metal solution (*see* Subheading 2.1, **item 1**), 1 ml 1% thiamine.
4. Solid MM, add 20 g agar to the MM described in Subheading 2.3.
5. For solid transformation medium using antibiotics (TMs), add 1 M sorbitol to the MM solid medium keeping the glucose.
6. For TM selecting for auxotrophies, replace glucose in MM solid medium (*see* Subheading 2.1, **item 4**) with 342.3 g sucrose.
7. For solid MM enriched with 5-fluoroorotic acid (5-FOA), we use 1.3 mg 5-FOA/ml for *Aspergillus* species and supplement with uridine and uracil at 10 mM.

2.2 Vectors for cas9 and sgRNA Expression

Currently, four CRISPR shuttle vectors (Fig. 2) are available for our system, each containing a different fungal selection marker (*see* **Note 8**).

2.3 Primers for CRISPR Plasmid Construction

1. For all CRISPR experiments, synthesize the following two primers:

   PgpdA-fwd: GGGTTTAA$\underline{\text{U}}$GCGTAAGCTCCCTAATTGGC
   TtrpC-rv: GGTCTTAA$\underline{\text{U}}$GAGCCAAGAGCGGATTCCTC

2. For each individual CRISPR experiments, synthesize the following two primers:

   sgRNA-x-fwd:
   AGTAAGC$\underline{\text{U}}$CGTC$X_1XXXXXXXXXXXXXXXXXXX_{20}$
   GTTTTAGAGCTAGAAATAGCAAGTTAAA

   sgRNA-x-rv:
   AGCTTAC$\underline{\text{U}}$CGTTTCGTCCTCACGGACTCATCAG$\underline{\text{X}_1\text{XXXXX}_6}$
   CGGTGATGTCTGCTCAAGCG

The row of italicized X's (1-20) in sgRNA-x-fwd is identical to the variable protospacer sequence. The row of underlined X's (1-6) in sgRNA-x-rv encodes the section of the hammerhead ribozyme that forms a hairpin in the unprocessed sgRNA transcript by base-pairing to the protospacer. Since this hairpin is essential for sgRNA release, the italicized nucleotide sequence X_1–X_6 needs to be identical to the underlined nucleotide sequence X_1–X_6. Underlined U's are used for sgRNA assembly by USER cloning. For other cloning methods, replace U's by T's. Bold letters represent the part of the primer that anneals to the template during PCR.

2.4 Reagents for PCR and Vector Construction

1. PCR materials for generation of sgRNA inserts: pFC334 template, Phusion HF PCR buffer (5×), dNTP mix (2 mM), DMSO, Pfu X7 polymerase [14] (*see* **Note 9**).
2. USER cloning materials: predigested vector, agarose-gel purified PCR fragments, appropriate buffer (e.g., Cutsmart 10×), USER enzyme.
3. Chemically competent *E. coli* cells and Luria broth (LB) solid medium containing ampicillin at 100 μg/ml.

2.5 Protoplasts and Reagents for Transformation and Spot Assay

1. Competent cells, e.g., in the form of protoplasts (*see* **Note 10**).
2. PCT: 50% w/vol PEG 3300-8000, 50 mM $CaCl_2$, 20 mM Tris, 0.6 M KCl. pH is adjusted with 2 N HCl to 7.5. Store at 4 °C.
3. Aspergillus Transformation Buffer (ATB): 1.2 M sorbitol; 50 mM $CaCl_2 \cdot 2H_2O$; 20 mM Tris; and 0.6 M KCl. pH is adjusted with 2 N HCl to 7.2.
4. Antibiotics for selection, e.g., hygromycin and bleomycin.

3 Methods

3.1 Determination of Antibiotic Concentrations in Transformation Experiments

To determine the concentration of antibiotics in transformation experiments for a given fungus, we recommend to perform spot assays (*see* Fig. 3). If the concentration is already known, proceed to Subheading 3.2.

1. Use a sterilized metal stamp to make a print indicating spotting positions on your favorite solid media (*see* **Note 11**). Use solid medium containing a wide range of drug concentrations from 0 μg/ml (control) to for example 100 μg/ml. Repeat assay until a conditions causing 10,000-fold reduction of growth is identified (*see* Fig. 3).
2. From a solution containing approximately 10^7 spores/ml ddH_2O, make six serial tenfold dilutions in ddH_2O in a microtiter plate.

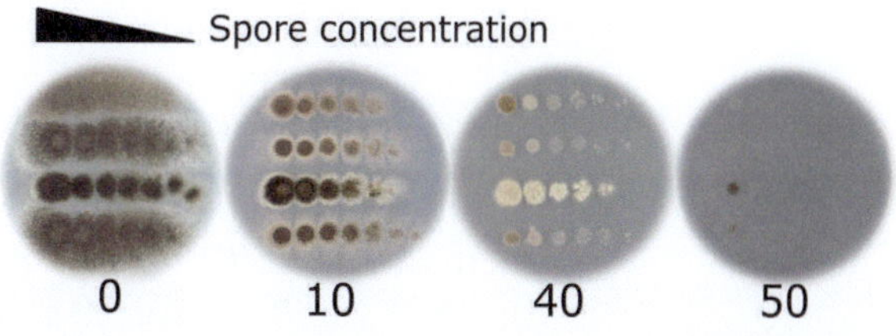

Fig. 3 Spot assay to determine sensitivity to the antibiotics hygromycin. The concentrations of hygromycin B used in this example are 0, 10, 40, and 50 μg/ml. This example shows that all strains are sensitive to the drug and that 50 μg/ml should be sufficient in transformation experiments

3. Spot 5 µl of spore solution to the plates in rows (*see* **Note 12**), starting from the lowest concentration to the highest.
4. Incubate at standard growth temperature for your species, and inspect the plates every day until concluding (*see* Fig. 3).

3.2 Preparation of CRISPR Vector Fragments for USER Cloning

The CRISPR vector set is originally designed for USER cloning. Move to Subheading 3.3, if an alternative approach is used.

1. Select a plasmid containing a marker that is suitable for the target fungus.
2. Digest the 5 µl of the relevant vector pFC330-333 (200–500 ng/µl) in a 100 µl working volume overnight at 37 °C with 10 U PacI restriction enzyme according to manufacturer's instructions.
3. Add additional 5 U PacI and 5 U Nt.BbvCI nicking enzyme for 3 h at 37 °C.
4. Heat-inactivate at 80 °C for 20 min.
5. Check restriction of plasmid digest by agarose-gel electrophoresis.
6. Purify vector fragments.

3.3 Generation of the Two DNA Insert Fragments Containing the sgRNA

Construction of each new CRISPR vector requires generation of two PCR reactions using the primers described in Subheading 2.3 (*see* Fig. 2). As template DNA for both PCR reactions, use an existing CRISPR vector containing an sgRNA, e.g., pFC334. PCR reaction 1 (*see* Subheading 2.4, **item 1**) amplifies fragment 1, which contains the promoter, *PgpdA*, and the major part of the HH ribozyme, by primers PgpdA-fwd and sgRNA-x-rv. PCR reaction 2 amplifies fragment 2, which contains the remaining part of the HH ribozyme, the sgRNA, the HDV ribozyme and the terminator, by primers sgRNA-x-fwd and TtrpC-rv. For uracil-containing primers, remember to use a polymerase that tolerates uracil (*see* **Note 9**). For the design of the protospacer, either use a script for the design, or handpick by inspecting sequences (*see* **Note 13**).

1. Standard PCR 50 µl mix for amplifying sgRNA fragments (uracil containing primers): Mix in ddH$_2$O the following solutions: 1 µl pFC334 template (<2 ng/µl), 10 µl Phusion HF buffer (5×), 5 µl dNTP mix (2 mM), 3 µl of primers (10 µM), 1.5 µl DMSO, 0.5 µl Pfu X7 polymerase (2 U/µl) (*see* **Note 9**).
2. Using PfuX7 as polymerase, we recommend for both PCR fragments 1 and 2: 98 °C for 2 min, followed by 35 cycles of touchdown PCR; 98 °C for 10 s, 62–52 °C for 20 s, 72 °C for 1 min. End the reaction with 72 °C for 5 min and cool to 12 °C.
3. Gel purification of PCR fragment 1 (~540 bp) and PCR fragment 2 (~400 bp).

3.4 USER Cloning

This section is specific for USER cloning. For alternative approaches *see* **Note 7**.

1. Make 10 μl USER cloning mix: 1 μl predigested vector (~5 ng/μl), 4 μl agarose-gel purified PCR fragment 1 (>20 ng/μl), 4 μl agarose-gel purified PCR fragment 2 (>20 ng/μl), 0.5 μl buffer for enzymes (e.g., Cutsmart 10×), 0.5 μl USER enzyme mix. For a negative control, add ddH_2O instead of PCR fragments.
2. Incubate the USER cloning mix for 35 min at 37 °C.
3. Transfer to room temperature for 25 min then to ice.
4. Simultaneously, thaw the chemically competent *E. coli* cells on ice for 10 min.
5. To the USER cloning mix, add very gently 50 μl chemically competent *E. coli* cells. Do not mix by pipetting.
6. Incubate on ice for 10 min.
7. Heat-shock for 1 min 15 s at 42 °C.
8. Incubate on ice for 5 min.
9. Plate cells on LB + ampicillin plates (100 μg/ml) and incubate overnight at 37 °C.
10. Purify plasmids from an appropriate amount of clones based on comparing to a negative control.
11. Analyze by restriction digestion, e.g., BspEI. This enzyme clearly shows whether the fragments have been correctly inserted into pFC330-333.
12. All cloned sgRNAs should be sequenced prior to use to check for sequence errors.

3.5 Transformation of Fungi—Using Antibiotics

1. In transformation experiments using either *hph* or *ble* as genetic markers; gently mix approximately 10^7 protoplasts (in 100 μl ATB) with 3 μg of Cas9/sgRNA vector. Use the equivalent amount of DNA if you are also using a gene-targeting vector.
2. Incubate on ice for 10 min.
3. To the DNA-protoplast mix, add 1 ml PCT solution, incubate for 15 min at room temperature.
4. Add *X* μg/ml hygromycin or bleomycin, as predetermined by the spot assay per 15 ml molten TMs 40–45 °C). Remember nutritional supplementation if the target gene for manipulation is a prototrophic gene.
5. Gently mix molten agar with DNA-protoplast-PCT mix.
6. Immediately pour into an empty 9 cm petri dish, and incubate for 24 h at optimal temperature.
7. Add an overlay of 15 ml TMs including the same amount of antibiotic.
8. Streak-purify candidate transformants prior to verification.

3.6 Transformation of Fungi—Using Auxotrophic Markers

1. In transformation experiments using nutritional markers such as *pyrG* or *argB*; gently mix approximately 10^7 protoplasts (in 100 μl ATB) with 3 μg of Cas9/sgRNA vector. Use the equivalent amount of DNA if you are also using a gene-targeting vector.
2. To the DNA–protoplast mix, add 150 μl PCT solution, and incubate for a minimum of 10 min at room temperature.
3. Add 250 μl ATB and plate on TM using the nutritional requirement.
4. Incubate until colonies appear and streak-purify on selective growth medium.

3.7 Evaluation of Cas9 Performance in Your Fungus

To evaluate CRISPR activity in a given fungus we recommend to introduce point mutations into genes where defects will cause visible phenotypes. In case, no gene is known to cause a visible phenotype in the fungus, mutate *pyrG* and select for mutations on MM containing uracil, uridine and 5-FOA. An example of a pigment-gene mutagenesis strategy is provided below.

1. Construct a CRISPR vector encoding Cas9/sgRNA that targets a gene involved in conidial pigmentation *wA/albA* and *yA* using protocols in Subheadings 3.2–3.4.
2. Transform your fungus (*see* Subheading 3.5) with the vector aiming at introducing mutations in your target gene by erroneous NHEJ repair, *see* Fig. 4.
3. Evaluate CRISPR efficiency by visual screening for conidia heads showing a mutant phenotype.
4. For strains with low CRISPR mutagenesis efficiency, restreak transformants on selective medium and examine whether additional CRISPR mutations accumulate as the function of cell divisions (*see* **Note 3**). If no additional mutagenesis occur, *see* **Notes 5** and **14**).
5. Streak-purify mutant strains (*see* **Note 15**).
6. PCR-amplify the mutated gene and sequence the gene.
7. Validate that the new mutations occurred at the location at the predicted Cas9/sgRNA cut site.
8. Since Cas9/sgRNA is not desirable in the final mutant strain, restreak validated mutant strains on nonselectable solid medium.
9. Grow colonies to a diameter of at least 2 cm and transfer spores from the periphery to new solid medium. Test the new isolates for the failure to grow on selective medium. The AMA1 plasmid is readily lost and most of the new isolates will likely be correct [8, 12].

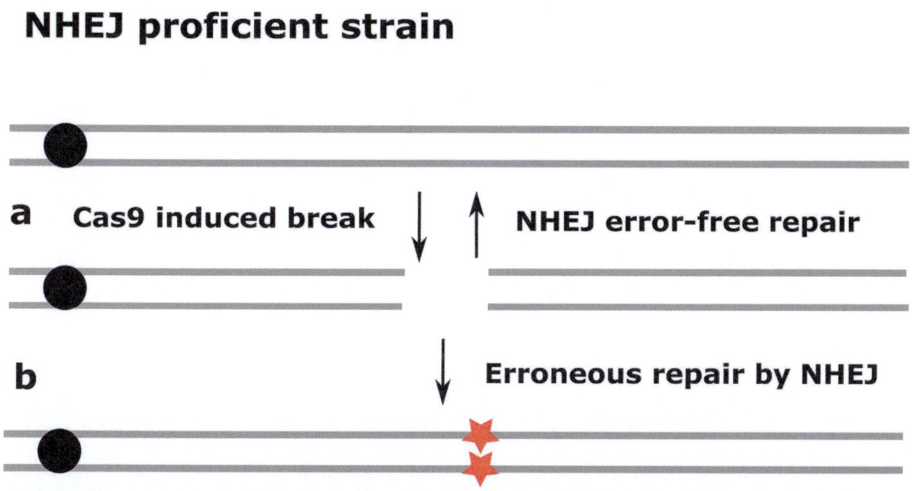

Fig. 4 The interplay of DSB formation by Cas9 and repair by NHEJ. First, Cas9 and the specific sgRNA makes the DSB and NHEJ repairs the break without error (**a**). Cas9 cuts again (**a**). This continuous until the protospacer no longer matches the sequence in the locus due to erroneous repair by NHEJ (**b**)

3.8 Establishing pyrG Auxotrophy

To facilitate genetic engineering it is favorable to mutate the gene encoding Orotidine-5′-phosphate decarboxylase encoding gene, *pyrG*. Using a heterologous *pyrG* in subsequent transformation experiments allows the subsequent selection in transformation experiments using, and it can be recycled for multiple engineering endeavors [15, 16].

1. Construct a CRISPR vector encoding Cas9/sgRNA that targets *pyrG* using protocols 3.2–3.4.
2. Transform your fungus (*see* Subheading 3.5) with the vector to induce NHEJ based mutagenesis at *pyrG*, *see* Fig. 4.
3. For transformation medium, remember to add supplements (*see* **Note 16**).
4. To identify *pyrG* mutant strains, harvest all spores from the transformation plate and dissolve in 500 μl water, *see* Fig. 5a. To avoid plates with too many mutants, make serial tenfold dilutions and add 100 μl of the diluted samples to 5-FOA enriched medium (Subheading 2.1, **item** 7) as shown in Fig. 5b. Incubate until colonies appear.
5. Pick single colonies (an additional round of restreaking on solid 5-FOA may be necessary) and transfer to solid medium without and with uridine and uracil to identify mutants, Fig. 5c.
6. Amplify the mutant *pyrG* by PCR and sequence the mutation to validate that occurred at the predicted Cas9/sgRNA cut site.

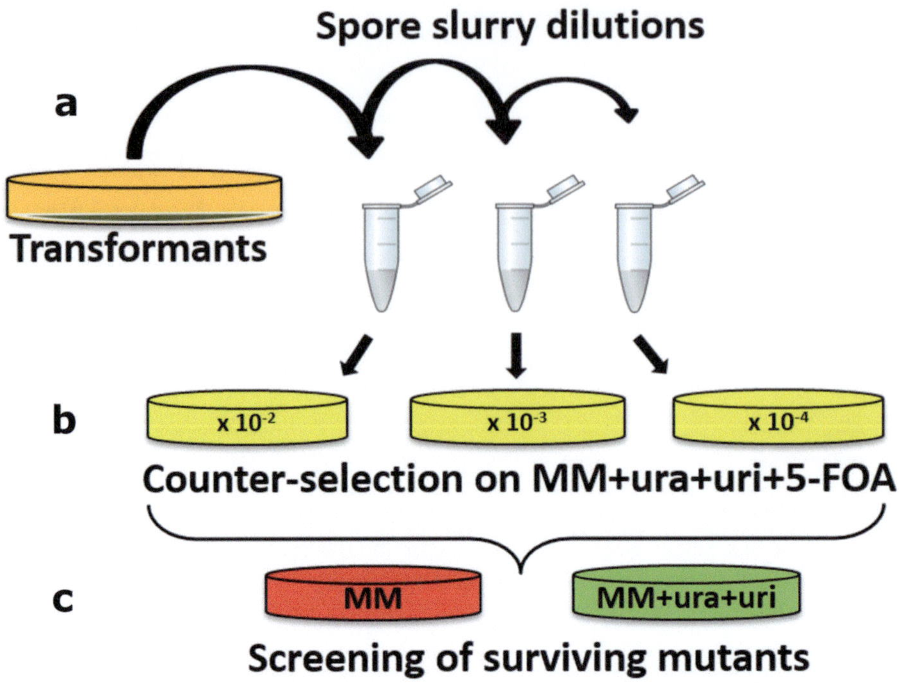

Fig. 5 Identification of Cas9/sgRNA mediated mutagenesis of *pyrG*. Harvest spores for the transformation plate (**a**). Make serial tenfold dilution and add 100 μl of the diluted samples to the 5-FOA enriched medium as shown (**b**). Transfer discrete single colonies to both selective and nonselective plates (**c**). Potential *pyrG* deficient strains should not grow on solid MM as marked in red, but on solid MM supplemented with uracil and uridine shown in green

3.9 Using CRISPR-Cas9 to Enhance Gene Targeting

Importantly, CRISPR-Cas9 can stimulate gene targeting allowing for a wider range of modifications (*see* **Note 1**). Cotransforming a conventional gene-targeting substrate with the episomal CRISPR-Cas9 system expressing an sgRNA specific for the gene target (*see* Fig. 6) will often yield a significantly higher gene-targeting frequency than what can be achieved in transformations with the gene-targeting substrate alone [5, 8].

1. Construct a circular gene-targeting substrate containing another genetic marker than the one harbored on the CRISPR vector (*see* **Note 17**). If the strain has only one marker, then select for the gene-targeting plasmid only.
2. Construct a CRISPR vector encoding Cas9 and an sgRNA targeting the gene of interest using protocols 3.2–3.4.
3. Cotransform into the protoplasts using 3 μg of each plasmid (*see* Subheading 3.5 or 3.6).
4. For strains with low CRISPR efficiency, it may be advantageous to plate the transformation mix on solid medium selecting for both the CRISPR plasmid and the gene-targeting substrate (*see* **Note 4**). For strains with high CRISPR efficiency, it would be advantageous to select only for the gene-targeting substrate to loose unnecessary Cas9/sgRNA as fast as possible.

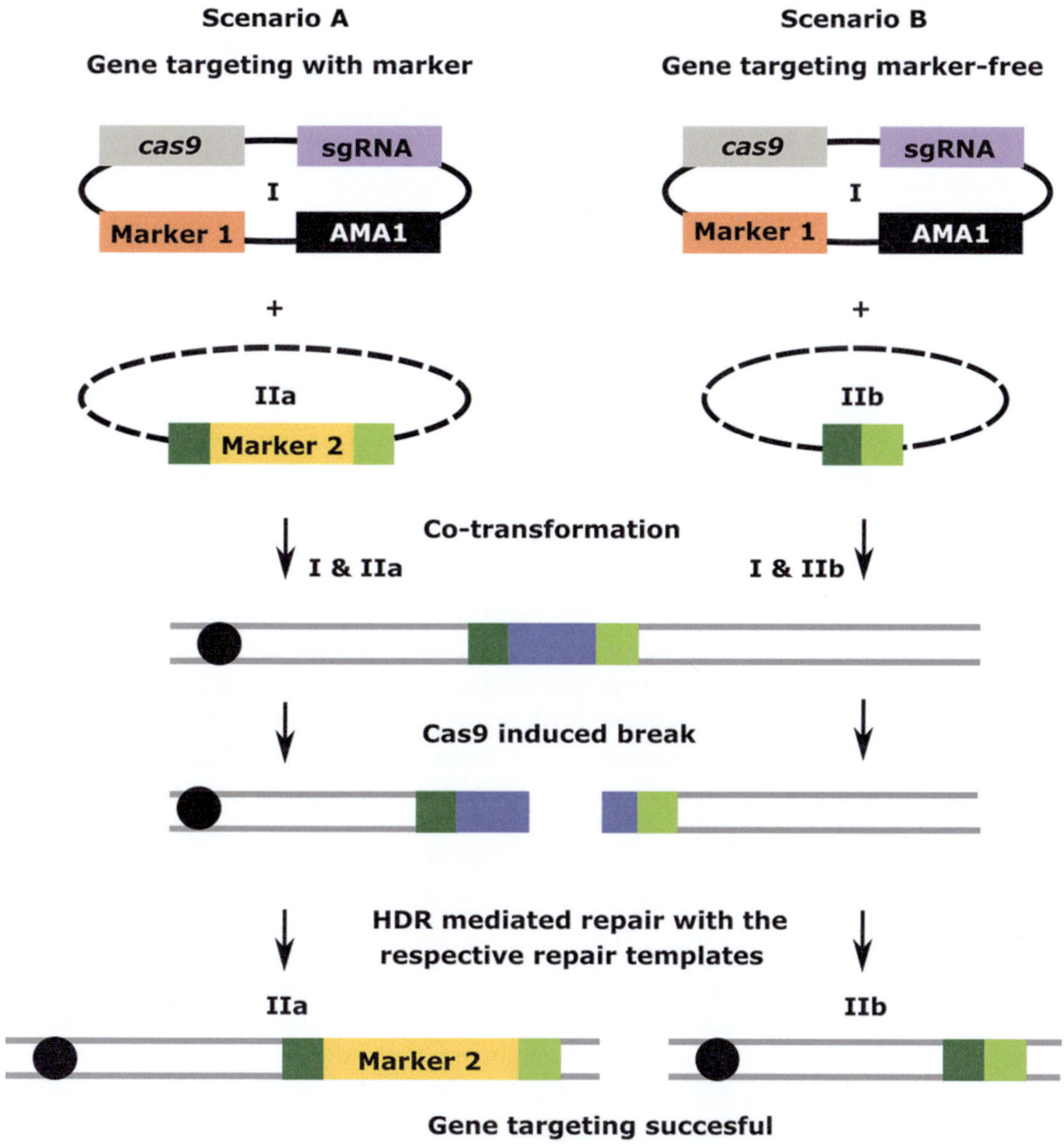

Fig. 6 CRISPR-Cas9 mediated gene targeting. Two scenarios, A and B, for gene targeting are shown. Scenario A employs a gene targeting substrate, vector IIa, containing a selectable marker; and scenario B employs aa marker-free gene-targeting substrate, vector IIb. The same Cas9/sgRNA construct, vector I, is used to stimulate gene targeting in both scenarios

5. Transformants should be streak-purified using selection.
6. Run diagnostic PCRs and Southern blots to verify true homokaryotic transformants.
7. In cases where selection for the CRISPR is maintained throughout the experiment, the CRISPR plasmid should be lost before further characterization of the mutant strain.

3.10 Marker-Free Gene Editing

CRISPR-Cas9 can also be utilized to facilitate marker-free gene editing based on repair by HR (*see* **Note 18**). This allows for cleaner introduction of point mutations, promoter replacement, in-frame protein tagging and multiplexing etc.

1. Assemble the gene-targeting substrate without adding the genetic marker.
2. Use Subheading 3.9, **step 3**, followed by selection only with the CRISPR plasmid.
3. Continue as in Subheading 3.9, **steps 4–7**.

4 Notes

1. Gene targeting facilitates numerous engineering possibilities including creation of full gene deletions, gene insertions, introduction of specific point mutations, extending genes in frame with sequences encoding epitope tags or fluorescent proteins as well as promoter and terminator swaps [3–5, 8].
2. Some sequences are known to be low on G dinucleotides, e.g., centromere and telomere sequences, and therefore these are examples of sequences with low editing potential.
3. Erroneous repair of DNA DSBs by NHEJ can be harnessed to introduce targeted mutations, typically small deletions or insertions, which disrupt the reading frame of a gene in question without the need to insert a selection marker [5]. Importantly, DNA DSBs are often repaired correctly by NHEJ to restore the original sequence. For some fungal species/loci where NHEJ repair fidelity is high, it is therefore necessary to express *cas9* and sgRNA constitutively and propagate the transformed fungus for several generations. Together, this will increase the frequency of CRISPR induced mutations in the mycelium exploiting that mutated loci are not substrates for the Cas9/sgRNA nuclease.
4. Insertion of a gene-targeting substrate by HR is an infrequent event in many filamentous fungi [1]. However, specific DNA DSBs introduced by Cas9 in the target sequence stimulate HR-based gene targeting as the gene-targeting substrates serve as a repair template for sealing the break. If the breaks are repaired by using information on the sister chromatid, the broken sequence is restored to wild-type and can therefore be cleaved once again by the Cas9/sgRNA endonuclease setting the stage for a second chance of gene targeting.
5. It may be necessary to use species specific promoters to achieve sufficient expression levels of the *cas9* and sgRNA genes.
6. If the AMA1 sequence does not support episomal plasmid propagation, it may be necessary to integrate *cas9* into the genome. The sgRNA component can be made in vitro and transformed into the host [17]. Alternatively, both *cas9* and

the sgRNA genes can be integrated into the genome. In this case, at least one of the genes should be controlled by an inducible promoter.

7. Use USER cloning for CRISPR vector construction, but vector construction can also be performed by other flexible methods like SLIC, Gibson DNA assembly, In-fusion, or SLiCE [18–20].
8. The plasmids pFC330-334 can be retrieved from Addgene.
9. Use a polymerase that tolerates uracil in the primer. We use the noncommercial and proofreading PfuX7 polymerase [14], but the commercially available Phusion U (Life Technologies) can also be used.
10. A suitable protocol for protoplastation or an equivalent method to enable uptake of DNA in the strain needs to be in place. If not, we recommend using a protocol that works for a closely related fungus.
11. Adjust the pH in the medium for optimal drug functionality.
12. If possible, include a strain/species with a known tolerance to the drug concentrations as a control.
13. Protospacer design for multiple species is facilitated by the OPTIMUS script [5]. Single protospacer design is also possible from online resources. Depending on the purpose of the sgRNA, there are different requirements for protospacer design. For simple NHEJ mediated mutagenesis, use a protospacer matching a sequence early in the reading frame of the target gene to reduce the size of the truncated protein and thereby the chance that it may retain any activity. For gene deletions this is not as critical.
14. In case the CRISPR-Cas9 system in a specific fungus does not seem to be functional, or have very low efficiency, try testing several protospacer sequences.
15. Occasionally, heterokaryons may arise from multinucleate protoplasts. In this case, additional purification of the colonies is necessary.
16. If *pyrG* is used as an auxotrophic marker, species in the nidulans section often require both uracil and uridine to supplement *pyrG* auxotrophy, whereas species in section Nigri only require uridine.
17. For classical gene-targeting experiments, a linear DNA substrate is usually preferred since it is more recombinogenic than in the circular form [21]. However, with CRISPR technology Cas9 introduces a DSB at the target locus to stimulate recombination. It may therefore be advantageous to use a circular over a linear gene-targeting substrate to avoid formation of false positives resulting from ectopic integrations via the NHEJ pathway.
18. For marker-free gene-targeting it may be advantageous to eliminate a gene in the NHEJ pathway to increase the efficiency.

References

1. Krappmann S (2007) Gene targeting in filamentous fungi: the benefits of impaired repair. Fungal Biol Rev 21:25–29. https://doi.org/10.1016/j.fbr.2007.02.004
2. Doudna JA, Charpentier E (2014) The new frontier of genome engineering with CRISPR-Cas9. Science 346:1258096. https://doi.org/10.1126/science.1258096
3. Sander JD, Joung JK (2014) CRISPR-Cas systems for editing, regulating and targeting genomes. Nat Biotechnol 32:347–355. https://doi.org/10.1038/nbt.2842
4. Jinek M, Chylinski K, Fonfara I, Hauer M, Doudna JA, Charpentier E (2012) A programmable dual-RNA-guided DNA endonuclease in adaptive bacterial immunity. Science 337:816–821. https://doi.org/10.1126/science.1225829
5. Nødvig CS, Nielsen JB, Kogle ME, Mortensen UH (2015) A CRISPR-Cas9 system for genetic engineering of filamentous fungi. PLoS One 10:e0133085. https://doi.org/10.1371/journal.pone.0133085
6. Weber J, Valiante V, Nødvig CS, Mattern DJ, Slotkowski R, Mortensen UH, Brakhage AA (2017) Functional reconstitution of a fungal natural product gene cluster by advanced genome editing. ACS Synth Biol 20:62–68. https://doi.org/10.1021/acssynbio.6b00203
7. Wenderoth M, Pinecker C, Voß B, Fischer R (2017) Establishment of CRISPR/Cas9 in *Alternaria alternata*. Fungal Genet Biol 101:55–60
8. Nielsen ML, Isbrandt T, Rasmussen KB, Thrane U, Hoof JB, TO L, Mortensen UH (2017) Genes linked to production of secondary metabolites in *Talaromyces atroroseus* revealed using CRISPR-Cas9. PLoS One 12:e0169712
9. Kalderon D, Roberts BL, Richardson WD, Smith AE (1984) A short amino acid sequence able to specify nuclear location. Cell 39:499–509
10. Lanford RE, Butel JS (1984) Construction and characterization of an SV40 mutant defective in nuclear transport of T antigen. Cell 37:801–813
11. Gao Y, Zhao Y (2014) Self-processing of ribozyme-flanked RNAs into guide RNAs in vitro and in vivo for CRISPR-mediated genome editing. J Integr Plant Biol 56:343–349
12. Gems D, Johnstone IL, Clutterbuck AJ (1991) An autonomously replicating plasmid transforms *Aspergillus nidulans* at high frequency. Gene 98:61–67
13. Nour-Eldin HH, Geu-Flores F, Halkier BA (2010) USER cloning and USER fusion: the ideal cloning techniques for small and big laboratories. Methods Mol Biol 643:185–200
14. Nørholm MHH (2010) A mutant Pfu DNA polymerase designed for advanced uracil-excision DNA engineering. BMC Biotechnol 10:21
15. Alani E, Cao L, Kleckner N (1987) A method for gene disruption that allows repeated use of *URA3* selection in the construction of multiply disrupted yeast strains. Genetics 116:541–545
16. d'Enfert C (1996) Selection of multiple disruption events in *Aspergillus fumigatus* using the orotidine-5′-decarboxylase gene, *pyrG*, as a unique transformation marker. Curr Genet 30:76. https://doi.org/10.1007/s002940050103
17. Liu R, Chen L, Jiang Y, Zhou Z, Zou G (2015) Efficient genome editing in filamentous fungus *Trichoderma reesei* using the CRISPR/Cas9 system. Cell Discov 1:15007. https://doi.org/10.1038/celldisc.2015.7
18. Gibson DG, Young L, Chuang R, Venter JC, Hutchison CA III, Smith HO (2009) Enzymatic assembly of DNA molecules up to several hundred kilobases. Nat Methods 6:343–345
19. Li MZ, Elledge SJ (2012) SLIC: a method for sequence- and ligation-independent cloning. Methods Mol Biol 852:51–59
20. Zhang Y, Werling U, Edelmann W (2012) SLiCE: a novel bacterial cell extract-based DNA cloning method. Nucleic Acids Res 40:1–10
21. Rothstein RJ (1983) One-step gene disruption in yeast. Methods Enzymol 101:202–211

Chapter 12

Evolutionary Adaptation to Generate Mutants

Ronald P. de Vries, Ronnie Lubbers, Aleksandrina Patyshakuliyeva, Ad Wiebenga, and Isabelle Benoit-Gelber

Abstract

In this chapter we describe a method to generate mutants of filamentous fungi using their genomic plasticity and rapid adaptability to their environment. This method is based on spontaneous mutations occurring in relation to improved growth of fungi on media by repeated inoculation resulting in adaptation of the strain to the condition. The critical aspect of this method is the design of the selective media, which will depend strongly on the phenomenon that will be studied. This method is advantageous over UV or chemical random mutagenesis as it results in a lower frequency of undesired mutations and can result in strains that combined with (post)genomic approaches can enhance our understanding of the mechanisms driving various biological processes. In addition, it can be used to obtain better strains for various industrial applications. The method described here is specific for sporulating fungi and has so far not yet been tested for nonsporulating fungi.

Key words Evolution, Adaptation, Mutants, Screening, Selection

1 Introduction

Microorganisms have a strong ability to adapt rapidly to different environmental conditions. This ability can be further developed as has been previously shown in studies to improve yeast strains for different biotechnological applications [1, 2]. Similarly, adaptive evolution was applied for *Aspergillus nidulans*, which resulted in phenotypic changes and higher fitness [3–5]. Recently, we applied this method in *Aspergillus niger* to improve cellulase production [6] and in *Aspergillus oryzae* to improve inulinase production [7]. An analysis of the spontaneous mutation rate measurement in filamentous fungi has been conducted with *Aspergillus nidulans* by a replica plating technique, illustrating the relevance of the method for organisms that grow as mycelia. Since it is an easy way of obtaining nucleus samples through the spores and therefore determine mutation rate per nucleus instead of per generation as for unicellular organisms [8]. Differences in spontaneous mutation frequencies as a function of environmental stress was assessed in *A. niger* and

Ronald P. de Vries et al. (eds.), *Fungal Genomics: Methods and Protocols*, Methods in Molecular Biology, vol. 1775, https://doi.org/10.1007/978-1-4939-7804-5_12, © Springer Science+Business Media, LLC, part of Springer Nature 2018

Penicillium lanosum and showed a strong correlation with the environmental conditions [9].

The method itself is rather straight forward, but the critical aspect is the development of the screen for the phenotype of interest. This method works best when selecting for improved growth as this will ultimately result in the improved strain becoming dominant in the mixture. For example in the case of the cellulase mutant [6], selection was based on improved growth on cellulose as the sole carbon source, while inulin was used in the *A. oryzae* study [7]. We also recently isolated mutants with improved tolerance to aromatic compounds (Lubbers et al., unpublished results). More details on this are given in the first section of the Methods below.

The method has been designed for fungi that sporulate on synthetic media and therefore is only applicable for those species in its current form. With certain modifications it may also be applicable to fungi that are propagated by mycelial fragments.

2 Materials

All harvesting and inoculations should be done in a sterile environment (e.g., laminar or circular flow cabinet) and under stringent sterile conditions to avoid contamination of the evolved mixture with other spores. As the exact media required depends on the screen used, only generic descriptions are given in this method. The number of adaptive evolution replicates is left to the appreciation of the user although we recommend a minimum of three paralleled replications for further statistical accuracy.

1. Spore plates of the species of interest. Prepare spore plates of the starting strain on media that is suitable for the species of interest. Ideally, this medium should not be selective for a specific phenotype, but be a rather rich medium that support good growth.
2. Selective medium for the evolution generation and selection of the final strains. This medium should be a minimal salt medium, such as that described for *A. niger* [10], when selection is based on improved growth on a certain carbon source. The medium may be adapted if a different nutrient is used for the selection (e.g., nitrogen, phosphate, iron). When selection is based on tolerance against a toxic compound it may sometimes be better to use a richer medium (e.g., by adding peptone or yeast extract), to ensure that the fungus can tolerate the toxic compound better.
3. ACES buffer: 10 mM N-(2-acetamido)-2-aminoethanesulfonic acid + 0.02% Tween 80, pH 6.1–7.5.
4. Microscope and hemocytometer/cell counter to determine the concentration of spores.

3 Methods

3.1 Screening for Improved Growth on a Specific Carbon Source

1. Growth on the desired carbon source should be significantly limited and ideally should also have limited sporulation.
2. It may in some cases be impossible to use a certain carbon source for a specific species, and several species/strains may need to be tested to identify the best combination. For example for improved growth on inulin *A. niger* was not suitable as it already grew very well, but *A. oryzae* had strongly restricted growth, offering options for improvement.
3. Determine which concentration of the carbon source is most suitable for the species, taking into account that carbon catabolite repression should be avoided as much as possible. For *Aspergillus*, typically monosaccharides and oligosaccharides should be used at 25 mM and polysaccharides at 1% final concentration, but this may differ per species.
4. Harvest spores from spore plates of your species of interest using 10 ml ACES buffer. Dilute the suspension 10- to 100-fold and determine the spore concentration.
5. Plate 5–100 μl (depending on the abundance of growth of the species) of a 10^6 spores/ml suspension on the selection plate and spread them evenly using a spatula.
6. Incubate the plates until there is a layer of mature spores on the selection plate and harvest the spores.
7. Count the spore suspension and plate 5–100 μl of a 10^6 spores/ml dilution of this suspension evenly on new selection plates.
8. Repeat **steps 6** and 7 until significant growth improvement can be observed.
9. From this final spore suspension, dilute to a final concentration of 10^2 spores/ml and plate 100 μl on a new selection plate.
10. Select the colonies that grow best and purify these by streaking them two times on selection plates.
11. Prepare spore plates of the evolved strain(s) and its/their original parent, harvest them and prepare dilutions containing 10^6 spores/ml, 10^5 spores/ml, 10^4 spores/ml and 10^3 spores/ml.
12. On a square Petri dish with selection media, plate 2 μl of each dilution in a horizontal row and put the different strains vertically underneath each other.
13. Incubate the strains to evaluate the difference in growth with the parent.
14. Select the strains with the most significant change and analyze these further by physiological testing, genome resequencing, transcriptomics, proteomics, and/or metabolomics.

3.2 Screening for Improved Tolerance to a Toxic Compound

1. The selection medium should be supplemented with glucose or another good carbon source at 25 mM—1% final concentration to avoid growth effects due to carbon limitation.
2. Make plates with this medium and a range of concentrations of the toxic compound. Determine at which concentration growth is strongly reduced, but not absent.
3. Prepare a first set of selection plates which contains this concentration of the toxic compound.
4. Harvest spores from spore plates of your species of interest using 10 ml ACES buffer. Dilute the suspension 10- to 100-fold and determine the spore concentration.
5. Plate 5–100 μl (depending on the abundance of growth of the species) of a 10^6 spores/ml suspension on the selection plate and spread them evenly using a spatula.
6. Incubate the plates until there is a layer of mature spores on the selection plate and harvest the spores.
7. Count the spore suspension and plate 5–100 μl of a 10^6 spores/ml dilution of this suspension evenly on new selection plates, in which the concentration of the toxic compound is 50% higher than in the previous selection plate.
8. Repeat **steps 6** and 7 until no growth is available anymore or until the desired concentration of the toxic compound is reached.
9. From this final spore suspension, dilute to a final concentration of 10^2 spores/ml and plate 100 μl on a new selection plate.
10. Select the colonies that grow best and purify these by streaking them two times on selection plates.
11. Prepare spore plates of the evolved strain(s) and its/their original parent, harvest them and prepare dilutions containing 10^6 spores/ml, 10^5 spores/ml, 10^4 spores/ml, and 10^3 spores/ml.
12. On a square petri dish with selection media, plate 2 μl of each dilution in a horizontal row and put the different strains vertically underneath each other. For selection media, use the original concentration of the toxic compound, the final concentration and a concentration that is in between them.
13. Incubate the strains to evaluate the difference in growth with the parent.
14. Select the strains with the most significant change and analyze these further by physiological testing, genome resequencing, transcriptomics, proteomics, and/or metabolomics.

References

1. Gonzalez-Ramos D, Gorter de Vries AR, Grijseels SS, van Berkum MC, Swinnen S, van den Broek M, Nevoigt E, Daran JM, Pronk JT, van Maris AJ (2016) A new laboratory evolution approach to select for constitutive acetic acid tolerance in *Saccharomyces cerevisiae* and identification of causal mutations. Biotechnol Biofuels 9:173
2. Wisselink HW, Toirkens MJ, Wu Q, Pronk JT, van Maris AJ (2009) Novel evolutionary engineering approach for accelerated utilization of glucose, xylose, and arabinose mixtures by engineered *Saccharomyces cerevisiae* strains. Appl Environ Microbiol 75:907–914
3. Schoustra S, Punzalan D (2012) Correlation of mycelial growth rate with other phenotypic characters in evolved genotypes of *Aspergillus nidulans*. Fungal Biol 116:630–636
4. Schoustra SE, Bataillon T, Gifford DR, Kassen R (2009) The properties of adaptive walks in evolving populations of fungus. PLoS Biol 7:e1000250
5. Schoustra SE, Punzalan D, Dali R, Rundle HD, Kassen R (2012) Multivariate phenotypic divergence due to the fixation of beneficial mutations in experimentally evolved lineages of a filamentous fungus. PLoS One 7:e50305
6. Patyshakuliyeva A, Arentshorst M, Allijn IE, Ram AFJ, de Vries RP, Gelber IB (2016) Improving cellulase production by *Aspergillus niger* using adaptive evolution. Biotechnol Lett 38:969–974
7. Culleton H, Majoor E, McKie VA, de Vries RP (2016) Evolutionary adaptation as a tool to generate targeted mutant strains as evidence by increased inulinase production in *Aspergillus oryzae*. In: de Vries RP, Benoit Gelber I, Andersen MR (eds) *Aspergillus* and *Penicillium* in the post-genomic era. Caister Academic Press, Norfolk, UK, pp 189–196
8. Baracho MS, Baracho IR (2003) An analysis of the spontaneous mutation rate measurement in filamentous fungi. Genet Mol Biol 26:83–87
9. Lamb BC, Mandaokar S, Bahsoun B, Grishkan I, Nevo E (2008) Differences in spontaneous mutation frequencies as a function of environmental stress in soil fungi at "Evolution Canyon," Israel. Proc Natl Acad Sci U S A 105:5792–5796
10. de Vries RP, Burgers K, van de Vondervoort PJI, Frisvad JC, Samson RA, Visser J (2004) A new black *Aspergillus* species, *A. vadensis*, is a promising host for homologous and heterologous protein production. Appl Environ Microbiol 70:3954–3959

Part II

Data Processing and Analysis

Chapter 13

Genome Assembly

Alicia Clum

Abstract

Genome assembly uses sequence similarity to go from sequencing reads to longer contiguous sequences (contigs). Scaffolds are contigs linked together by gaps where the order and orientation of the contigs is known but the exact sequence connecting two contigs is unknown, represented by Ns which estimate the gap length. Here we describe recommendations for genome assembly for different sequencing technologies, describe organelle assembly, and review how to perform assembly quality control.

Key words Assembly, de Bruijn graph, k-mer, String graph, Overlap–layout–consensus, Ploidy, Sequencing, Genomics, Contigs, Scaffolds

1 Introduction

Genome assembly is an important process for several reasons. One is to reduce the complexity of the data, making downstream analysis such as annotation and comparative genomics more computationally feasible. Another is to span the length of genes or gene clusters to understand biological function. The longer the assembled contigs are, the more the information you have about the order and relative orientation of genes. Reconstructing the consensus for internal transcribed spacer (ITS) is useful for taxonomic identification [1] and generating phylogenetic trees.

There are several challenges for sequencing and assembly of fungi. Fungal genomes cover a large spread of repeat content and genome size (Fig. 1). Frequently ploidy, the number of sets of chromosomes in a cell, is unknown prior to sequencing and assembly. Fungi can be haploid, diploid, or multinucleated, and separation of alleles may be challenging [2]. Many assemblers discard overrepresented sequences such as repeats, or assembly results are fragmented when sequencing errors accumulate to sufficient depth as to appear real. Since reads representing mitochondria occurs at a higher coverage than main chromosomes they can require a separate assembly. Endosymbionts, while biologically interesting, can

Ronald P. de Vries et al. (eds.), *Fungal Genomics: Methods and Protocols*, Methods in Molecular Biology, vol. 1775,
https://doi.org/10.1007/978-1-4939-7804-5_13,

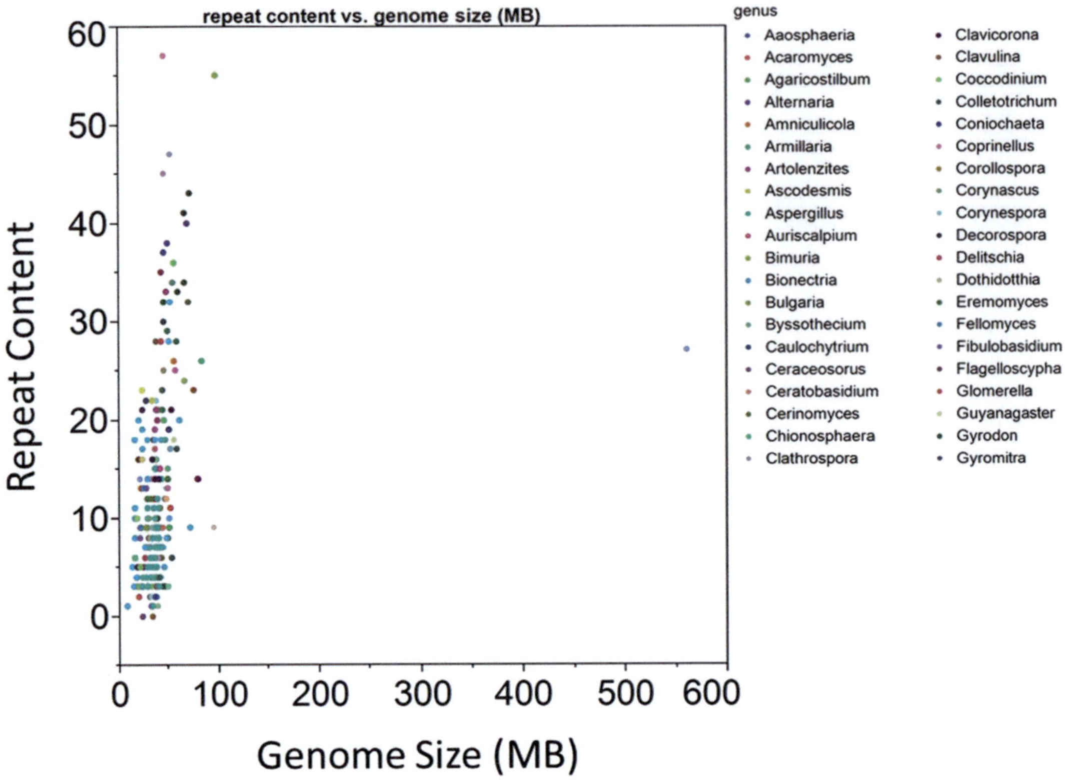

Fig. 1 Genome size in MB and percent repeat content at a k-mer of 25 for taxonomically diverse range of fungi

confuse assemblers expecting a single organism at relatively uniform coverage.

Assembly methods depend on sequencing approaches which include short-read Illumina and long-read PacBio or Oxford Nanopore. The protocols in this chapter use the following assembler types: de Bruijn, overlap–layout–consensus (OLC), and string graph assemblers. Most short-read assemblers are de Bruijn assemblers which break sequencing reads into overlapping k-mers. k-mers are all possible substrings of length k for a given sequence. Selecting an optimal k-mer size is a balancing act; too short and small repeats will not be resolved and more memory will be required, and too long and k-mers may contain sequencing errors. OLC assemblers do all-vs-all pairwise alignments between reads which do not scale well with millions of short reads but have seen a resurgence in popularity in recent years because of long read technologies. String graphs also use reads, discarding alignments that can be inferred from transitivity. For a review of assembly algorithms see *Sequence assembly demystified* [3]. It is recommended that anyone using these software tools reviews any relevant literature specific to the software and the type of assembly algorithm that is used for best practices, description of parameters and outputs. Publications and manuals for all assemblers discussed here are provided in the references.

2 Materials

Genome assembly requires access to compute servers or a personal computer with 120–250 Gb of memory. For de Bruijn graph assembly, memory scales with genome size, so more memory may be needed for fungi larger than 100 Mb. We will discuss different assembly methods depending on the sequencing technology. For Illumina sequencing, we start with a 270 bp insert Illumina fragment library, sequenced 2 × 150 bp such that the ends of the reads overlap, sequenced to several hundred fold coverage in fastq format. Required software and databases includes CASAVA [4], BBTools [5], Velvet [6, 7], NCBI RefSeq [8], bwa [9], MegaBLAST [10], ALLPATHS-LG [11, 12], wgsim [13], and VelvetOptimiser [14].

For PacBio we target a hundred fold coverage of a 10 kb or larger library, yielding reads several kilobases or longer in length in fasta and HDF5 format. Far few total number of reads are needed because of their length. As such, OLC and string graphs are favored here with memory scaling with the number of input reads. For de novo assembly, when possible, we recommend using a long read technology as it will produce a more contiguous and complete assembly (Fig. 2). Required software includes SRMT Portal [15], Falcon [16, 17], Celera [18, 19], and SAMtools [20].

For assembly QC required software includes MegaBLAST, NCBI databases (nt, RefSeq bacteria, RefSeq archaea, RefSeq fungi, RefSeq mitochondrion), UNITE [21], ESTmapper [22], and CEGMA [23].

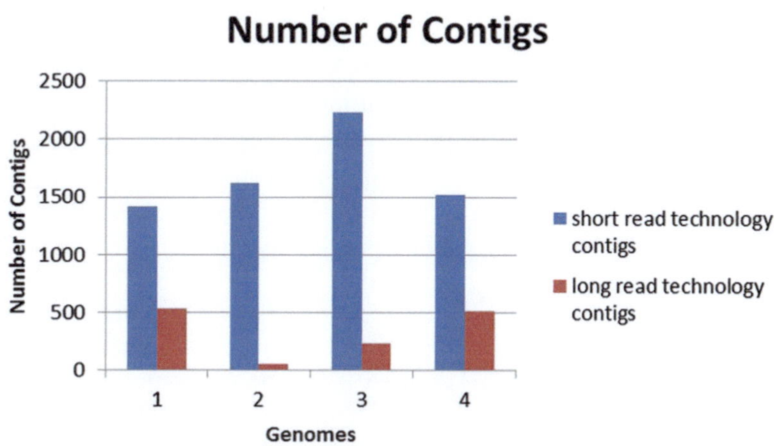

Fig. 2 Number of contigs for the same genome comparing a short read technology approach to a long read technology approach

3 Methods

3.1 Assembly of Illumina Short Reads

Once basecalling and demultiplexing is done with CASAVA we adapter trim data that passes Illumina's chastity filtering. Trimming rather than discarding sequences that matches adapter sequence can retain more data if you some of your inserts are shorter than the read length. You will need to know the adapter sequences for whatever library prep was used to generate the data.

1. Here is an example of command used for trimming using BBDuk, part of BBtools:

   ```
   bbduk.sh ktrim=r minlen=41 mink=11 tbo k=23 hdist=1
   hdist2=1 in1=my_data.fastq ref=adapters.fasta
   out1=temp.fastq.gz
   ```

   We trim the bases to the right (ktrim) of the sequences that match the adapter file using a k-mer size of 23 (k), dropping down to a minimum k-mer (mink) of 11 for additional sensitively at the end of reads with a hamming distance of 1 for both (hdist, hdist2). Hamming distance denotes the distance between two strings. We discard reads and their pair if the read length (minlen) of either drops below 41. Additionally we trim adapters based on where paired reads overlap (tbo).

2. Next, we filter to remove Phi X, which is standardly spiked in to calibrate the basecaller, any other synthetic spike-ins, and any remaining adapters using BBDuk:

   ```
   bbduk.sh maq=13 trimq=0 qtrim=r maxns=0 minlen=41
   minlenfraction=0.33 k=25 hdist=1 in1=temp.fastq.gz
   out1=temp2.fastq.gz outm=synth1.fq.gz
   ref=synthetic_contams.fasta
   ```

   We use a k-mer of 25 (k) with a hamming distance of 1 (hdist), trim bases to the right (qtrim) of a quality score of 0 (trimq), remove any reads which have a minimum average quality (maq) below 13 after quality trimming, and remove reads with ambiguous bases (maxns).

3. Once the data is preprocessed the next step is to separate organellar from the nuclear genome reads. Our approach here is to subsample two million reads and assemble with Velvet. The resulting assembly is aligned to NCBI RefSeq mitochondrion database with a minimum identity of 80% to identify organellar contigs. A secondary assembly is performed with Velvet using cov_cutoff, max_coverage, and exp_cov cutoffs defined from the coverages associated with the contigs previously identified as organelle. These variables signify the minimum, maximum and expected coverage respectively. Read pairs providing linking support between the assembled contigs are identified by aligning the original input fastq to the assem-

bled contigs with bwa using "mem -t 16". The paired reads are used in conjunction with NCBI alignment results to RefSeq mitochondrion to identify trusted organelle contigs. Nuclear genome 18S ribosomal elements are identified by alignment to NCBI nt database with MegaBLAST with a minimum percent identity of 80% and excluded from the list. An enriched set of organelle reads is then created from the original input fastq reads by k-mer matching with BBduk, using defaults, against the resulting whitelist of organelle contigs. Those that do not match the organelle contigs are output into a separate nonOrganelle fastq for downstream assembly.

4. For organelle genome assembly at time of writing ALLPATHS-LG was our preferred assembler of choice for fungal data although it is no longer being actively developed. ALLPATHS-LG requires an unamplified overlapping fragment library and a long mate-pair library. In the absence of a long mate-pair library we simulate 25× of either 1,000 bp or 3,000 bp from the Velvet organelle assembly using wgsim, details provided in the next section. We have an estimated genome size from the prior step so we use that to assemble 125× of the enriched organelle matching read set together with 25x simulated long mate-pairs. Detailed parameters for setting up ALLPATHS-LG parameters are provided in the next section.

5. For the nuclear genome assembly at time of writing an ALLPATHS-LG was our preferred assembler. Reads are subsampled to 20 million reads and assembled with VelvetOptimiser which iterates of k-mers to pick an optimal one.

```
VelvetOptimiser2.pl --s 61 --e 97 --i 4 --t 4 --f "-
shortPaired -fmtAuto my_fastq.gz" --o "-ins_length
250
-min_contig_lgth 500"
```

We use a minimum k-mer of 61 and a max of 97 with a step of 4, an insert length of 250 bp, and a minimum contig length of 500 bp.

If long mate-pair data is not available wgsim is used to simulate ~25× of 2 × 100 bp 3,000±300 bp inserts which is assembled along with the fragment data using ALLPATHS-LG (*see* **Note 1**).

```
wgsim -e 0 -1 100 -2 100 -r 0 -R 0 -X 0 -d 3000 -s 300
-N 4000000 contigs.fa 3000.read1.fastq 3000.read2.fastq
```

Here -e specifies the base error rate, -1 and -2 are specify read lengths -r specifies rate of mutation, -R specifies fraction of indels, -X specifies the probability an indel is extended -d specifies the distance of the insert, -s is the standard deviation, and -N is the number of read pairs.

ALLPATHS-LG requires creating a library and groups csv file. The library csv file contains the following information: library_name, project_name, organism_name, type,paired, frag_size, frag_stddev, insert_size, insert_stddev, read_orientation, genomic_start, genomic_end. The groups csv file contains the following information: group_name, library_name, file_name. We use the assembled genome size from VelvetOptimiser as the input estimated genome size for ALLPATHS-LG. The first three scripts below prepare the data with **step 4** running the actual assembly. See references for location of full ALLPATHS-LG manual.

```
CacheLibs.pl
CACHE_DIR=/projects/genome/allpaths/my_cache_direc-
tory
IN_LIBS_CSV=in_libs.csv ACTION=ADD
CacheGroups.pl
CACHE_DIR=/projects/genome/allpaths/my_cache_direc-
tory
PICARD_TOOLS_DIR=/path/to/picard/toools
IN_GROUPS_CSV=in_groups.csv ACTION=ADD
CacheToAllPathsInputs.pl
CACHE_DIR=/projects/genome/allpaths/my_cache_direc-
tory
DATA_DIR=/projects/genome/allpaths/run_assembly
GENOME_SIZE=40000000 PLOIDY=2 GROUPS="list groups
from
above" FRAG_COVERGE=125 JUMP_COVERAGE=25
RunALLPATHS-LG PRE=/projects REFERENCE_NAME=genome
DATA_SUBDIR=allpaths/run_assembly
RUN=run.150x_125_25
```

3.2 Assembly of Long Reads

1. For illustrative purposes we use P6/C4 PacBio RS II long read data. As with Illumina, removing sequencing adapters and control sequences are required prior to assembly. We use the default RS_Filter protocol provided in SMRT Portal to removed adapters with a small custom modification to additionally remove their synthetic control sequence. This uses a minimum subread length of 50 bp, minimum polymerase read quality of 75, and a minimum polymerase read length of 50 bp. To remove control sequences manually with blasr from SMRT Portal:

```
blasr reads.{fasta,bas.h5}.fasta 2kb_Control.fasta -
unaligned my_cleaned_reads.fasta
```

2. Next, we generate an initial assembly of the data with Falcon. Falcon is a string graph assembler that can scale to large genome sizes and was developed to handle diploid genomes. *See* Fig. 3 for an example fungal config. The example config provided is compatible with a Univa Grid Engine (UGE) scheduler and specifies parameters for coverage cutoffs, length

```
[General]
# list of files of the initial subread fasta files
input_fofn = input.fofn

input_type = raw

# The length cutoff used for seed reads used for initial mapping
length_cutoff = 3000

# The length cutoff used for seed reads used for pre-assembly
length_cutoff_pr = 3000

# Cluster queue setting, -P added by falcon_asm
sge_option_da = --pe pe_slots 8
sge_option_la = -pe pe_slots 2
sge_option_pda = --pe pe_slots 8
sge_option_pla = -pe pe_slots 2
sge_option_fc = -pe pe_slots 24
sge_option_cns = -pe pe_slots 8

# concurrency setting
pa_concurrent_jobs = 8
cns_concurrent_jobs = 8
ovlp_concurrent_jobs = 8

# overlapping options for Daligner
pa_HPCdaligner_option =  -v -dal4 –t16 -e.70 -l1000 -s1000
ovlp_HPCdaligner_option = -v -dal4 -t832-h60 -e.96 -l500 -s1000

pa_DBsplit_option = -x500 -s400
ovlp_DBsplit_option = -x500 -s400

# error correction consensus option
falcon_sense_option = --output_multi --min_idt 0.70 --min_cov 4 --local_match_count_threshold 2 --max_n_read 200 --n_core 6

# overlap filtering options
overlap_filtering_setting = --max_diff 100 --max_cov 100 --min_cov 5 --bestn 10
```

Fig. 3 A sample Falcon configuration file

cutoffs, database sizes, etc. See references for URL of Falcon github repository with additional example configs, detailed parameter explanations and information about outputs. The Falcon assembler is launched with the following command:

```
fc_run.py run.cfg
```

3. As part of the assembly process, long reads are error-corrected to become >99% accurate [24], known as preassembly reads. We take these preassembly reads and use k-mer depth and GC content to enrich for candidate mitochondrial reads, followed by machine learning to identify and separate mitochondrial preassembly reads. These reads are subsequently assembled by Celera. For more information see the mitochondria reconstruction Subheading 3.3.
4. A second Falcon assembly is generated using the preassembly reads after mitochondrial reads have been removed from the dataset. The bulk of the compute is the read vs. read comparison,

so rerunning post-preassembly runs quickly. The output is a fasta files containing primary contigs which represent the primary haplotype and associate contigs which represent the alternative haplotype where there are structural variants.

5. Finally Quiver from SRMT Portal should be run on all contigs. Quiver is a tool to improve the final consensus accuracy. All scripts listed below come with SMRT portal:

   (a) The reference should be indexed using referenceUploader:

```
referenceUploader -c -n "reference" -p
/my/directory/location -f /my/draft/assembly.fasta --
saw="sawriter -blt 8 -welter" --samIdx="samtools faidx"
```

   The –c flag specifies a new reference should be created with –n for name. –saw and –samIdx are flags that get passed to SAMtools.

   (b) Align raw data using pbalign. This requires a file with a list of the raw data (input.fofn), reference locations, and alignment parameters. See SMRT Portal documentation for full parameter explanations. Example command:

```
pbalign "input.fofn" "/my/directory/location/reference"
"/my/dir/location/output.cmp.h5" --seed=1 --
minAccuracy=0.75 --minLength=50--concordant --
algorithmOptions="-useQuality" --algorithmOptions=' -
minMatch 12 -bestn 10 -minPctIdentity 70.0' --
hitPolicy=randombest --tmpDir=$TMPDIR --forQuiver
```

   (c) Chemistry needs to be loaded to the alignment file using the loadChemistry.py command where input.fofn is a file listing the raw HDF5 files (*see* **Note 2**). Example command:

```
loadChemistry.py input.fofn output.cmp.h5
```

   (d) variantCaller.py with –algorithm=quiver is what outputs the corrected consensus. The final assembly fasta in this example is polished_assembly.fasta. Example command:

```
variantCaller.py -P
$SEYMOUR_HOME/analysis/etc/algorithm_parame-
ters/2014-
09/ --algorithm=quiver /my/dir/location/output.cmp.h5 -
r /my/dir/location/reference/sequence/reference.fasta -
o corrections.gff -o polished_assembly.fasta -o
polished_assembly.fastq.gz
```

SEYMOUR_HOME is an environmental variable from SMRT Portal, –P specifies algorithm parameters, and –r specifies the reference and –o specifies output files.

3.3 Reconstructing Mitochondria

There are several reasons we separate out and reassemble mitochondria. Mitochondrial reads generally occur at much higher depth than the main genome. Some assemblers intentionally exclude regions of high depth. Higher abundance of these reads can mean that sequencing errors can accumulate and start to look real, complicating the assembly graph and causing assembly fragmentation. For both these reason separating and reassembly this data can produce more contiguous and complete results. Lastly, mitochondria can use a different genetic code than the nuclear genome [25, 26] so it is preferable for annotation if the consensus sequences are provided separately. There are some rules that generally hold true that we can use to identify these areas. For cultured isolates, mitochondrial reads occur at higher sequencing depth than the nuclear data and are generally lower GC content than the main chromosomes. Other features such as tetranucleotide frequency can be used or Hidden Markov Models (HMM) can be generated for conserved mitochondrial genes like cytochrome c oxidase I (COX1) [27]. Frequently there is enough similarity to already sequenced mitochondria that BLAST hits to RefSeq mitochondria can be used to confirm suspect contigs. Note that some fungi lack mitochondria [28] or could be lost during DNA extraction or library size selection.

3.4 Assembly QC

The purpose of assembly QC is to confirm the organism identity, check for contamination, evaluate if there is sufficient coverage, and determine if the library type and sequencing technology used produce an assembly that meets your scientific needs. Some of these methods may provide redundant information, with the hope that at least one method will catch problems. Here we use the input reads, the contigs generated from the assembly, transcriptome data if available, and Sanger sequence of the ITS.

1. The ITS has been formally proposed as the primary fungal barcode marker [1]. We use MegaBLAST with an identity of 90% to compare the Sanger ITS sequenced region to what is assembled, as well as to UNITE which is a publicly available database of ribosomal DNA ITS sequences. We also MegaBLAST with an identity of 90% to various other NCBI databases such as nt, RefSeq bacteria, archaea, fungi, and mitochondrion to confirm that there are no hits to mitochondria in the nuclear assembly or to contaminants and to confirm organism's identity. Example MegaBLAST command:

```
blastn -task megablast -perc_identity 90 -evalue 1e-30
-dust yes -num_threads 8 -query polished_assembly.fasta
-subject /mydatabases/UNITE
```

2. If available, we map a 1% subsample of the transcriptome reads or the entire transcriptome assembly to the genome assembly. For mapping transcriptome reads we use bbmap.sh with default

parameters and a minimum scaffold size of 500 bp. To map the transcriptome assembly we use ESTmapper using a 90% identity and 85% coverage cutoff. If either of these is below 90%, further investigation is warranted. Good mapping would confirm that RNA and DNA are from the same organism and help assess assembly completeness. Example command to map transcriptome reads as follows:

```
bbmap.sh ref= polished_assembly.fasta minscaf=500
in=transcriptome_reads.fastq.gz
out=mapped_transcriptome.sam > mapping.stdout 2 >
mapping.stderr
```

3. Another way to assess completeness is using Core Eukaryotic Genes Mapping Approach (CEMGA) or Benchmarking Universal Single-Copy Orthologs (BUSCO) [29]. CEGMA is based on conserved protein families among eukaryotes. BUSCO is based on near-universal single-copy orthologs from OrthoDB v9 [30]. CEGMA is unfortunately no longer supported. Multiple copies of genes which are expected to be present in only one copy may indicate contamination. For cultured isolates, anything below 90% should be investigated further. Example CEGMA command:

```
cemga --genome polished_assembly.fasta
```

4. To look for contamination we break the contigs into 5000 bp pieces and map the reads back using a short or long read aligner and MegaBLAST each piece using 90% identity to RefSeq fungi, bacteria and archaea. Then a plot for each taxonomic level is made plotting the coverage vs GC content, colored by taxonomic hit. Frequently contaminates are a different GC content or coverage, so unless a sample is contaminated by a closely related organism at the same coverage plotting this information can be a very simple visual for identifying contamination (Fig. 4).
5. We also recommend calculating the tetranucleotide frequencies and generating a principal component analysis plot. Tetranucleotide frequencies are highly conserved across a genome [31]. Therefore, contigs which belong to different organisms should cluster separately even if their GC content is similar.
6. Other figures than can be useful visuals are k-mer histograms or plotting contig length vs. GC content. To make a k-mer histogram plots we use kmercountexact.sh from BBtools with default parameters.

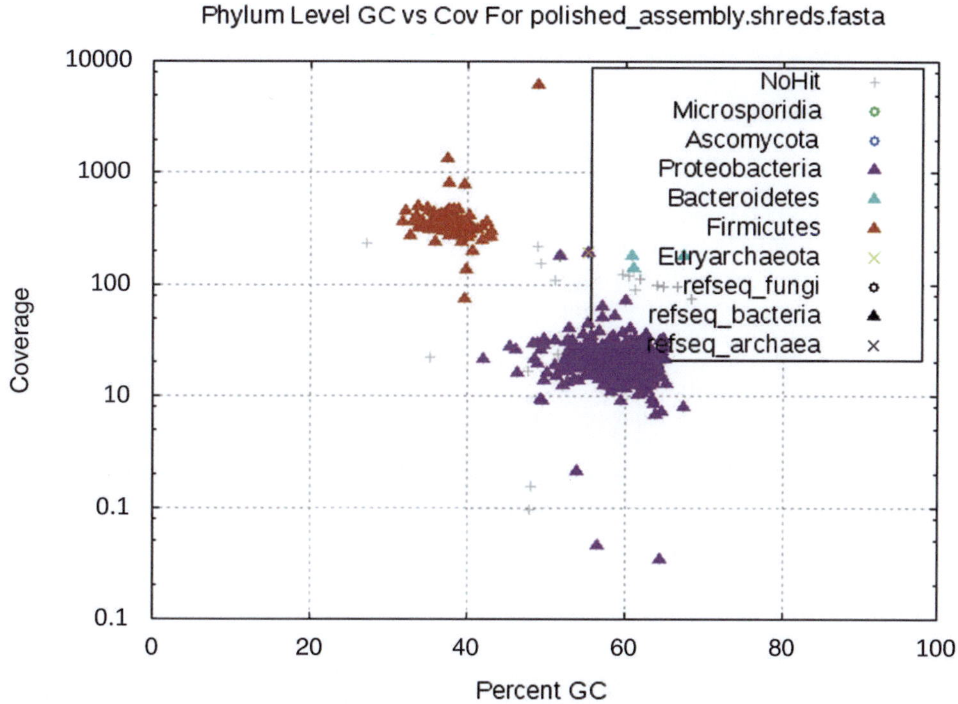

Fig. 4 An example of identifying the presence of multiple organisms by plotting coverage vs. GC content overlaid with taxonomic hits

4 Notes

1. We no longer generate Illumina mate-pair data but ALLPATHS-LG and several other de Bruijn assemblers can accommodate this data which can improve scaffolding and resolve some repeats.
2. Quiver has specific training models for each chemistry version, so results will be suboptimal if this step is skipped.

Acknowledgment

The work conducted by the US Department of Energy Joint Genome Institute, a DOE Office of Science User Facility, is supported under Contract No. DE-AC02-05CH11231.

References

1. Schoch CL, Seifert KA, Huhndorf S, Robert V, Spouge JL, Levesque CA, Chen W, Fungal Barcoding Consortium, Fungal Barcoding Consortium Author List (2012) Nuclear ribosomal internal transcribed spacer (ITS) region as a universal DNA barcode marker for fungi. Proc Natl Acad Sci U S A 109(16):6241–6246. https://doi.org/10.1073/pnas.1117018109
2. Roper M, Ellison C, Taylor JW, Glass NL (2011) Nuclear and genome dynamics in multinucleate ascomycete fungi. Curr Biol 21(18):R786–R793. https://doi.org/10.1016/j.cub.2011.06.042
3. Nagarajan N, Pop M (2013) Sequence assembly demystified. Nat Rev Genet 14(3):157–167. https://doi.org/10.1038/nrg3367
4. Illumina. http://support.illumina.com/sequencing/sequencing_software/casava/documentation.html. Accessed 30 Nov 2016
5. Joint Genome Institute (1997) BBTools. http://jgi.doe.gov/data-and-tools/bbtools/. Accessed 30 Nov 2016
6. Zerbino DR, Birney E (2008) Velvet: algorithms for de novo short read assembly using de Bruijn graphs. Genome Res 18(5):821–829. https://doi.org/10.1101/gr.074492.107
7. The European Bioinformatics Institute. https://www.ebi.ac.uk/~zerbino/velvet/Manual.pdf. Accessed 30 Nov 2016
8. O'Leary NA, Wright MW, Brister JR, Ciufo S, Haddad D, McVeigh R, Rajput B, Robbertse B, Smith-White B, Ako-Adjei D, Astashyn A, Badretdin A, Bao Y, Blinkova O, Brover V, Chetvernin V, Choi J, Cox E, Ermolaeva O, Farrell CM, Goldfarb T, Gupta T, Haft D, Hatcher E, Hlavina W, Joardar VS, Kodali VK, Li W, Maglott D, Masterson P, McGarvey KM, Murphy MR, O'Neill K, Pujar S, Rangwala SH, Rausch D, Riddick LD, Schoch C, Shkeda A, Storz SS, Sun H, Thibaud-Nissen F, Tolstoy I, Tully RE, Vatsan AR, Wallin C, Webb D, Wu W, Landrum MJ, Kimchi A, Tatusova T, DiCuccio M, Kitts P, Murphy TD, Pruitt KD (2016) Reference sequence (RefSeq) database at NCBI: current status, taxonomic expansion, and functional annotation. Nucleic Acids Res 44(D1):D733–D745. https://doi.org/10.1093/nar/gkv1189
9. Li H, Durbin R (2009) Fast and accurate short read alignment with Burrows-Wheeler transform. Bioinformatics 25(14):1754–1760. https://doi.org/10.1093/bioinformatics/btp324
10. Boratyn GM, Camacho C, Cooper PS, Coulouris G, Fong A, Ma N, Madden TL, Matten WT, McGinnis SD, Merezhuk Y, Raytselis Y, Sayers EW, Tao T, Ye J, Zaretskaya I (2013) BLAST: a more efficient report with usability improvements. Nucleic Acids Res 41(Web Server issue):W29–W33. https://doi.org/10.1093/nar/gkt282
11. Gnerre S, Maccallum I, Przybylski D, Ribeiro FJ, Burton JN, Walker BJ, Sharpe T, Hall G, Shea TP, Sykes S, Berlin AM, Aird D, Costello M, Daza R, Williams L, Nicol R, Gnirke A, Nusbaum C, Lander ES, Jaffe DB (2011) High-quality draft assemblies of mammalian genomes from massively parallel sequence data. Proc Natl Acad Sci U S A 108(4):1513–1518. https://doi.org/10.1073/pnas.1017351108
12. The Broad Institute. http://software.broadinstitute.org/allpaths-lg/blog/?page_id=12. Accessed 30 Nov 2016
13. GitHub wgsim. https://github.com/lh3/wgsim/. Accessed 30 Nov 2016
14. GitHub VelvetOptimiser. https://github.com/tseemann/VelvetOptimiser. Accessed 30 Nov 2016
15. PACBIO®. http://www.pacb.com/products-and-services/analytical-software/smrt-analysis/. Accessed 30 Nov 2016
16. Chin CS, Peluso P, Sedlazeck FJ, Nattestad M, Concepcion GT, Clum A, Dunn C, O'Malley R, Figueroa-Balderas R, Morales-Cruz A, Cramer GR, Delledonne M, Luo C, Ecker JR, Cantu D, Rank DR, Schatz MC (2016) Phased diploid genome assembly with single-molecule real-time sequencing. Nat Methods 13(12):1050–1054. https://doi.org/10.1038/nmeth.4035
17. Github. https://github.com/PacificBiosciences/FALCON/wiki/Manual. Accessed 30 Nov 2016
18. Myers EW, Sutton GG, Delcher AL, Dew IM, Fasulo DP, Flanigan MJ, Kravitz SA, Mobarry CM, Reinert KH, Remington KA, Anson EL, Bolanos RA, Chou HH, Jordan CM, Halpern AL, Lonardi S, Beasley EM, Brandon RC, Chen L, Dunn PJ, Lai Z, Liang Y, Nusskern DR, Zhan M, Zhang Q, Zheng X, Rubin GM, Adams MD, Venter JC (2000) A whole-genome assembly of Drosophila. Science 287(5461):2196–2204
19. Sourceforge. http://wgs-assembler.sourceforge.net/wiki/index.php?title=Main_Page. Accessed 30 Nov 2016
20. Li H, Handsaker B, Wysoker A, Fennell T, Ruan J, Homer N, Marth G, Abecasis G, Durbin R, Genome Project Data Processing Subgroup (2009) The sequence alignment/

map format and SAMtools. Bioinformatics 25(16):2078–2079. https://doi.org/10.1093/bioinformatics/btp352

21. Koljalg U, Larsson KH, Abarenkov K, Nilsson RH, Alexander IJ, Eberhardt U, Erland S, Hoiland K, Kjoller R, Larsson E, Pennanen T, Sen R, Taylor AF, Tedersoo L, Vralstad T, Ursing BM (2005) UNITE: a database providing web-based methods for the molecular identification of ectomycorrhizal fungi. New Phytol 166(3):1063–1068. https://doi.org/10.1111/j.1469-8137.2005.01376.x
22. Xue W, Lee W-J, Tseng C-W (2005) ESTmapper: efficiently aligning DNA sequences to genomes. IPDPS 7(8):196a. https://doi.org/10.1109/IPDPS.2005.204
23. Parra G, Bradnam K, Korf I (2007) CEGMA: a pipeline to accurately annotate core genes in eukaryotic genomes. Bioinformatics 23(9):1061–1067. https://doi.org/10.1093/bioinformatics/btm071
24. Chin CS, Alexander DH, Marks P, Klammer AA, Drake J, Heiner C, Clum A, Copeland A, Huddleston J, Eichler EE, Turner SW, Korlach J (2013) Nonhybrid, finished microbial genome assemblies from long-read SMRT sequencing data. Nat Methods 10(6):563–569. https://doi.org/10.1038/nmeth.2474
25. Osawa S, Jukes TH, Watanabe K, Muto A (1992) Recent evidence for evolution of the genetic code. Microbiol Rev 56(1):229–264
26. Jukes TH, Osawa S (1993) Evolutionary changes in the genetic code. Comp Biochem Physiol B 106(3):489–494
27. Aguileta G, de Vienne DM, Ross ON, Hood ME, Giraud T, Petit E, Gabaldon T (2014) High variability of mitochondrial gene order among fungi. Genome Biol Evol 6(2):451–465. https://doi.org/10.1093/gbe/evu028
28. Alexopolous CJ, Mims CW, Blackwell M (2004) Introductory mycology, 4th edn. Wiley, Hoboken, NJ. ISBN 0-471-52229-5
29. Simao FA, Waterhouse RM, Ioannidis P, Kriventseva EV, Zdobnov EM (2015) BUSCO: assessing genome assembly and annotation completeness with single-copy orthologs. Bioinformatics 31(19):3210–3212. https://doi.org/10.1093/bioinformatics/btv351
30. Kriventseva EV, Tegenfeldt F, Petty TJ, Waterhouse RM, Simao FA, Pozdnyakov IA, Ioannidis P, Zdobnov EM (2015) OrthoDB v8: update of the hierarchical catalog of orthologs and the underlying free software. Nucleic Acids Res 43(Database issue):D250–D256. https://doi.org/10.1093/nar/gku1220
31. Noble PA, Citek RW, Ogunseitan OA (1998) Tetranucleotide frequencies in microbial genomes. Electrophoresis 19(4):528–535. https://doi.org/10.1002/elps.1150190412

Chapter 14

Fungal Epigenomics: Detection and Analysis

Stephen J. Mondo, Rita C. Kuo, and Vasanth R. Singan

Abstract

Across Eukaryota, DNA modifications play an important role in regulation of gene expression. While 5-methylcytosine (5mC) has been explored in depth, other modifications such as 6-methyladenine (6 mA) have historically been overlooked, in part due to technical difficulties in collecting/analyzing these data. However, recent technological advances have enabled exploration of these marks with much greater detail and on a larger scale. In this chapter, we discuss multiple methods for identifying and analyzing both 5mC and 6 mA across fungi.

Key words Epigenomics, Fungi, 6-Methyladenine, 5-Methylcytosine, SMRT-analysis, Bisulfite-sequencing, 6 mA-immunoprecipitation followed by sequencing (6 mA-IP-seq)

1 Introduction

Owing to their small and comparatively simple genomes, Fungi represent one of the most attractive kingdoms for exploration of eukaryotic genomics and gene regulation. In Fungi (and most eukaryotes), 5mC is a well-known suppressor of gene expression, primarily of transposons [1]. However, explorations into the role of 6 mA in eukaryotic gene regulation has only recently begun [2–5]. While bisulfite sequencing has a long history of success for high confidence identification of methylated cytosines [6], high-throughput characterization of 6 mA in eukaryotes is largely accomplished through use of PacBio sequencing and 6 mA-immunoprecipitation followed by sequencing (6 mA-IP-seq) [7]. Each of these methods and caveats is discussed below.

2 Materials

Library preparation for bisulfite sequencing, PacBio sequencing, and 6 mA-IP-Seq all require separate materials, which are listed individually below.

Ronald P. de Vries et al. (eds.), *Fungal Genomics: Methods and Protocols*, Methods in Molecular Biology, vol. 1775, https://doi.org/10.1007/978-1-4939-7804-5_14, © Springer Science+Business Media, LLC, part of Springer Nature 2018

2.1 Materials for Bisulfite-Seq Library Preparation

1. Reagents and kits: NEB Ultra II DNA Library Prep (New England Biolabs), Kapa HiFi Uracil + (Kapa Biosystems), Agencourt AMPure XP beads (Backman Culture Life Sciences), EZ DNA Methylation-Lightening Kit (Zymo Research), Buffer EB (Qiagen Inc.), TE Buffer, NEBNext® Multiplex Methylated Oligos for Illumina® (Methylated Adaptor Index Primers Set, New England Biolabs).
2. Equipment: Covaris LE220 and Covaris Micro Tube (Covaris Inc.).

2.2 Materials for PacBio Library Preparation

1. Reagents and kits: Covaris g-TUBES (Covaris), PacBio SMRTbell Template Prep Kit (Pacific Biosciences), PacBio AMPure Beads (Pacific Biosciences), DNF-464 High Sensitivity Large Fragment 50 kb Analysis Kit (Advanced Analytical).
2. Equipment: Eppendorf Mini Spin Plus (Eppendorf), Fragment Analyzer (Advanced Analytical).

2.3 Materials for 6 mA-IP and Library Preparation

1. Reagents and kits: Agencourt AMPure XP beads (Backman Culture Life Sciences), Pierce™ Protein A Magnetic Beads (Thermo Fisher Scientific), Anti-6 mA antibody (ABE572, Millipore), NEBNext Multiplex Oligos and PCR primers for Illumina (New England Biolabs; You can replace this with IDT Illumina adapters), NEBNext Ultra II DNA Library Prep Kit (New England Biolabs), TE Buffer, bovine serum albumin (BSA).
2. 5× IP buffer: 50 mM Tris–HCl (pH 7.4), 750 mM NaCl, and 0.5% Igepal CA-630.
3. Equipment: Covaris LE220 and Covaris micorTUBEs (Covaris Inc.).

3 Methods

3.1 Bisulfite-Sequencing for Detecting of Cytosine Methylation

This protocol utilizes NEB Ultra II DNA Library Prep kit for Illumina adapter ligation and Zymo EZ DNA Methylation-Lightening Kit for bisulfite conversion. The procedures are optimized based on the manufacturers' protocol and are designed to produce Bisulfite-Seq with target size around 500 bp for sequencing on the Illumina HiSeq platform using a 2 × 150 recipe. Starting DNA amount is 1 μg. Methylated adapters are required. You can also add 0.02% of synthetic DNA spike-in to verify bisulfite conversion via PCR prior sequencing.

1. DNA fragmentation (Optional: Add 0.1% of **unmethylated** lambda DNA as a Spike-In control before DNA fragmentation). Bring 1 μg of DNA to 50 μL using TE buffer and

sonicate to 500 bp in a microTUBE using a Covaris LE220 (peak power 450, duty factor 10%, cycle burst 200, 80 s, temperature 4 °C).

2. Confirm the size of the sheared DNA with TapeStation or Bioanalyzer (*see* **Note 1**).
3. Purify DNA with 0.65× Ampure beads (32.5 μL, *see* **Note 2**), elute DNA in 50 μL of EB Buffer.
4. For end repair, A-tailing and adaptor ligation using NEBNext Ultra II Kit, add 7 μL of Ultra II End Prep Reaction Buffer and 3 μL of Ultra II End Prep Enzyme Mix to the fragmented DNA. Pulse-vortex, quickly spin, and incubate the mixture on a thermocycler at 20 °C for 30 min, 65 °C for 30 min, and 4 °C for 5 min.
5. Add 2.5 μL of Methylated Illumina adapter (10 μM) to the reaction and mix well, then add 1 μL of the Ligation Enhancer and 30 μL of Ligation Master Mix and incubate the mixture (93.5 μL in total) in a thermocycler at 20 °C for 30 min.
6. Purify DNA with 0.9× Ampure beads (84 μL) and then elute DNA with 21 μL of EB Buffer. Use NanoDrop or Qubit to quantify DNA. Total amount of DNA should be less than 1 μg (optimal range for conversion is 200–500 ng).
7. To perform Bisulfite Conversion with Zymo EZ DNA Methylation-Lightening Kit, add 130 μL of Lightening Conversion Reagent to 20 μL of adapter ligated DNA then place the PCR tube in a thermocycler for the following steps: 99 °C for 8 min, 54 °C for 60 min, and 4 °C for storage up to 20 h.
8. Clean up DNA using the provided columns. Add 600 μL of M-Binding Buffer to the bisulfite converted DNA, mix well and add the mixture to the column. Centrifuge at 10,000 × *g* for 30 s, then discard flow-through.
9. Add 100 μL of M-Wash Buffer to the column and centrifuge at max speed for 30 s.
10. Add 200 μL of L-Desulphonation Buffer to the column and incubate the reaction at room temperature for 20 min and then centrifuge at max speed for 30 s.
11. Wash the column twice with 200 μL of M-Wash Buffer.
12. Place the column into a new tube and centrifuge one more time at max speed for 1 min to eliminate any buffer residual on the column. Place the column into a new tube, add 27 μL of M-Elution Buffer to the column, close the cap and incubate at room temperature for 15 min.
13. Centrifuge at max speed for 1 min.
14. Optional: Conversion QC using Spike-In (If Spike-In DNA was added prior DNA fragmentation, otherwise move to **step 17**).

15. Prepare 2 PCR reactions. Add the following component to each reaction: **Reaction 1:** 1 μL of Bisulfite converted DNA, 2 μL of wild type primer mix (25 μM, *see* **Note 3**), 9.5 μL of water, and 12.5 μL of 2× Kapa HiFi Uracil. **Reaction 2:** 1 μL of bisulfite-converted DNA, 2 μL of bisulfite primer mix (25 μM, *see* **Note 3**), 9.5 μL of water, and 12.5 μL of 2× Kapa HiFi Uracil.
16. Place the tube on a thermocycler and perform the following steps: 98 °C for 45 s, then 35 cycles of 98 °C for 15 s, 60 °C for 30 s, and 72 °C for 30 s. Incubate for 72 °C for 1 min, then hold at 4 °C.
17. Use Bioanalyzer or TapeStation to verify the size of the PCR amplicon. If the DNA has been converted successfully, PCR with wild type primer should have **no** amplification and PCR with bisulfite primer should have amplification with a peak around the size that you designed for PCR.
18. Library amplification, clean up and QC. Add 2 μL of Illumina Primer Mix (25 μM) and 25 μL of 2× Kapa HiFi Uracil + to the converted DNA. Place the tube on a thermocycler and perform the following steps: 98 °C for 45 s, followed by 10 cycles of 98 °C for 15 s, 60 °C for 30 s, and 72 °C for 30 s. Lastly, hold at 72 °C for 1 min, then hold at 4 °C.
19. Purify library 2 times with 0.9× Ampure beads and elute library with 28 μL of EB Buffer after the second purification. Use Bioanalzyer or TapeStation to confirm size, quality, and quantity of library.

Following library preparation, quantify the library using qPCR. Then prepare samples for sequencing on the Illumina HiSeq sequencing platform, following a 2 × 151 indexed run recipe. Please *see* Chapter 1 for Illumina sequencing preparation.

3.2 Analysis of Bisulfite-Sequencing Data

A plethora of tools exist for analyzing bisulfite-seq data and this protocol specifically discuss two software packages, bbtools (for preprocessing; https://sourceforge.net/projects/bbmap/) and bismark (for methylation analysis; [8]).

1. Preprocessing. Raw fastq files from Illumina sequencers need to be processed to trim adapter sequences and low quality sequences. Additionally, spike-in sequences and contaminants should be removed. BBDuk within the bbtools package is used to initially trim adapters. A fasta file containing all adapters used with sequencing is provided as a reference to bbduk. Bbduk uses a kmer-based trimming routine to trim adapters from the reads.
2. Trim adaptors from raw reads using bbduk. An example command and parameters used are described as follows:

```
bbduk.sh in=raw.fastq.gz out=trimmed.fastq.gz
ref=adapters.fa k=23 ktrim=r minlen=51 mink=11
hdist=1
```

Parameters: `k=23`, Kmer length used for finding adapters/contaminants. Contaminants/adapters shorter than k will not be found. `ktrim=r`, trim reads to the right, to remove bases matching reference kmers. `minlen=51`, reads shorter than 51 bp after trimming will be discarded. `mink=11`, look for shorter kmers at the reads tops down to 11 bp. `hdist=1`, maximum hamming distance for ref. kmers (substitutions only).

3. BBDuk is used as second pass to trim for low quality sequences. An example command and parameters used are described as follows:

```
bbduk.sh in=trimmed.fastq.gz out=filtered.fastq.
gz qtrim=r trimq=6 minlength=51 hdist=1
```

Parameters: `qtrim=r`, trim reads to the right, to remove bases below quality score. `trimq=6`, regions with average quality score below 6 will be trimmed.

4. The filtered fastq file is used for subsequent methylation analysis. Detailed help/usage guides for bbduk can be obtained by executing

```
bbduk.sh -help
```

5. Methylation analysis. This protocol describes methylation analysis using Bismark v0.16.3. To prepare the genome for bisulfite-seq analysis, it first needs to be prepared for mapping. This includes bisulfite conversion and indexing using bowtie (*see* **Note 4**):

```
bismark_v0.16.3/bismark_genome_preparation --bow-
tie1 --single_fasta --genomic_composition --verbose
/PATH/TO/genome_ref
```

Parameters: `--bowtie1`, use bowtie1. `--single_fasta`, input is a single fasta formatted file. `--genomic_composition`, calculate frequency of all mono and dinucleotides across the reference genome. `--verbose`, provide detailed information while running. `/PATH/TO/genome_ref`, location of the folder containing your reference genome sequence. Bisulfite converted sequences and bowtie indexes are created within the same folder.

6. Alignment. The next step aligns reads to the genome reference and calls methylation. Sequence reads are fully bisulfite-converted into forward (C→T) and reverse reads (G→A) and then aligned to bisulfite-converted genome. Uniquely aligned reads are reported from the four alignments processes and then compared to normal genomic sequence to infer methyla-

tion states at all cytosine positions. The alignments are reported in BAM/SAM format. A summary of the alignment results is provided as a text file. Example command line:

```
bismark_v0.16.3/bismark -X 1000 --bowtie1 -n 1 -l
50 /PATH/TO/genome_ref -1 read_1.fastq -2 read_2.
fastq
```

Parameters: `-X 1000`, the maximum insert size for valid paired-alignments set to 1000 bp. `--bowtie1`, Use bowtie1 for mapping. `-n 1`, maximum number of mismatches permitted in the seed set to 1. `-l 50`, the seed length is set to 50, i.e., the number of bases of high quality end of the read to which the `-n` ceiling applies.

7. Deduplication. This post-processing step removes redundant alignments to control for PCR amplification biases. A new deduplicated BAM file is created as part of this step. Example command line:

```
bismark_v0.16.3/deduplicate_bismark --bam read_1_
bismark_pe.bam
```

Parameters: `--bam`, the input bam file for deduplication.

8. Methylation Extraction (*see* **Note 5**). This is an optional step that extracts methylation information from the alignments. Deduplicated BAM file may be provided as input for the extractor. This routine provides a lot of additional information including context-specific methylation statistics, filtering options, and creation of bedGraph and coverage files. Example command line:

```
bismark_v0.16.3/bismark_methylation_extractor --mul-
ticore 16 --bedGraph --CX --genome_folder /PATH/
TO/genome_ref --cytosine_report read_1_bismark_
pe.deduplicated.bam
```

Parameters: `--bedGraph`, write methylation output into a sorted bedGraph file. `--CX`, sorted bedGraph file contains information on every single cytosine that was covered in the experiment. `--cytosine_report`, produce a genome-wide cytosine report for all cytosines in the genome.

3.3 PacBio Sequencing for Characterization of Adenine DNA Methylation

Due to the kinetics of DNA sequencing using the PacBio platform, a wealth of epigenomic information can be extracted from DNA, making it an attractive tool not only for generation of high quality genome assemblies, but also for simultaneous characterization of the methylome. The protocol for PacBio library preparation requires 1–5 μg of high molecular weight gDNA to prepare libraries for PacBio sequencing, which can be utilized to detect 6 mA. These procedures were adapted from Pacific Bioscience's protocol. The average size of the library should be around 10 kb.

A 6-h movie is recommended for sequencing and at least a 50×–coverage effort is required to obtain high confidence data.

1. DNA fragmentation using g-TUBEs (*see* **Note 6**): adjust DNA concentration to 100 ng/μL (5 μg of DNA in 50 μL of buffer), then load the sample to the cap of g-TUBE and centrifuge the g-TUBE using an Eppendorf MiniSpin Plus centrifuge for 60 s at 2,029 × *g*.
2. Reverse the g-TUBE and repeat spin for 60 s at 5500 rpm and collect the sample from the g-TUBE cap.
3. Perform a QC using Fragment Analyzer or a pulse gel (*see* **Note 7**).
4. Purify and size select DNA with 0.45× Ampure PB beads (e.g., add 22.5 μL of beads to 50 μL of DNA). Elute DNA in 39 μL of EB and transfer 38 μL to a new tube (*see* **Note 8**).
5. ExoVII digestion: add 5 μL of Damage Repair Buffer, 5 μL ATP High, 0.5 μL of DNA+, 0.5 μL of dNTP, and 1 μL of EXoVII to 38 μL of sheared DNA for 50 μL of total reaction volume. Mix the reaction by flicking the tube, quick spin and incubate at 37 °C for 15 min.
6. Repair damaged DNA by adding 2 μL of Damage Repair Mix into the ExoVII treated DNA. Mix the reaction by flicking the tube, followed by a quick spin, then incubate at 37 °C for 30 min and return the reaction to room temperature for 3 min.
7. Next, perform end repairing by adding 2.5 μL End Repair Mix to the DNA. Mix the reaction by flicking the tube, then incubate the reaction at 25 °C for 5 min. Purify the DNA with 0.45× Ampure PB beads and elute DNA in 32 μL of EB Buffer (*see* **Note 8**).
8. Adapter Ligation and Exonuclease digestion: Add 1 μL of PacBio Blunt Adapter to the end-repaired DNA (*see* **Note 9**). Mix the reaction, then add 4 μL of 10× Template Prep Buffer, 2 μL of ATP Lo, and 1 μL of Ligase to 33 μL of adapter–DNA mixture for 40 μL of total reaction volume. Mix the reaction again by flicking the tube and incubate at 25 °C overnight, then 65 °C for 10 min and return the reaction to room temperature for 3 min.
9. Add 1 μL of Exonuclease III and Exonuclease VII. Mix the reaction by flicking the tube, then incubate at 37 °C for 1 h, then return the reaction to 4 °C and proceed with bead purification immediately.
10. Purify the library with 0.45× Ampure PB beads twice (*see* **Notes 8** and **9**) and elute the final library in 20 μL of EB Buffer. Use Fragment Analyzer and Qubit to assess quality and to quantify the library. Library yield should be >10% of your

DNA input. After the library has passed QC, anneal PacBio Sequencing primer to the SMRTbell template library and bind the PacBio polymerase to the template. Use at least a 6-h movie for sequencing to obtain high coverage of consensus reads. The loading concentration should be optimized as it will significantly affect your data quality. Please *see* Chapter 1 for preparing samples for PacBio sequencing.

3.4 SMRT-Analysis of PacBio Data to Detect Methylated Adenines

After sequencing, the methylome can be analyzed through mapping raw read data back to the resulting assembly. For high confidence calling of modified adenines, PacBio recommends a minimum of 25× per-strand coverage. PacBio has developed tools for analyzing these data, which include prediction of modifications on all four nucleotides (*see* **Note 10**). Detection of methylated bases is accomplished through the use of the PacBio SMRT-portal, an open source workbench for analysis of PacBio data (for more information, see http://www.pacb.com/products-and-services/analytical-software/smrt-analysis).

1. Log in and upload genome assembly to SMRT portal. Click Import and Manage → Manage Reference Sequences. At the bottom right hand corner, select "New…" and input info on the Name (filename), Organism (Organism name), and ploidy, then browse to select reference genome for upload.
2. Design your job. Once upload of reference genome is complete, select "Design Job" at the top left of the screen and click "Create New." From protocols, select RS_Modification_Detection and the reference genome uploaded in **step 1** from the Reference pull-down. You can further modify parameters of the run by pressing the "…" button to the left of the protocol pulldown if desired. Select the SMRT cells which correspond to your particular sequencing run and move them to the "SMRT Cells in this job" panel by clicking the right-facing arrow.
3. Launch the analysis. Input a job name (*see* **Note 11**) and start the run by clicking "Start." This will launch the modification detection pipeline. You can monitor your job by selecting the "monitor jobs" tab, clicking on the run of interest, and pressing "Open." This will give you information on the progress of the pipeline, and results of steps already completed (*see* **Note 12**).
4. Once the modification detection pipeline has completed, it will produce two outputs relevant to DNA modifications: modifications.csv and modifications.gff. The csv file contains information on every single base in your genome assembly including probability that the base is modified, the modification ratio, coverage, etc. As this information is stranded, each base has two rows of output, one row for each direction. The modifications.gff file contains info on each base showing significant evi-

dence of modification (mQV > 20, roughly equivalent to $p \leq 0.05$), in gff format. For more details on what these files contain and how these data are calculated, see https://github.com/PacificBiosciences/Bioinformatics-Training/wiki/Methylome-Analysis-Technical-Note.

5. To increase stringency for calling methylated adenines, we recommend further filtering your results by using a minimum per-strand coverage cutoff (at least 15×) and increased mQV minimum of 25 (roughly equivalent to $p \leq 0.01$), as well as a maximum coverage cutoff which should be determined on a per-sample basis (*see* **Note 13**).
6. For quick viewing of 6 mA in a genome browser such as IGV [9], you can collect all 6 mA modifications using the built-in unix command `grep`. Example command line:

```
grep m6A modifications.gff > outfile.gff
```

Parameters: *m6A*, the search string to look for within the provided infile. `modifications.gff`, the file within which you are searching. >, instead of reporting to screen (stdout), save to a file. outfile.gff, the filename to save your results. Replace "`outfile.gff`" with your desired filename, then load that into IGV as a new track (Fig. 1).

3.5 6 mA Detection with IP-Sequencing

6 mA-IP was modified from the previous published 6 mA RNA-IP protocol [7, 10]. This protocol is to prepare libraries for sequencing on the Illumina MiSeq platform using a 1 × 50 run recipe.

1. DNA fragmentation: Dilute 10 μg of genomic DNA (*see* **Note 14**) to 110 μL using TE buffer. Sonicate in 55 μL to 150 bp in 2 microTUBEs using a Covaris LE220 (peak power 450, duty factor 30, cycle burst 1000, 600 s). Verify the size of the sheared DNA (*see* **Note 15**). Combine samples together into a lobind 1.5 mL tube, purify sheared DNA using 1:1 Ampure beads and elute DNA in 50 μL of EB Buffer.
2. End repair, A-tailing and adaptor ligation using NEBNext UltraII kit: add 7 μL of Ultra II End Prep Reaction Buffer and 3 μL of End Prep Enzyme Mix to 50 μL of sheared DNA. Mix the reaction well, then incubate the mixture in a thermocycler at 20 °C for 30 min, 65 °C for 30 min, and 4 °C for 5 min.
3. Add 2.5 μL of Illumina adapter (50 μM) to the end prep mixture (60 μL) and mix well, then add 1 μL of the Ligation Enhancer and 30 μL of Ligation Master Mix to the end prep mixture and incubate the mixture in a thermocycler at 20 °C for 30 min.
4. Purify DNA with 0.9× Ampure PB beads and elute DNA in 50 μL of EB Buffer.

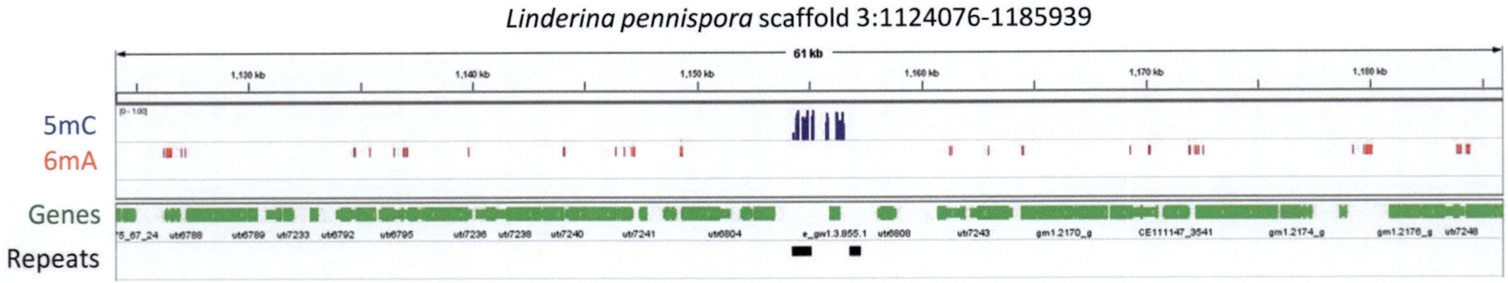

Fig. 1 Epigenomic marks found in *Linderina pennispora.* The blue track shows 5mC marks detected using the described methods for bisulfite sequencing (Subheading 3.1) and analysis (Subheading 3.2). Similarly, the red track shows 6 mA detected using PacBio sequencing (Subheading 3.3) and analysis (Subheading 3.4). Genes are shown in green and repeats in black

5. Adapter ligated DNA denaturing: incubate samples at 95 °C for 10 min and chill on ice immediately. **Important!** Save 5 ng of the single stranded, unchipped DNA as a control library.
6. Prepare Protein A beads by adding 50 μL of Protein A beads to a lobind tube and wash **twice** using cold IP Buffer (10 mM Tris–HCl (pH 7.4), 150 mM NaCl, 0.1% Igepal CA-630) by vortex.
7. Add 50 μL of IP buffer to the beads, then transfer 40 μL of the beads to a new tube for preblocking the beads with BSA (**step 8**). Keep the remaining 10 μL for preclearing DNA (**step 10**).
8. Preblock Protein A beads by transferring 40 μL of the Protein A beads to a 1.5 mL lobind tube and adding 460 μL IP buffer and bovine serum albumin (20 μg/μL) to the beads. Incubate the beads at 4 °C for 6 h.
9. Wash the beads twice with 1 mL IP buffer by vortex and quickly spin, then add 50 μL of IP buffer to the beads.
10. Preclear DNA: Add the denatured DNA to the 10 μL of Protein A beads, then add IP buffer to the tube to bring up the volume to 2 mL and incubate at 4 °C for 2 h.
11. Place the tube with the precleared DNA on a magnetic stand and transfer the clear liquid to a new 2 mL tube.
12. 6 mA antibody–DNA hybridization: Add 1 μg of anti-6 mA antibody (*see* **Note 16**, ABE572, Millipore) to the precleared DNA, then incubate the DNA–6 mA antibody mixture at 4 °C for 4 h (*see* **Note 16**).
13. Immunoprecipitation: Add 50 μL of the preblocked beads to the DNA–6 mA antibody mixture and incubate the mixture at 4 °C for 2 h (*see* **Note 17**).
14. Bead clean up (*see* **Note 18**). Wash the beads with 1 mL of cold IP buffer four times by using magnetic stand and inverting the tube (do not vortex). Wash the beads two more times with 1 mL of cold TE buffer. Discard supernatant, then add 23 μL of nuclease-free water to the beads.
15. One bead PCR amplification: Add 2 μL of Illumina PCR Primer Mix, 25 μL Kapa PCR Ready Mix to 23 μL of the bead–DNA complex, then PCR-amplify with the following program: 98 °C for 3 min, followed by 10–15 cycles (*see* **Note 19**) of 98 °C for 30 s, 65 °C for 30 s, 72 °C for 15 s. Hold at 72 °C for 5 min then incubate at 4 °C.
16. Purify the library with 0.9× Ampure Beads twice and elute the library with 28 μL of EB Buffer. Use Bioanalzyer confirm size, quality and quantity of library. Optional: Perform qPCR to evaluate IP efficiency prior sequencing. Following library preparation, quantify the library using qPCR. Then prepare samples for sequencing on the Illumina MiSeq sequencing platform, using a 1 × 50 run recipe. Please *see* Chapter 1 for Illumina sequencing preparation.

3.6 Analysis of 6 mA-IP-Seq Data

1. Filter reads following steps listed in **step 1** of Subheading 3.2.
2. Map 6 mA IP-seq treatment and control reads to the reference genome using your preferred aligner. We recommend BWA [11] using default parameters. Start by indexing your reference genome. Example command line:

```
bwa index reference.fasta
```

3. Run alignments for treatment and control reads. Example command lines:

   (a) `bwa mem -t 10 reference.fasta IP_seq_treatment > IP_treatment.sam`

   (b) `bwa mem -t 10 reference.fasta IP_control_treatment > IP_control.sam`

   Parameters: `-t`, the number of threads use. `reference.fasta`, your reference genome to align reads to.

4. Use samtools view to convert sam to bam file. Example command lines:

   (a) `samtools view -bS IP_treatment.sam > IP_treatment.bam`

   (b) `samtools view -bS IP_control.sam > IP_control.bam`

   Parameters: `-bS`, convert sam to bam.

5. Use macs2 [12] to call peaks. Example command line:

```
macs2 callpeak -t IP_treatment.bam -c IP_control.
bam -f BAM -g genomeSize(bp) -n   outfile_name -q
0.01 -nomodel
```

   Parameters: `-t`, treatment filename. `-c`, control filename. `-f`, filetype. `-g`, genome size (in basepairs). `-n`, output prefix. `-q`, q-value threshold for determining significant peaks. `-nomodel`, specify this parameter if you want to skip model building (*see* **Note 20**).

6. macs2 will produce several outputs, including an Excel formatted spreadsheet (outfile_name_peaks.xls) which contains information on all 6 mA enriched regions detected, fold enrichment, qvalue, location, etc. Optionally, add the following to command line arguments if you want macs2 to produce bedGraph (`-B` or `--bdg`) or wiggle (`-W` or `--wig`) formatted files of your results, which may be useful for visualization.

4 Notes

1. The average size should be around 500 bp. TapeStation and Bioanalyzer are convenient, but it is important to follow manufacturer's instruction to obtain a more accurate result. If you

are using electrophoresis for size estimation, make a 1.2% agarose gel and use 50 ng of DNA for electrophoresis.

2. The ratio of Ampure Beads to DNA affects the size of purified DNA. In general, we use a lower ratio (e.g., 0.65×) to select larger fragments. The size of sheared DNA can vary among different species and instruments. You can optimize DNA size by adjusting the bead–DNA ratio if the average size of sheared DNA is not around 500 bp.
3. If your sample contains control DNA, two different primer sets (i.e., wild type vs. bisulfite-converted primers) are required to confirm efficiency of bisulfite conversion. When you design the primers, keep the length of the primer between 20 bp and 40 bp and the length of amplicon below 300 bp. A free online tool from Zymo Research is available (http://www.zymoresearch.com/tools/bisulfite-primer-seeker).
4. Organelle (mitochondria/chloroplast) sequences can be included as part of the genome reference. The bisulfite-conversion efficiency can be estimated by calculating the ratio of methylated to unmethylated C's in the organelle sequences. Typically organelles are not methylated and the methylation ratio can provide a direct readout of the bisulfite-conversion efficiency. Alternatively if synthetic DNA spike-in was added, ratio of methylated to unmethylated C's in the spike-in reads can be used to estimate bisulfite-conversion efficiency.
5. Two additional scripts, `bismark2report` and `bismark2summary` provide the results in HTML and text formats. `bismark2report` generates HTML summary for alignments and optionally if methylation extractor and deduplication results are available these are included with the HTML report. `bismark2summary` collates and summarizes all results in the run folder into a single large table and also a HTML to graphically visualize the summary statistics.
6. It is important to obtain as high quality DNA as possible before you start. You do not need to shear your sample if DNA is already degraded, but heavily degraded DNA is not recommended. Here are some indicators for high quality DNA:

   (a) The ratio of 260/280 should be close to 1.8 and 260/230 should be close to 2.0. If the ratio of 260/230 is off, there might be contaminants or inhibitors in the sample, which will affect ligation and polymerase efficiency during sequencing.

   (b) DNA should have little smear. If you observe a smear lower than 1 kb, an abundance of short reads will appear in your sequencing data.

(c) It is important to minimize freeze–thaw cycles and avoid harsh vortex during DNA extraction.

7. Shear one more time if the average size is larger than 20 kb as the data is more reliable with consensus reads. If the average size is too large, there will be less consensus reads.
8. Library yields are affected by bead incubation and DNA elution time. Fifteen minutes of bead binding and 20 min of elution are recommended. Incubate the mixture by shaking on a mixer at 350 rpm at room temperature. Mix the reaction by flicking the tube to avoid more DNA shearing during library prep.
9. To avoid adapter dimers, it is important to (1) add adapter to the reaction before adding ligase and (2) invert the tube a couple times to remove adapter dimers from the tube during bead washing steps.
10. While the kinetic signature of methylated adenines is fairly distinct, modification detection of other bases is much more challenging and results should be interpreted with caution if you are not using amplified template DNA as a control. In addition, if a genome has extremely low levels of a particular modification, the signal to noise ratio of SMRT-analysis may be very low and therefore multiple replicates/aggressive downstream filtering methods should be used to boost confidence in results.
11. To up ram and max number of hours, append to the job name relevant info separated by ##. For example to use 9 G ram and 192 h run time, append ##9G##192 h to the job name. Depending on the genome size and number of SMRT-cells used, modification detection can take a long time, so it is best to allot as much run time as possible.
12. The modification detection pipeline was designed for small genomes, so depending on your genome size completing modification detection may be challenging. If this occurs, try reducing the number of SMRT-cells used (as long as coverage is ≈25× per strand, this is sufficient to detect 6 mA). If your genome is very repetitive, you may also need to hard mask repeats prior to this step. If you plan to mask repeats, first try to complete analysis using a reduced set of SMRT cells—it is important to check whether there is any signal of 6 mA in these regions before you decide whether or not to mask.
13. A maximum coverage cutoff is recommended but not necessary. In genomes with extremely low overall levels of adenine methylation, filtering to remove highly covered sights may be advisable, as false positives are much more likely to be detected at these locations. An appropriate way to detect a maximum

coverage cutoff is to use the built-in R boxplot function and remove any sites covered greater than upper limit.

14. The amount of input DNA depends on abundance of 6 mA in the genome. If 6 mA is symmetric and abundant, the amount of DNA can be reduced to 1 μg.
15. The optimum average size for 6 mA IP-Seq is about 150 bp. IP efficiency will decrease if DNA size is too large.
16. Performing an antibody-titration experiment (e.g., add 0.1, 0.5, 1, and 2 μg of antibody to the reaction) is sometimes useful to optimize IP efficiency. The incubation time can be increased if library yield is low after 15 PCR cycles.
17. Incubation time can be adjusted. For example, you can increase the incubation time to overnight for immunoprecipitation if your IP efficiency is low.
18. It is important to wash Protein A beads thoroughly to remove unwanted DNA from the sample after immunoprecipitation.
19. Take 1 μL of PCR products and run an HS Bioanalyzer chip after 10 cycles of PCR. If amplicons are observed and the concentration is higher than 2 ng/μL, proceed with bead purification, otherwise, continue PCR up to 15 cycles.
20. If `-nomodel` is specified, macs2 will set this value at ½ of the fragment size. `-nomodel` is recommended for histone and genome modification data [12]. However, if you choose to build the shifting model, you may need to tune the `-mfold` parameter for your particular dataset.

Acknowledgments

The work conducted by the US Department of Energy Joint Genome Institute, a DOE Office of Science User Facility, is supported under Contract No. DE-AC02-05CH11231.

References

1. Zemach A, McDaniel IE, Silva P et al (2010) Genome-wide evolutionary analysis of eukaryotic DNA methylation. Science (New York, NY) 328:916–919
2. Fu Y, Luo G-Z, Chen K et al (2015) N6-Methyldeoxyadenosine marks active transcription start sites in chlamydomonas. Cell 161:879–892
3. Zhang G, Huang H, Liu D et al (2015) N6-methyladenine DNA modification in Drosophila. Cell 161:893–906
4. Greer EL, Blanco MA, Gu L et al (2015) DNA methylation on N6-adenine in C. elegans. Cell 161:868–878
5. Mondo SJ, Dannebaum RO, Kuo RC et al (2017) Widespread adenine N6-methylation of active genes in fungi. Nat Genet 49(6):964–968
6. Krueger F, Kreck B, Franke A et al (2012) DNA methylome analysis using short bisulfite sequencing data. Nat Methods 9:145–151

7. Dominissini D, Moshitch-Moshkovitz S, Salmon-Divon M et al (2013) Transcriptome-wide mapping of N(6)-methyladenosine by m(6)A-seq based on immunocapturing and massively parallel sequencing. Nat Protoc 8:176–189
8. Krueger F, Andrews SR (2011) Bismark: a flexible aligner and methylation caller for bisulfite-seq applications. Bioinformatics 27:1571–1572
9. Robinson JT, Thorvaldsdóttir H, Winckler W et al (2011) Integrative genomics viewer. Nat Biotechnol 29:24–26
10. Dominissini D, Moshitch-Moshkovitz S, Schwartz S et al (2012) Topology of the human and mouse m6A RNA methylomes revealed by m6A-seq. Nature 485:201–206
11. Li H (2013) Aligning sequence reads, clone sequences and assembly contigs with BWA-MEM, arXiv:1303.3997 [q-bio]
12. Zhang Y, Liu T, Meyer CA et al (2008) Model-based analysis of ChIP-Seq (MACS). Genome Biol 9:R137

Chapter 15

Fungal Genome Annotation

Sajeet Haridas, Asaf Salamov, and Igor V. Grigoriev

Abstract

The term "genome annotation" includes identification of protein-coding and noncoding sequences (e.g., repeats, rDNA, and ncRNA) in genome assemblies and attaching functional information (metadata) to these annotated features. Here, we describe the basic outline of fungal nuclear and mitochondrial genome annotation as performed at the US Department of Energy Joint Genome Institute (JGI).

Key words Genome, Annotation, Gene prediction, Pipeline, Functional annotation

1 Introduction

Genome annotation consists of three main steps:

1. Identifying noncoding features of the genome that do not code for proteins.
2. Identifying protein coding genes, generally referred to as gene prediction.
3. Attaching biological information (functional annotation) to these genome features, for example, pfam domains, repeat classes, putative gene functions, and descriptive names.

In practice, the vast majority of annotation efforts are to identify protein-coding genes in the genome and to assign biologically meaningful names and functions to these genes.

The quality of automated annotation of genomes is highly dependent on the quality of the assembly and the availability of associated data such as RNA and protein sequences from the organism in question or a close relative. Annotation of genomes is a complex process with many input and output files and interdependent procedures. Frequently, these operations are combined into a single annotation pipeline which feeds appropriate inputs to the underlying software tools and keeps track of the output files. Many sequencing centers like the Joint Genome Institute (JGI) and the Broad Institute have developed specialized pipelines to run

Ronald P. de Vries et al. (eds.), *Fungal Genomics: Methods and Protocols*, Methods in Molecular Biology, vol. 1775, https://doi.org/10.1007/978-1-4939-7804-5_15,

on large compute clusters [1–3]. The protocol described here is based on the JGI Genome Annotation pipeline approximated using readily available software tools (*see* **Note 1**).

2 Materials

Several pieces of software will be required for the annotation. You will need to install these programs on a Unix-like operating system using the documentation included with each one. Some of the most popular ones include (*see* **Note 1**) the following:

1. RepeatMasker (http://repeatmasker.org) to identify known repeats in the genome.
2. RepeatScout [4] and RepeatModeler (http://www.repeatmasker.org/RepeatModeler.html) to identify novel repeats in the genome.
3. BLAT [5] (https://genome.ucsc.edu/FAQ/FAQblat.html) and BLAST [6] to align transcripts and protein sequences to the genome.
4. A short read aligner like BWA [7] (http://bio-bwa.sourceforge.net) or BBmap (https://sourceforge.net/projects/bbmap/) to align transcriptome reads to the genome.
5. Ab initio gene modelers like GeneMark [8] (http://exon.gatech.edu/GeneMark/), SNAP [9] (http://korflab.ucdavis.edu/software.html), Augustus [10] (http://bioinf.uni-greifswald.de/augustus/), and Fgenesh [11] (http://www.softberry.com).
6. Evidence-based gene modelers like GeneWise [12] (http://www.ebi.ac.uk/~birney/wise2/), PASA [13] (http://pasapipeline.github.io), and Fgenesh+ (http://www.softberry.com).
7. A filtering pipeline like EVidenceModeler [14] (EVM, https://evidencemodeler.github.io) or Maker [15] (http://www.yandell-lab.org/software/maker.html) to filter and consolidate the results of the multiple genome predictors.
8. Databases like the NCBI nonredundant protein database (nr), RefSeq (https://www.ncbi.nlm.nih.gov/refseq/), UniProt/Swiss-Prot (http://www.uniprot.org) to identify homologs and impute function to the predicted genes.
9. Any number of specialized tools to add functional annotation to the predicted genes such as pfam domains [16] (http://pfam.xfam.org), signal peptides [17] (http://www.cbs.dtu.dk/services/SignalP/), and EC assignment [18] (e.g., http://priam.prabi.fr).

The annotation will require several input files including the genome assembly FASTA file (*see* **Notes 2** and **3**). Depending on the type of annotation, several additional input files will be required like mRNA or protein sequences from this or a related organism to predict protein-coding genes.

3 Methods

3.1 Identify Noncoding Genome Features

The genome encodes several kinds of noncoding features including repeats, tRNA and other ncRNA, and rDNA. Each of these can be identified using specific software tools. Of these, identification of repeats and masking them is critical to successful identification of good quality protein-coding genes (*see* **Note 4**).

The following three-step process masks repeats and transposable elements (TE) in genome sequence using RepeatModeler and RepeatMasker using 4 compute threads (-pa 4). Additional steps below identify other noncoding sequences in the genome assembly and can be skipped without affecting the downstream prediction of protein-coding genes.

1. Build a database of the genome FASTA file for RepeatModeler to run:

```
$ /path/to/RepeatModeler/BuildDatabase -name myGenome -engine ncbi myGenome.fasta
```

2. Run RepeatModeler.

```
$ RepeatModeler -engine ncbi -pa 4 -database myGenome
```

Since this step will run for a long time, users may want to consider running this using nohup, screen, or a job submission system like qsub, and capturing the stdout and stderr into log files. Once the RepeatModeler run is complete, the identified repeats will be in the folder RM_*some_name* as

```
consensi.fa.classified.
$ /path/to/RepeatMasker/util/queryRepeatDatabase.pl -species fungi > fungi_repeats.lib
```

Add these newly identified repeat sequences to the RepeatMasker library and create a custom library for this genome:

```
$ cat fungi_repeats.lib consensi.fa.classified >
myGenome.custom.repeat.lib
```

3. Now, run RepeatMasker on the genome using the custom library. This should also be run using nohup, screen, or some job submission system like qsub so that it can run to completion.

```
$ RepeatMasker -engine crossmatch -lib myGenome.cus-
tom.repeat.lib -pa 4 -no_is myGenome.fasta
```

The "-no_is" in the above command skips the bacterial insertion element check. You can choose not to use this option.

4. Predict tRNAs using tRNAscan-SE [19] (http://lowelab.ucsc.edu/tRNAscan-SE/):

```
$ tRNAscan-SE -o myGenome.tRNAscan.results myGen-
ome.fasta
```

The predicted tRNAs are listed in the output file set with the –o option.

5. Predict other noncoding elements. snoSeeker [20] can be used to identify snoRNA. If specialized sequencing for other ncRNA (like miRNA) was performed, reads can be aligned to the genome and tools like miRDeep [21] or miRanalyzer [22] can be used to identify these features in the genome. Alternately, users can identify known ncRNA homologs using ERPIN [23] and miRAlign [24]. Tools for identification of novel ncRNAs are still in development and remain highly experimental, often with high error rates. In the absence of accepted standards and high reliance on experimental evidence to validate these annotations, these techniques are beyond the scope of this guide. A general method is to use Infernal (http://eddylab.org/infernal/) to predict all noncoding RNAs which have corresponding covariance models in the RFAM database:

```
$ cmsearch -tblout output.file --cut_ga Rfam.cm
myGenome.fasta
```

The `-tblout` option puts the output in the file `output.file` in a tabular format which is easy to parse. Users may choose to output the full format output (default) in readable form.

3.2 Identify Protein Coding Genes

Due to intron–exon structure of eukaryotic genes, gene prediction in eukaryotes is one of the most challenging parts of the genome annotation. We recommend using several gene prediction approaches to combine different lines of evidence used for annotation: ab initio, homology-based, and transcriptome-based.

1. **Ab initio gene prediction**. All ab initio gene finding tools, e.g., GeneMarkHMM, FGENESH, Augustus, SNAP, and GlimmerHMM [25] have to be individually trained using the masked genome as described in the previous section. Here is an example of executing self-training GenMark v4.32 running on 16 compute threads:

```
$ gmes_petap.pl --ES --fungus --cores 16 --sequence
myGenome.fasta
```

2. **Homology-based gene prediction**. Homology-based gene prediction is done by mapping proteins from other organisms to the genome of interest. For example, GeneWise can use a large database like nr, uniref90, or UniProt/Swiss-Prot to generate homology-based gene models (*see* **Note 5**). The command below uses GeneWise v2.2 to search both DNA strands and produce gene models in the gff3 output format:

```
$ genewise protein.database.fasta myGenome.fasta -
both -gff
```

3. **Transcriptome-based gene prediction**. RNA-Seq data can be used in two different ways. RNA-seq reads can be mapped to the genome to predict transcripts and generate a gff3 file using the cufflinks suite as described by Trapnell et al. [26]. However, you may want to restrict the max-intron-length parameter to about 1–2 kb (from the 300 kb default) because average fungal introns are about 60 bp.

   The second approach involves mapping RNA-Seq assemblies to the genome and then building gene models from these aligned transcripts. You can see a complete walkthrough of the procedure using PASA and an explanation of the parameters at http://pasapipeline.github.io.

```
$ Launch_PASA_pipeline.pl -c alignAssembly.config -C -
R -g myGenome.fasta -t transcripts.fasta.clean -T -u
transcripts.fasta -f FL_accs.txt -USE_GMAP
```

4. **Inspect gene models**. Visually inspect the gene predictions from the different modelers by loading them into a genome viewer (like IGV, http://software.broadinstitute.org/software/igv/) to make sure that most models from the different modelers are similar to each other at the same locus for several random loci (Fig. 1). If one or more modelers produce models that are significantly different from the others, identify and correct the source of the error. For example, if the models generated by GlimmerHMM are significantly different from Augustus and SNAP (which are similar to each other), this could point to poor training of GlimmerHMM. In this case, the GlimmerHMM predictions should be either dropped from downstream processing or retrained and rerun until congruence with other modelers is achieved (also *see* **Note 6**).

5. **Select best models**. Since using multiple gene predictors creates several alternative gene models for every locus, we would like to select or construct from existing models the best model for each locus. At the JGI we use a scoring filtering procedure where every model is evaluated by transcriptome and homology support. There are several publicly available tools like EVidenceModeler to identify the best model at each locus as the first draft automated predicted gene set for this genome.

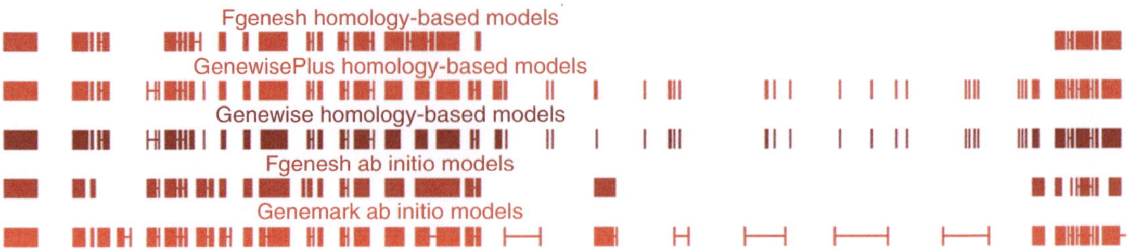

Fig. 1 Screenshot of the JGI genome browser in MycoCosm (http://jgi.doe.gov/fungi) showing the aberrant behavior of GeneMark compared to other gene predictors in this genome due to poor training. The bad short models from GeneWise are due to the badly curated protein database

Each model is given a particular weight based on the preponderance of evidence and a model is chosen for each locus based on the "winner take all" strategy. This set is further filtered after functional annotation (*see* below). In our experience, we have come to rely on a few broad criteria for weighing the fitness of a model. The weight given to particular model sources should be balanced against the probability of it being real or spurious (*see* **Notes 6–8**).

This produces the first automated approximation of the proteome. At this point, perform a sanity check on the data by comparing it to related genomes as a quality control exercise (*see* **Notes 9** and **10**).

3.3 Functional Annotation

Functional annotation of non-protein coding genome features is usually concurrent with their identification. Predicted proteins can be functionally annotated using a wide variety of tools depending on the user's requirements. The three general approaches include (1) characterization of protein sequence parts such as domains, (2) detecting similarity to already characterized protein sequences, and (3) annotation according to existing classification schemes such as EuKaryotic Orthologous Groups (KOG [27]), Gene Ontology (GO http://www.geneontology.org), Kyoto Encyclopedia of Genes and Genomes (KEGG [28] http://www.genome.jp/kegg). Many of these tools have online servers, but it may be more efficient for multiple genomes to use a local installation.

Some of the most popular domain or protein sequence feature predictors include the following:

1. hmmscan (http://hmmer.org) to identify pfam domains in predicted proteins. Genes encoding proteins harboring known TE domains should be removed from the predicted gene set because these are traditionally not included in the organism's gene set.

```
$ hmmscan -domtblout output_filename --cut_ga
Pfam-A.hmm myProteins.fasta
```

2. signalP (http://www.cbs.dtu.dk/services/SignalP/) to identify signal peptides suggesting protein secretion in predicted

genes. The presence of signal peptides can also serve as evidence for a valid gene model, which could be especially important for small single exon predictions where error (false prediction) rates are high.

```
$ signalp myProteins.fasta > output_filename
```

3. TMHMM (http://www.cbs.dtu.dk/services/TMHMM/) predicts transmembrane domains, useful to identify membrane-bound proteins:

```
$ tmhmm myProteins.fasta > output_filename
```

4. Psort (http://psort.hgc.jp/) predicts cellular localization of proteins:

```
$ runWolfPsortSummary fungi < myProteins.fasta >
output_filename
```

5. InterProScan [29] offers a collection of functional and structural protein domains. It is available from EMBL-EBI (http://www.ebi.ac.uk/interpro/download.html) and can be run using:

```
$ interproscan.sh -i myProteins.fasta -b output_
filename -f gff3
```

   You can use the option "-f tsv" if you prefer a tab-separated text file output rather than a gff3. The –b option automatically adds file extension based on the type defined by –f.

6. Protein alignments to NCBI nr, SwissProt (http://www.expasy.org/sprot/), or UniProt using blast can serve as the first approximation of protein function. Based on blast hits to specialized databases, you can perform targeted annotations such as the identification of peptidases using the MEROPS database (http://merops.sanger.ac.uk). Using these data, often a putative function can be assigned to over half of the predicted genes which can be used to provide a biologically meaningful descriptive name to the model.
7. Gene classification systems offer another way to annotate proteins and put these annotations in a comparative context. Some of these include: (1) Gene Ontology (http://www.geneontology.org) which assigns GO terms from one of three categories: Biological Process, Molecular Function, and Cellular compartment. Interpro and SwissProt hits are used to map gene ontology to predicted proteins. (2) KEGG for metabolic pathways. This assigns EC numbers (http://www.expasy.org/enzyme/) to the proteins and maps them to metabolic pathways. (3) KOG for eukaryotic clusters of orthologs, which also provides additional support for individual models.

Small models without functional annotations, especially single exon genes, may be spurious and may need to be removed unless there appear to be lineage-specific expansion of a novel gene family, which can be identified using tools like OrthoFinder [30] and OrthoMCL [31, 32]. A quality control check by comparison to related genomes is invaluable at this stage (*see* **Notes 11–13**).

3.4 Mitochondrial Genome Annotation

The mitochondrial genome of most organisms is highly conserved. The widely accepted view is that mitochondria evolved from an alpha-proteobacterial symbiont in the ancestor of all eukaryotes. Most mitochondria still retain many bacterial-type features such as its circular topology. Some mitochondria harbor multiple circular chromosomes (e.g., cucumber) and others have linear chromosomes (like some ciliates, cnidarians, and Chlamydomonas).

The mitochondrial genome has undergone massive reduction with many genes moving to the nuclear genome or their function being replaced by nuclear-encoded orthologs. The mitochondrial gene set is usually limited to known genes in the electron transport chain, two ribosomal RNAs, and several transfer RNAs. In spite of the limited set of fungal mitochondrial genes, the gene structures can be highly variable and complex. Several genes, especially *cox1*, usually have introns that harbor TE-like endonucleases. Some exons and introns can be very short (<10 bp) making their identification difficult. Due to these and other factors, accurate automated annotation of mitochondrial genomes has remained elusive.

The small size and limited gene set make the errors glaringly apparent. Many erroneous annotations and nonstandard gene names have been published and deposited into GenBank and RefSeq (Table 1), making their classification and identification a laborious manual process. The major steps in the annotation of mitochondrial genomes are:

1. Mitochondrial genes are transcribed polycistronically and cleaved by endonucleases at tRNAs. Therefore conceptually, the first step in mitochondrial genome annotation is the prediction of tRNAs. While several software packages exist for this, in our experience, tRNAscan-SE with organellar option (-O) works well. Other packages such as ARWEN [33] and RNAweasel [34] have also been reported to perform well on mitochondrial genomes.

```
$ tRNAscan-SE -O -o mito.tRNAscan.results mito.
fasta
```

2. A sizeable fraction of fungal mitochondrial protein-coding genes contain introns and they are usually of self-spliced type I or (rarely) of type II, which, unlike splicesomal introns, do not have conserved sequence motifs, and therefore present a challenge for their correct prediction. We use three methods for predicting mitochondrial protein-coding genes.

Table 1
Gene sizes (aa) of mitochondrial genes in RefSeq. Each of the 14 canonical protein coding genes (except for atp9) has over 4000 models in RefSeq. In addition, there are 1–300 models of many (>50) noncanonical mitochondrial genes. We looked at several (including some annotated as DNA or RNA polymerases) and found that these were erroneous annotations and spurious models

Gene	Min	Max	Median
cox1	65	778	516
cox2	13	1296	229
cox3	185	503	261
cob	252	537	380
nad1	54	386	322
nad2	135	923	347
nad3	66	567	116
nad4	77	580	459
nad4L	40	498	98
nad5	399	924	606
nad6	103	375	173
atp6	112	427	227
atp8	26	256	55
atp9	14	75	127

The first method is similar to prokaryotic gene finder algorithms, using translation code and protein-coding potential specific for mitochondrial genomes. The second and third methods use homology-based approaches to map intron-containing genes. For the second method, we use TBLASTN, which maps proteins to the genome without consideration of splice site consensuses and then, using a custom-made Perl script, refine the boundaries, preserving the reading frame. For the third method, we built HMMs for 14 core mitochondrial genes and use the GeneWise algorithm, with mitochondrial genetic code to predict them in the genome. Multiple predictions at particular loci can be filtered using custom scripts and manual inspection. These steps rely on the availability of a well curated and continuously updated library of good quality gene models. As seen in Table 1, standard databases like GenBank and RefSeq are unreliable for this step. Some curated databases such as the one by F.Lang (http://www.bch.umontreal.ca/

People/lang/FMGP/proteins.html) are good starting points for users to create such a database.

3. The ribosomal RNA genes are perhaps the most difficult mitochondrial genes to annotate due to their high levels of length variability. In our experience, Infernal (http://eddylab.org/infernal) with covariance models "LSU_rRNA_bacteria" and "SSU_rRNA_bacteria" from RFAM database [35] works well for fungal mitochondrial genomes.

```
$ cmsearch -tblout output.file --cut_ga covariance.
cm mito.fasta
```

4. Assembly of the circular mitochondrial genome is problematic, and its linear representation may sometimes show duplicate sequences at both ends. Such sequences may be genuine repeats in the genome sequence or artifacts of the assembly process. This can cause gene duplication or gene split across the ends. The presence of a truncated, duplicated or missing gene from the canonical set is potentially a sign of this. In such cases, resplitting the FASTA file at a new gene sparse location is recommended. The newly reconstituted FASTA sequence should be then reannotated using the steps outlined above.

4 Notes

Once the annotation is complete, it should be checked for accuracy and quality. The large number of poor models in GenBank is a testament to the lack of quality control in published genomes. Some of the most common errors that we notice are the presence of organellar scaffolds in genome assemblies, TE elements in the protein set, and incomplete or chimeric gene models. In order to generate a high-quality annotation, users should consider the following.

1. Many of the software tools used here are under active development. New version of the tools may have additional features and parameters. We have provided basic command line parameters that may change in later versions. Users should read documentation and help pages of the tools being used.
2. Use the highest possible quality assembly with a large L50 (the shortest contig length of the most contiguous 50% of the genome). A poor quality fragmented assembly will produce poor quality annotations.
3. Evaluate assembly quality using a completeness tool like CEGMA [36] or BUSCO [37]. These tools can suggest the upper limit of recoverable protein-coding genes from the assembly and can prod the user to assembly improvement if the numbers are below expectation.

4. Insufficient masking is usually more disastrous than overmasking. Undermasking can generate hundreds of spurious models, while overmasking might lose a few models that lie in or span these masked nonrepeats.
5. The quality of the protein database will have an effect on the homology-based models. Since protein conservation is restricted to active sites and those residues that affect structure and folding, it is often difficult to identify full-length gene models with homology evidence. Be aware of partial (incomplete) gene models. An ad hoc approach to extend the models in the 5′ and 3′ directions in order to identify the potential start and stop codons can affect functional annotation such as the prediction of secretion signals.
6. Compare selected model structure against mapping of RNA-Seq reads and assembly to evaluate gene models and intron–exon boundaries. In general, predicted gene models should closely match the mapping of RNA-Seq reads to the genome. A significant departure from this can point to poor structural annotation or an incorrect genome assembly.
7. Some fungal genomes are highly compact with closely spaced genes, often with overlapping UTRs. In these cases, RNA-Seq mapping sometimes produces chimeric models and noisy RNA-seq data exacerbates this problem. Using strand-specific RNA-Seq and comparison with high-quality annotations of related genomes can help mitigate such problems.
8. Gene calling for small models is highly error-prone. At the JGI, we predict genes >49aa and only keep models under 200aa if they have some additional evidence like the presence of pfam domains, signal peptides, transmembrane domains or similarity to proteins from a related genome with high quality gene predictions. These cutoffs are dependent on sequencing and assembly methods of genome and transcriptome.
9. Compare gene model statistics, such as a number of predicted genes, gene length distribution, and number and size of exons and introns, with other related genomes. This can point out errors in gene prediction.
10. Generate a phylogenetic tree using single copy orthologs (we use ~200) to confirm that the genome assembly is placed where expected as compared to its nearest relatives. This can help identify misidentified DNA source material or point to errors in its previous classification.
11. Compare functional annotation with that of related genomes such as the proportion of gene models with pfam domains. Like the previous point, this can also identify errors in structural and functional annotations.

12. Compare nr BLAST hits within each scaffold. A large proportion (>50%) of hits to other phyla can identify assembly artifacts. While bacterial and human contamination of samples and reads is well known, we have seen scaffolds showing hits to a wide variety of taxonomic lineages including amphibians, reptiles, and plants. The ability to extract DNA from a pure culture is not a sufficient safeguard from this, since contamination can occur at later stages which are often outside the control of the DNA producing laboratory.
13. After initial quality assessment of the annotated genome, it is often subject to a multi-tier process that includes an assessment by peers, community annotation, and GenBank review. While automated annotation is an important source of information, manual curation is still useful in many cases, despite a lack of published standards and user training/background variability. The main purpose of manual curation is to validate structure and function of individual genes based on available lines of evidence: similarity to mapped transcripts, genome conservation, and alignment with other proteins and domains. JGI Mycocosm (http://jgi.doe.gov/fungi) offers tools for community-based manual curation for every hosted genome and enables curators to add, remove, and modify both functional and structural annotations. Some popular tools for manual annotation include GenomeView (http://genomeview.org/) and Apollo (http://gmod.org/wiki/Apollo).

Acknowledgment

The work conducted by the US Department of Energy Joint Genome Institute, a DOE Office of Science User Facility, is supported under Contract No. DE-AC02-05CH11231.

References

1. Grigoriev IV, Nikitin R, Haridas S, Kuo A, Ohm R, Otillar R, Riley R, Salamov A, Zhao X, Korzeniewski F, Smirnova T, Nordberg H, Dubchak I, Shabalov I (2014) MycoCosm portal: gearing up for 1000 fungal genomes. Nucleic Acids Res 42(Database issue):D699–D704. https://doi.org/10.1093/nar/gkt1183
2. Haas BJ, Zeng Q, Pearson MD, Cuomo CA, Wortman JR (2011) Approaches to fungal genome annotation. Mycology 2(3):118–141. https://doi.org/10.1080/21501203.2011.606851
3. Kuo A, Bushnell B, Grigoriev IV (2014) Fungal genomics: sequencing and annotation. Adv Bot Res 70:1–52. https://doi.org/10.1016/b978-0-12-397940-7.00001-x
4. Price AL, Jones NC, Pevzner PA (2005) De novo identification of repeat families in large genomes. Bioinformatics 21(Suppl 1):i351–i358. https://doi.org/10.1093/bioinformatics/bti1018
5. Kent WJ (2002) BLAT – the BLAST-like alignment tool. Genome Res 12(4):656–664. https://doi.org/10.1101/gr.229202. Article published online before March 2002
6. Camacho C, Coulouris G, Avagyan V, Ma N, Papadopoulos J, Bealer K, Madden TL (2009) BLAST+: architecture and applications. BMC Bioinformatics 10:421. https://doi.org/10.1186/1471-2105-10-421
7. Li H, Durbin R (2010) Fast and accurate long-read alignment with burrows-wheeler

transform. Bioinformatics 26(5):589–595. https://doi.org/10.1093/bioinformatics/btp698

8. Ter-Hovhannisyan V, Lomsadze A, Chernoff YO, Borodovsky M (2008) Gene prediction in novel fungal genomes using an ab initio algorithm with unsupervised training. Genome Res 18(12):1979–1990. https://doi.org/10.1101/gr.081612.108
9. Korf I (2004) Gene finding in novel genomes. BMC Bioinformatics 5:59. https://doi.org/10.1186/1471-2105-5-59
10. Stanke M, Schoffmann O, Morgenstern B, Waack S (2006) Gene prediction in eukaryotes with a generalized hidden Markov model that uses hints from external sources. BMC Bioinformatics 7:62. https://doi.org/10.1186/1471-2105-7-62
11. Salamov AA, Solovyev VV (2000) Ab initio gene finding in Drosophila genomic DNA. Genome Res 10(4):516–522
12. Birney E, Clamp M, Durbin R (2004) GeneWise and Genomewise. Genome Res 14(5):988–995. https://doi.org/10.1101/gr.1865504
13. Haas BJ, Delcher AL, Mount SM, Wortman JR, Smith RK Jr, Hannick LI, Maiti R, Ronning CM, Rusch DB, Town CD, Salzberg SL, White O (2003) Improving the Arabidopsis genome annotation using maximal transcript alignment assemblies. Nucleic Acids Res 31(19):5654–5666
14. Haas BJ, Salzberg SL, Zhu W, Pertea M, Allen JE, Orvis J, White O, Buell CR, Wortman JR (2008) Automated eukaryotic gene structure annotation using EVidenceModeler and the program to assemble spliced alignments. Genome Biol 9(1):R7. https://doi.org/10.1186/gb-2008-9-1-r7
15. Holt C, Yandell M (2011) MAKER2: an annotation pipeline and genome-database management tool for second-generation genome projects. BMC Bioinformatics 12:491. https://doi.org/10.1186/1471-2105-12-491
16. Finn RD, Coggill P, Eberhardt RY, Eddy SR, Mistry J, Mitchell AL, Potter SC, Punta M, Qureshi M, Sangrador-Vegas A, Salazar GA, Tate J, Bateman A (2016) The Pfam protein families database: towards a more sustainable future. Nucleic Acids Res 44(D1):D279–D285. https://doi.org/10.1093/nar/gkv1344
17. Petersen TN, Brunak S, von Heijne G, Nielsen H (2011) SignalP 4.0: discriminating signal peptides from transmembrane regions. Nat Methods 8(10):785–786. https://doi.org/10.1038/nmeth.1701
18. Claudel-Renard C, Chevalet C, Faraut T, Kahn D (2003) Enzyme-specific profiles for genome annotation: PRIAM. Nucleic Acids Res 31(22):6633–6639
19. Lowe TM, Eddy SR (1997) tRNAscan-SE: a program for improved detection of transfer RNA genes in genomic sequence. Nucleic Acids Res 25(5):955–964
20. Yang JH, Zhang XC, Huang ZP, Zhou H, Huang MB, Zhang S, Chen YQ, Qu LH (2006) snoSeeker: an advanced computational package for screening of guide and orphan snoRNA genes in the human genome. Nucleic Acids Res 34(18):5112–5123. https://doi.org/10.1093/nar/gkl672
21. An J, Lai J, Lehman ML, Nelson CC (2013) miRDeep*: an integrated application tool for miRNA identification from RNA sequencing data. Nucleic Acids Res 41(2):727–737. https://doi.org/10.1093/nar/gks1187
22. Hackenberg M, Rodriguez-Ezpeleta N, Aransay AM (2011) miRanalyzer: an update on the detection and analysis of microRNAs in high-throughput sequencing experiments. Nucleic Acids Res 39(Web Server issue):W132–W138. https://doi.org/10.1093/nar/gkr247
23. Sebastian B, Aggrey SE (2008) Specificity and sensitivity of PROMIR, ERPIN and MIR-ABELA in predicting pre-microRNAs in the chicken genome. In Silico Biol 8(5–6):377–381
24. Wang X, Zhang J, Li F, Gu J, He T, Zhang X, Li Y (2005) MicroRNA identification based on sequence and structure alignment. Bioinformatics 21(18):3610–3614. https://doi.org/10.1093/bioinformatics/bti562
25. Majoros WH, Pertea M, Salzberg SL (2004) TigrScan and GlimmerHMM: two open source ab initio eukaryotic gene-finders. Bioinformatics 20(16):2878–2879. https://doi.org/10.1093/bioinformatics/bth315
26. Trapnell C, Roberts A, Goff L, Pertea G, Kim D, Kelley DR, Pimentel H, Salzberg SL, Rinn JL, Pachter L (2012) Differential gene and transcript expression analysis of RNA-seq experiments with TopHat and Cufflinks. Nat Protoc 7(3):562–578. https://doi.org/10.1038/nprot.2012.016
27. Koonin EV, Fedorova ND, Jackson JD, Jacobs AR, Krylov DM, Makarova KS, Mazumder R, Mekhedov SL, Nikolskaya AN, Rao BS, Rogozin IB, Smirnov S, Sorokin AV, Sverdlov AV, Vasudevan S, Wolf YI, Yin JJ, Natale DA (2004) A comprehensive evolutionary classification of proteins encoded in complete eukaryotic genomes. Genome Biol 5(2):R7. https://doi.org/10.1186/gb-2004-5-2-r7

28. Kanehisa M, Goto S, Hattori M, Aoki-Kinoshita KF, Itoh M, Kawashima S, Katayama T, Araki M, Hirakawa M (2006) From genomics to chemical genomics: new developments in KEGG. Nucleic Acids Res 34(Database issue):D354–D357. https://doi.org/10.1093/nar/gkj102
29. Quevillon E, Silventoinen V, Pillai S, Harte N, Mulder N, Apweiler R, Lopez R (2005) InterProScan: protein domains identifier. Nucleic Acids Res 33(Web Server issue):W116–W120. https://doi.org/10.1093/nar/gki442
30. Emms DM, Kelly S (2015) OrthoFinder: solving fundamental biases in whole genome comparisons dramatically improves orthogroup inference accuracy. Genome Biol 16:157. https://doi.org/10.1186/s13059-015-0721-2
31. Li L, Stoeckert CJ Jr, Roos DS (2003) OrthoMCL: identification of ortholog groups for eukaryotic genomes. Genome Res 13(9):2178–2189. https://doi.org/10.1101/gr.1224503
32. Fischer S, Brunk BP, Chen F, Gao X, Harb OS, Iodice JB, Shanmugam D, Roos DS, Stoeckert CJ Jr (2011) Using OrthoMCL to assign proteins to OrthoMCL-DB groups or to cluster proteomes into new ortholog groups. Curr Protoc Bioinformatics Chapter 6:Unit 6. 12 11–19. https://doi.org/10.1002/0471250953.bi0612s35
33. Laslett D, Canback B (2008) ARWEN: a program to detect tRNA genes in metazoan mitochondrial nucleotide sequences. Bioinformatics 24(2):172–175. https://doi.org/10.1093/bioinformatics/btm573
34. Gautheret D, Lambert A (2001) Direct RNA motif definition and identification from multiple sequence alignments using secondary structure profiles. J Mol Biol 313(5):1003–1011. https://doi.org/10.1006/jmbi.2001.5102
35. Nawrocki EP, Burge SW, Bateman A, Daub J, Eberhardt RY, Eddy SR, Floden EW, Gardner PP, Jones TA, Tate J, Finn RD (2015) Rfam 12.0: updates to the RNA families database. Nucl Acids Res 43(D1):D130–D137. https://doi:10.1093/nar/gku1063
36. Parra G, Bradnam K, Korf I (2007) CEGMA: a pipeline to accurately annotate core genes in eukaryotic genomes. Bioinformatics 23(9):1061–1067. https://doi.org/10.1093/bioinformatics/btm071
37. Simao FA, Waterhouse RM, Ioannidis P, Kriventseva EV, Zdobnov EM (2015) BUSCO: assessing genome assembly and annotation completeness with single-copy orthologs. Bioinformatics 31(19):3210–3212. https://doi.org/10.1093/bioinformatics/btv351

Chapter 16

Manual Gene Curation and Functional Annotation

Erin McDonnell, Kimchi Strasser, and Adrian Tsang

Abstract

No genome sequencing project is complete without structural and functional annotation. Gene models and functional predictions for these models can be obtained relatively easily using computational methods, but they are prone to errors. We describe herein the steps we use to manually curate gene models and functionally annotate them. Our approach is to examine each gene model carefully, and improve its structure if necessary, using a comprehensive set of experimental and computational data as evidence. Then, functional predictions are assigned to the gene models based on conserved protein domains and sequence similarities. We use stringent sequence similarity cutoffs and reviewed sequence-database records as external sources for our annotations. By methodically choosing which evidence to use for each annotation, we minimize the risk of adopting and assigning false predictions to the gene models.

Key words Whole-genome annotation, Manual gene curation, Manual functional assignment, Gold-standard genome resource

1 Introduction

Having a well-curated and well-annotated gene set for a genome of interest is extremely important. The information can be used to guide research, reexamine existing data, and better understand the organism under study. The dataset can also serve as a "gold-standard" reference for other genomes. The falling cost of DNA sequencing has led to a dramatic increase in the number of genomes being sequenced. Automated gene prediction pipelines are often used to identify gene models in the sequenced genomes. While gene-prediction algorithms are powerful and continuously improving, they still have limitations. Missed genes, false predictions, and merged or split gene models lead to errors in the final gene set. Functional predictions are automatically assigned to gene models and this, too, can be problematic, as it can lead to misannotations and the risk of propagating errors to other gene collections. Although manually reviewing gene models and their functional predictions is time-intensive, it is necessary if one wants to generate an accurate and reliable dataset. Here we describe a method

Ronald P. de Vries et al. (eds.), *Fungal Genomics: Methods and Protocols*, Methods in Molecular Biology, vol. 1775, https://doi.org/10.1007/978-1-4939-7804-5_16,

which is intended to streamline the task of whole genome annotation. It has been refined with the experience gained in the annotation of the genome of *Aspergillus niger* strain NRRL 3 (http://gbrowse.fungalgenomics.ca/), but it is applicable to other genomes across different organisms.

Using our strategy, the electronic gene models are displayed along the genome in parallel with experimental evidence and computational predictions including: strand-specific RNA-seq data, de novo assembled transcripts, peptides identified by protein mass spectrometry analyses, ChIP-seq peaks obtained using anti-histone H3K4me3 antibodies, intron positions and polarities predicted from mapped RNA-seq reads, orthologous gene predictions, and conserved protein domains. All of the information is examined together to assess the likelihood and most probable structure of a gene model. A strict protocol, which considers both sequence similarity to reviewed proteins and to conserved protein domains, is then used to assign a functional prediction to the gene model in a conservative manner. Groups of enzymes and gene models with similar domains are named using a controlled vocabulary. Finally, the evidence used to assign the functional annotation is given, along with standardized evidence codes, providing a confidence level to the annotation.

2 Materials

- SnowyOwl gene prediction pipeline [1].
- Gbrowse [2] or similar genome browser.
- Genome annotation editing tool, Apollo [3].

3 Methods

3.1 Preparation of Genome Browser

1. Load the genome assembly onto Gbrowse or a similar genome browser that supports FASTA, gff3, and bam files. Map and display the experimental evidence and predictions below onto the genome. The evidence and prediction should be displayed in parallel tracks such that all of the information pertaining to a genomic region can be viewed simultaneously.
2. Map and display computationally predicted gene models. This provides a starting set of gene models for the curator(s) to work with. Each gene model will be reviewed using the other tracks to assess its likelihood and structural accuracy. Evidently, the more accurate the starting set of predicted gene models is, the faster the manual process will be. There are several predictors that exist including Fgenesh [4], Fgenesh+ (www.softberry.com), and GeneWise [5] which are used in a pipeline by

the Joint Genome Institute for predicting genes in fungal genomes [6]. The HMM-based predictor GeneMark [7] and the genome annotation pipeline MAKER [8] are also suitable programs. Another pipeline that has been shown to predict highly dependable gene models in fungal genomes is called SnowyOwl [1] (*see* **Note 1**). The program provides two sets of predicted gene models: "Accepted" models that are highly probable and "Imperfect" models that do not meet the requirements for acceptability but still have some gene evidence. Both must be reviewed.

3. Map and display strand-specific RNA-seq reads. The short reads from transcripts/mRNAs are mapped onto the genome and indicate loci that are likely to contain expressed genes. The strand onto which they map specifies the orientation of the gene (forward or reverse). They mark intron and exon boundaries clearly and help to distinguish neighbouring genes from one another based on the level of read coverage and strand specificity (Fig. 1) (*see* **Note 2**).
4. Map and display de novo assembled transcripts. Protein-coding regions of transcripts deduced from de novo assembly of RNA reads provide strong support for a gene model. They can be extended or truncated at the ends but still indicate that a model is likely present at the mapped loci. The tools Trinity [9] and MEGAHIT [10] can be used for the assembly of RNA-seq reads.
5. Display GC content. GC content tends to be lower in intergenic regions (Fig. 2). Predicted models found in regions with low GC content are suspicious. A long stretch of genomic DNA sequence with low GC content may indicate a centromeric region.
6. Display six-frame translations. Displaying the six translation frames for the entire genome assembly can draw attention to loci that potentially contain genes. A noticeably long stretch of amino acid sequence with no stop codon(s) may indicate an open reading frame (Fig. 3).
7. Display predicted introns. Introns can be inferred from gaps in the mapped, strand-specific RNA-seq reads. The polarity is indicated and is based on the standard intron/exon boundary consensus sequences 5′-GT, AG-3′, the less common 5′-GC, AG-3′, or the rare 5′-AT, AC-3′ junctions (Fig. 3) (*see* **Note 3**).
8. Display mapped peptides. Proteins can be isolated from the culture filtrate and/or cells, trypsinized, and then virtually sequenced using mass spectrometry. The peptides identified are then mapped to the genome and used to support a gene model (Fig. 4) (*see* **Note 4**).

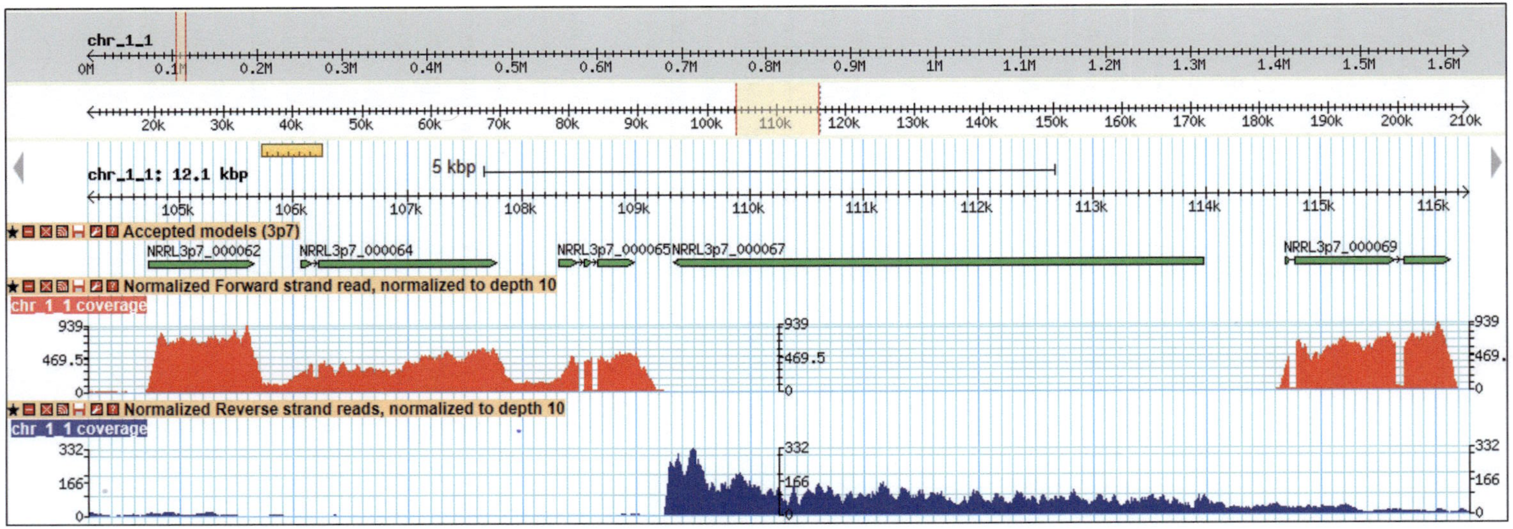

Fig. 1 Strand-specific RNA-seq reads. Track 1: predicted gene models, track 2: forward strand RNA-seq reads, track 3: reverse strand RNA-seq reads

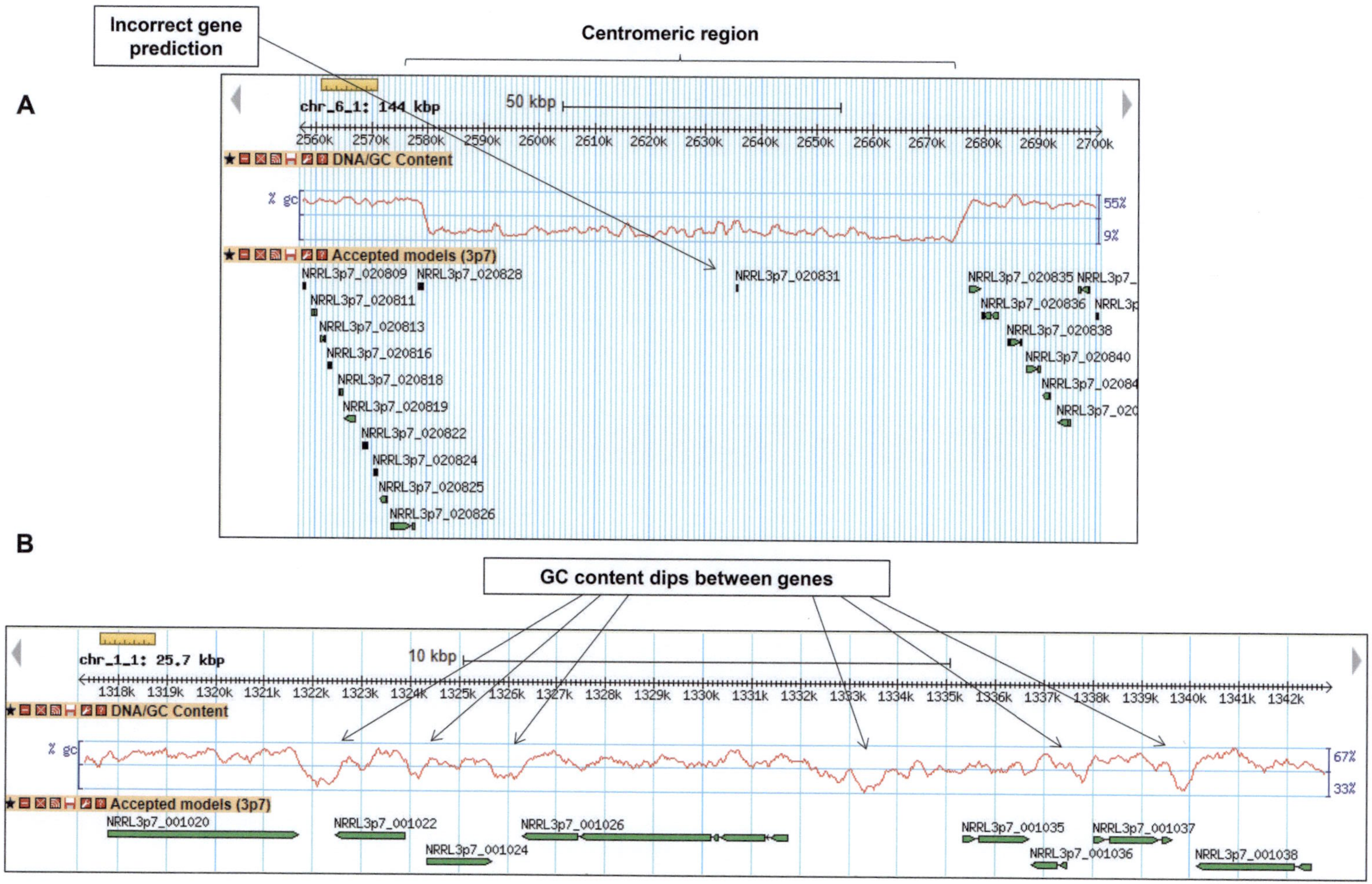

Fig. 2 GC content. Panels **a** and **b**. Track 1: Percentage of GC content along the genomic sequence, track 2: predicted gene models

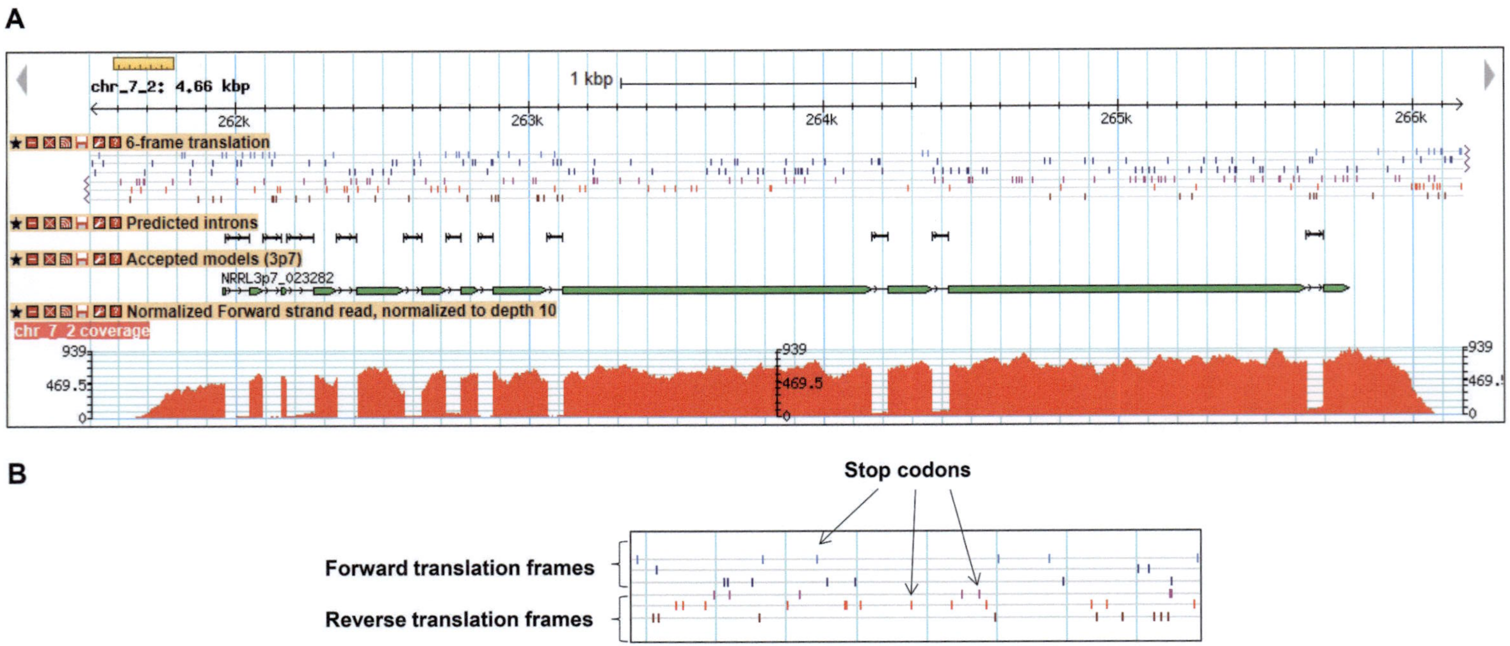

Fig. 3 Six-frame translations and introns. Panel **a**. Track 1: six-frame translation of the genome sequence, track 2: predicted introns, track 3: predicted gene model, track 4: forward strand RNA-seq reads. Panel **b**. Close-up of the six-frame translation track. The top three lines (blue) represent the three forward translation frames (left to right). The bottom three lines (red) represent the three reverse translation frames (right to left). A bar indicates a stop codon. The single-letter amino acid sequence is displayed when viewing a window of 400 bp or less

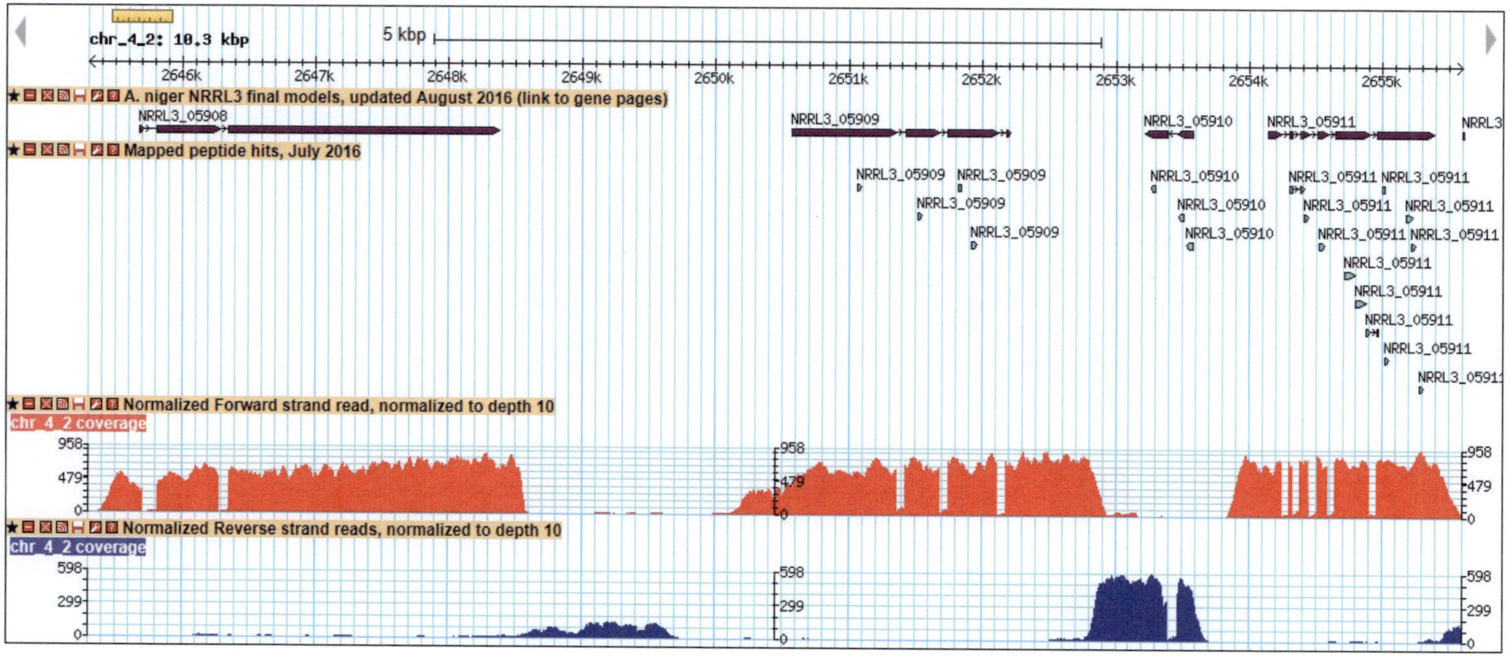

Fig. 4 Mapped peptides from proteomic analyses. Track 1: final gene models, track 2: mapped peptides, track 3: forward strand RNA-seq reads, track 4: reverse strand RNA-seq reads. Some gene models will not have mapped peptides

9. Map and display ChIP-seq reads (chromatin immunoprecipitation followed by sequencing). Trimethylation of lysine 4 on histone 3 (H3K4me3) is an epigenetic modification. H3K4me3 is enriched in genes undergoing active transcription, especially near the start of the gene. ChIP-seq can help identify transcriptionally active genes. Regions of the genome containing the H3K4me3 modification are captured using an anti-H3K4me3 antibody; the DNA is isolated, sequenced, and mapped onto the genome [11]. The results are visualized as peaks in Gbrowse. Read peaks near the 5′ end of gene models are often observed (Fig. 5).
10. Display conserved protein domains. The predicted set of gene models should be scanned for the presence of known functional domains or signatures. If a domain is detected within a translated gene model, this suggests that it has functional or structural similarity to other known proteins. Pfam [12], a collection of protein families supported by multiple sequence alignments and hidden Markov models, and InterProScan [13], a consortium that allows one to search for protein signatures from over a dozen signature databases simultaneously, are useful resources to use (Fig. 6). Moreover, the six-frame translations of the entire genome sequence should be run against Pfam. This is helpful for finding genes missing in the predicted set. Observing where the domains map may also help to elucidate the gene structures (*see* **Note 5**).
11. Map and display external predictions. For some species of interest, a set of predicted gene models may already exist. Even if the predictions were computed using a different genome assembly or are from a different strain, they could be useful for locating genes. The Aspergillus Genome Database (AspGD) [14], the MycoCosm portal from the Joint Genome Institute [15], and FungiDB [16] are databases that house genome assemblies and predicted gene model sets for fungal species. Bear in mind that external gene model predictions may not fully map onto the genome, and only part of a gene may be displayed in the track, because of weak similarity in parts of the gene model. Moreover, some predictions may roughly map onto the genome but may not be feasible because of errors in the external model or sequence differences in the genome assemblies that were used to make the predictions (Fig. 7).
12. Display orthologs. Gene models from closely related species can provide further support that a gene is present. Similar to the external predictions, orthologous gene models may not always fully map onto the genome (Fig. 8).
13. Display alternate models. The SnowyOwl pipeline generates a huge set of possible gene models. It uses these as input for the selection process that results in the Accepted and Imperfect

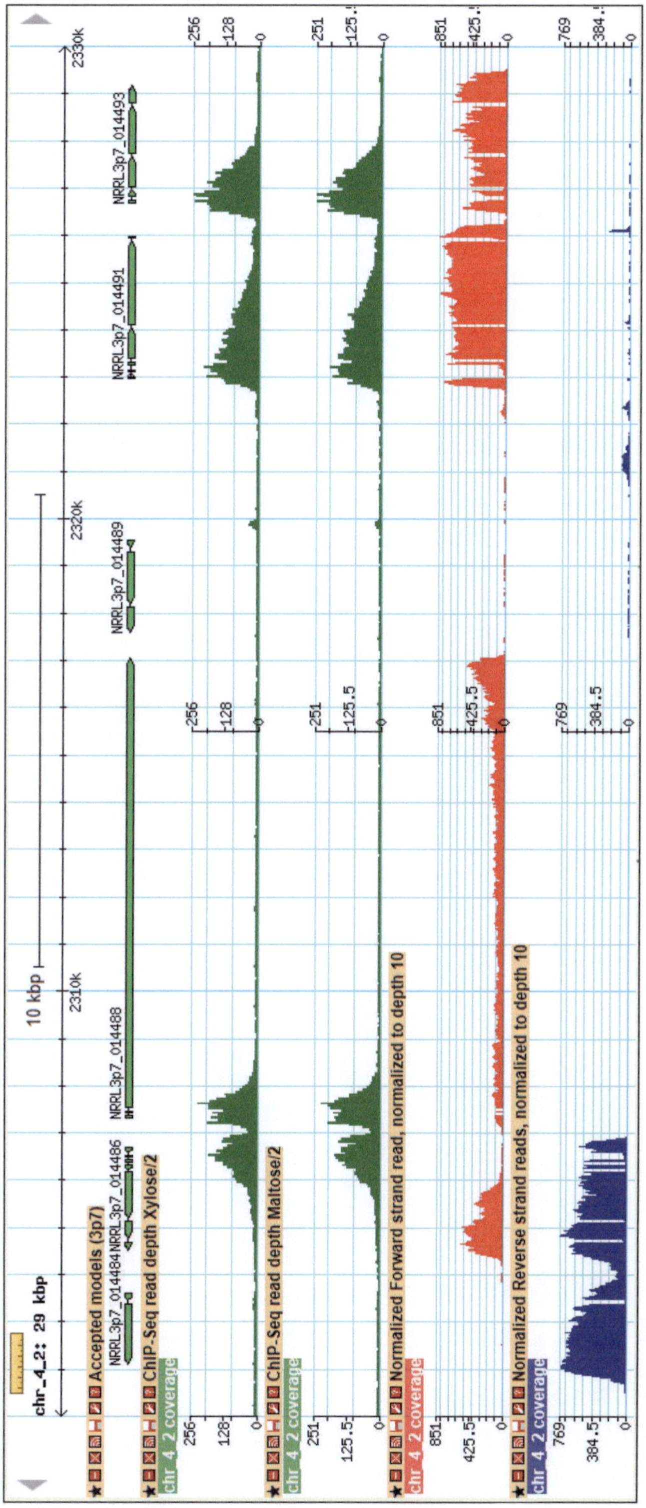

Fig. 5 ChIP-seq peaks viewed in Gbrowse. Track 1: predicted gene models, track 2: ChIP-seq reads from a sample with xylose as a carbon source, track 3: ChIP-seq reads from a sample with maltose as a carbon source, track 4: forward strand RNA-seq reads, track 5: reverse strand RNA-seq reads

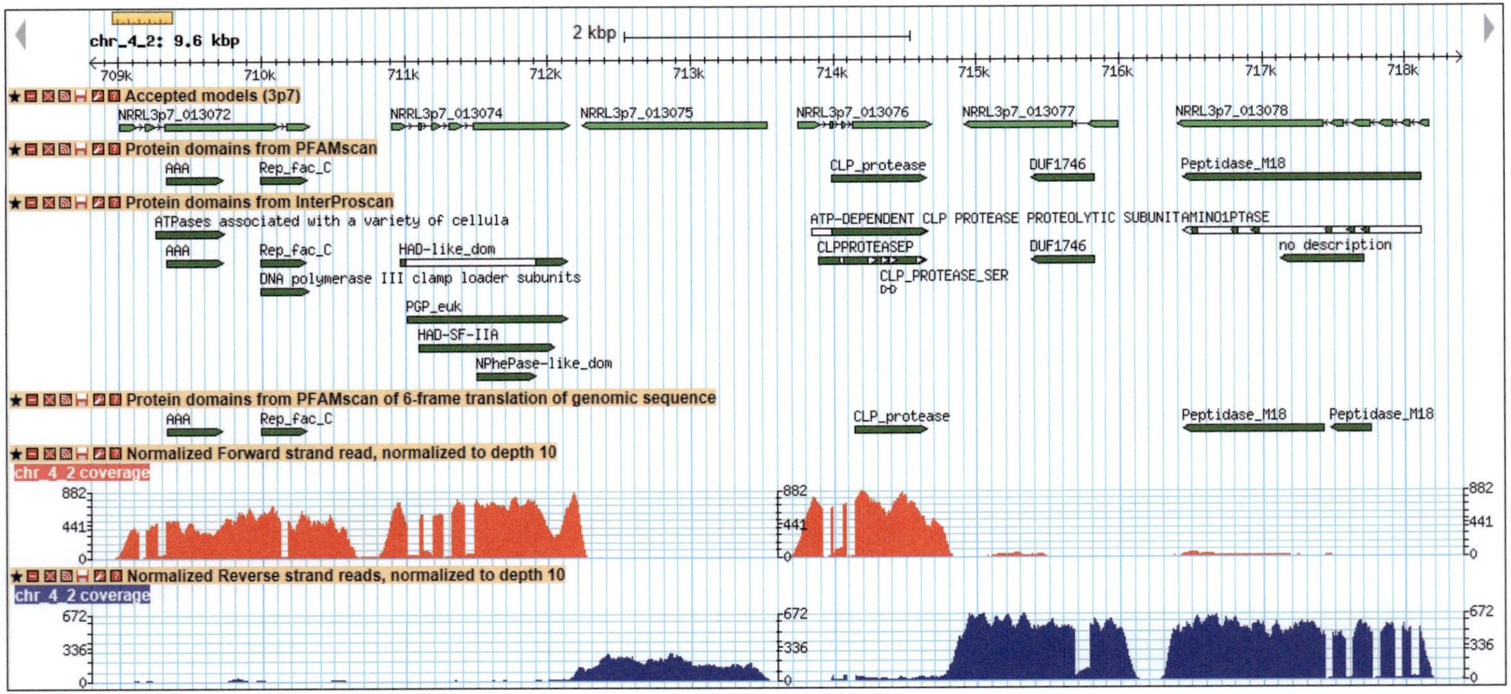

Fig. 6 Conserved protein domains. Track 1: predicted gene models, track 2: domains detected using Pfam [12], track 3: domains detected using InterProScan [13], track 4: conserved protein domains predicted from six-frame translation of the genome sequence, track 5: forward strand RNA-seq reads, track 6: reverse strand RNA-seq reads

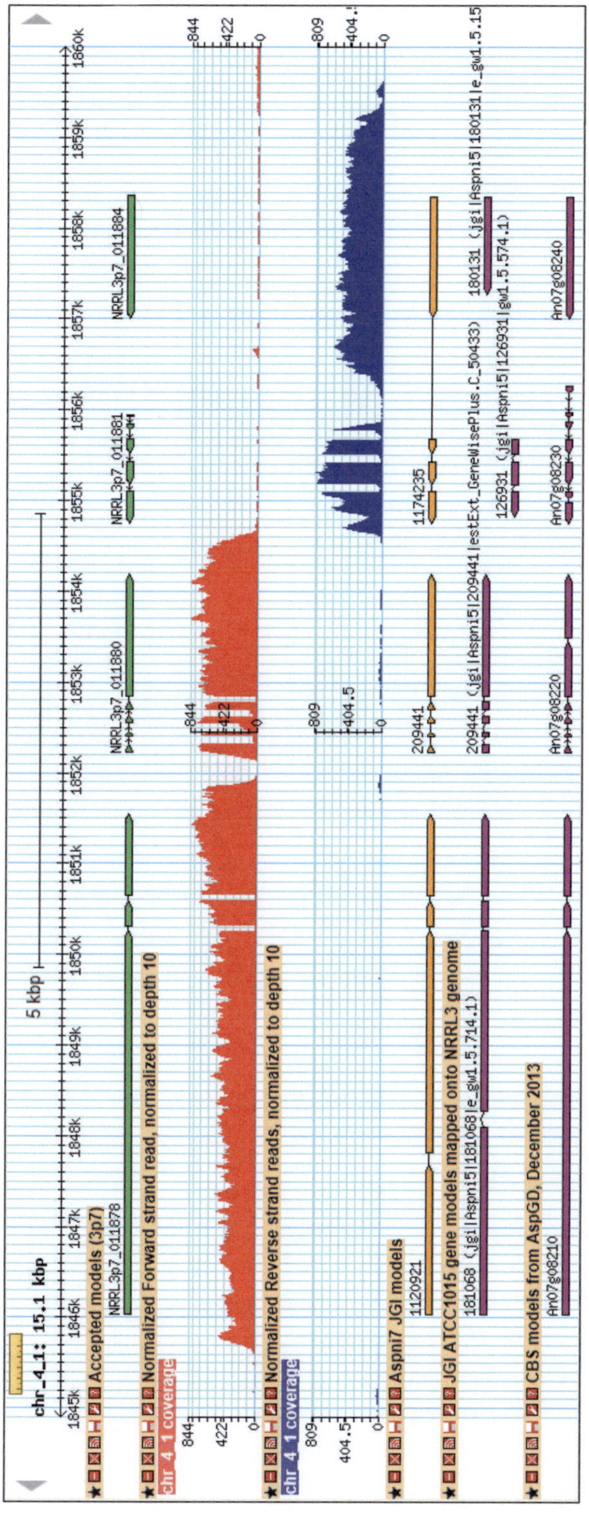

Fig. 7 External predictions. Track 1: predicted gene models, track 2: forward strand RNA-seq reads, track 3: reverse strand RNA-seq reads, track 4: mapped predicted gene models from the Joint Genome Institute for *A. niger* strain ATCC 1015 version 4.0 mapped onto the NRRL 3 genome [15], track 5: mapped predicted gene models from *A. niger* strain ATCC 1015 version 3.0 [17], track 6: mapped predicted gene models from *A. niger* strain CBS 513.88 [18]

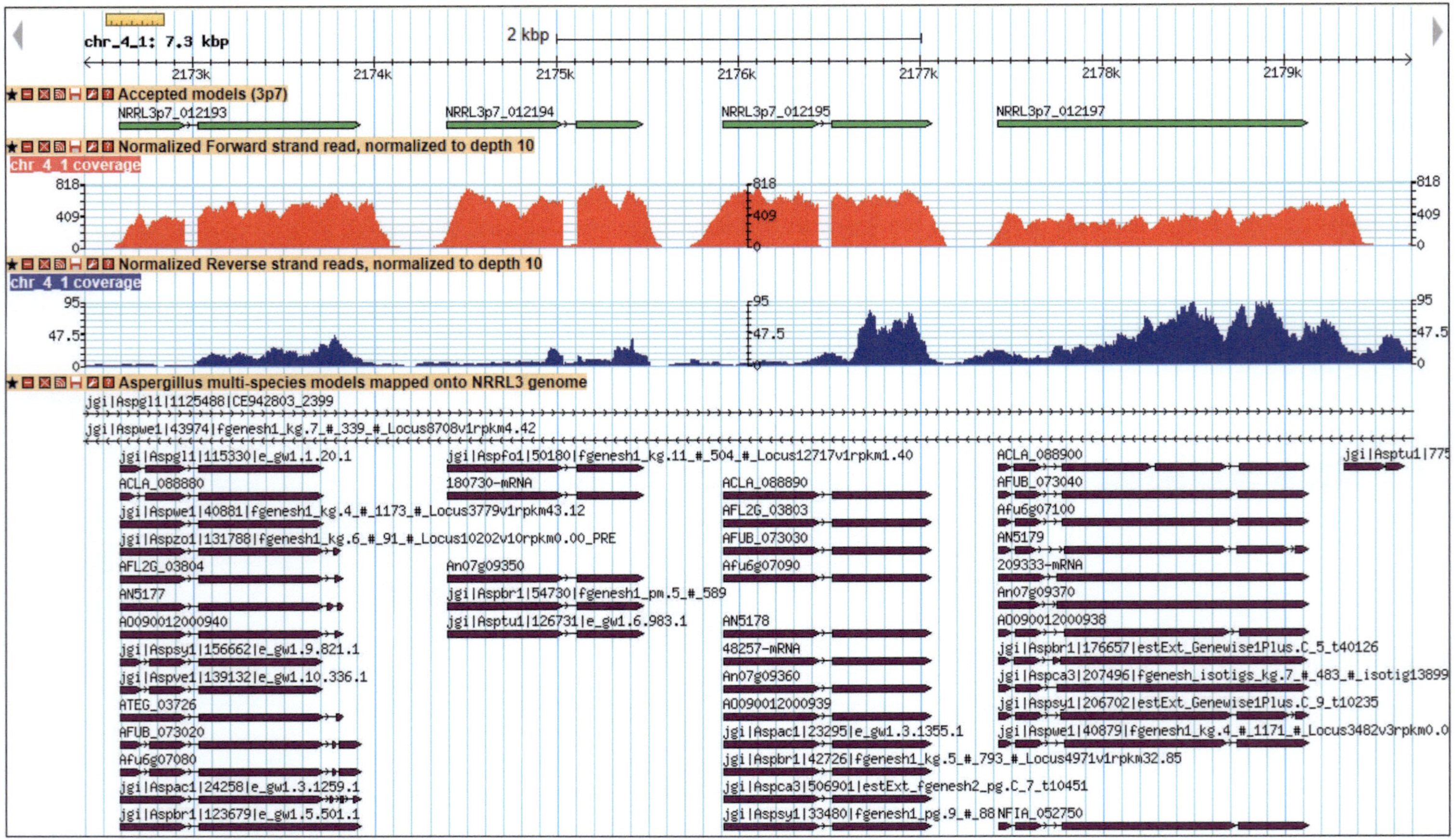

Fig. 8 Orthologous gene models mapped onto the genome of *A. niger* strain NRRL 3. Track 1: predicted gene models, track 2: forward strand RNA-seq reads, track 3: reverse strand RNA-seq reads, track 4: mapped predicted gene models from the Joint Genome Institute for all species in the *Aspergillus* genus [15]

models. It is worthwhile to display all the predicted gene models in a separate track so that they can readily be selected as alternate models or added in by the curator. Note that all of these models are structurally possible, unlike the mapped orthologs and external predictions. GeneMark [7] is another highly accurate gene predictor. Including the GeneMark predictions in a separate track can be informative as well (Fig. 9).

3.2 Manual Curation of Gene Models

1. Perform BLASTX with the predicted gene models against NCBI's nonredundant database [19]. Save the top ten hits and alignment details. These results provide a wide-reaching search for similar proteins from phylogenetically distant species and can be helpful when reviewing predicted gene models that are not supported by experimental evidence. They may also help to determine the structure of a gene (e.g., the BLASTX results show that the beginning or end of the protein is missing or that it is partly out of frame).
2. Navigate through the genome assembly with all of the evidence tracks opened (*see* **Note 6**). Inspect every chromosome/scaffold from beginning to end (*see* **Note 7**). As predicted gene models are happened upon, they should be examined (instructions below). Although the predicted gene models are of particular interest, an eye must be kept out for models that were missed by the prediction pipeline. When viewed in combination, the different types of mapped evidence and predictions should provide a strong indication of the likelihood and structure of each gene model.
3. Use a Microsoft Excel worksheet to store the curator's decisions (Fig. 10). For each predicted gene model in the Accepted set, a decision of "accept," "change," or "demote" should be provided. For each Imperfect model, the decision to "promote," "change," or "demote" must be specified. If "change" is selected, an alternate model must be provided in the adjacent column. Following this strict terminology is important for successfully implementing the decisions downstream (*see* **Note 8**). For models that were missed by the prediction pipeline, add a line into the Excel sheet between the appropriate neighboring gene models and assign the new model a temporary ID consisting of the preceding model's ID followed by an "a".
4. Pool each curator's decisions and perform a second round of review for the gene models where conflicting decisions were made.
5. Use the genome annotation editing tool, Apollo [3], to manually create or edit gene models in cases where there are no available models with the desired structure. An Apollo user guide with detailed instructions for creating and altering gene

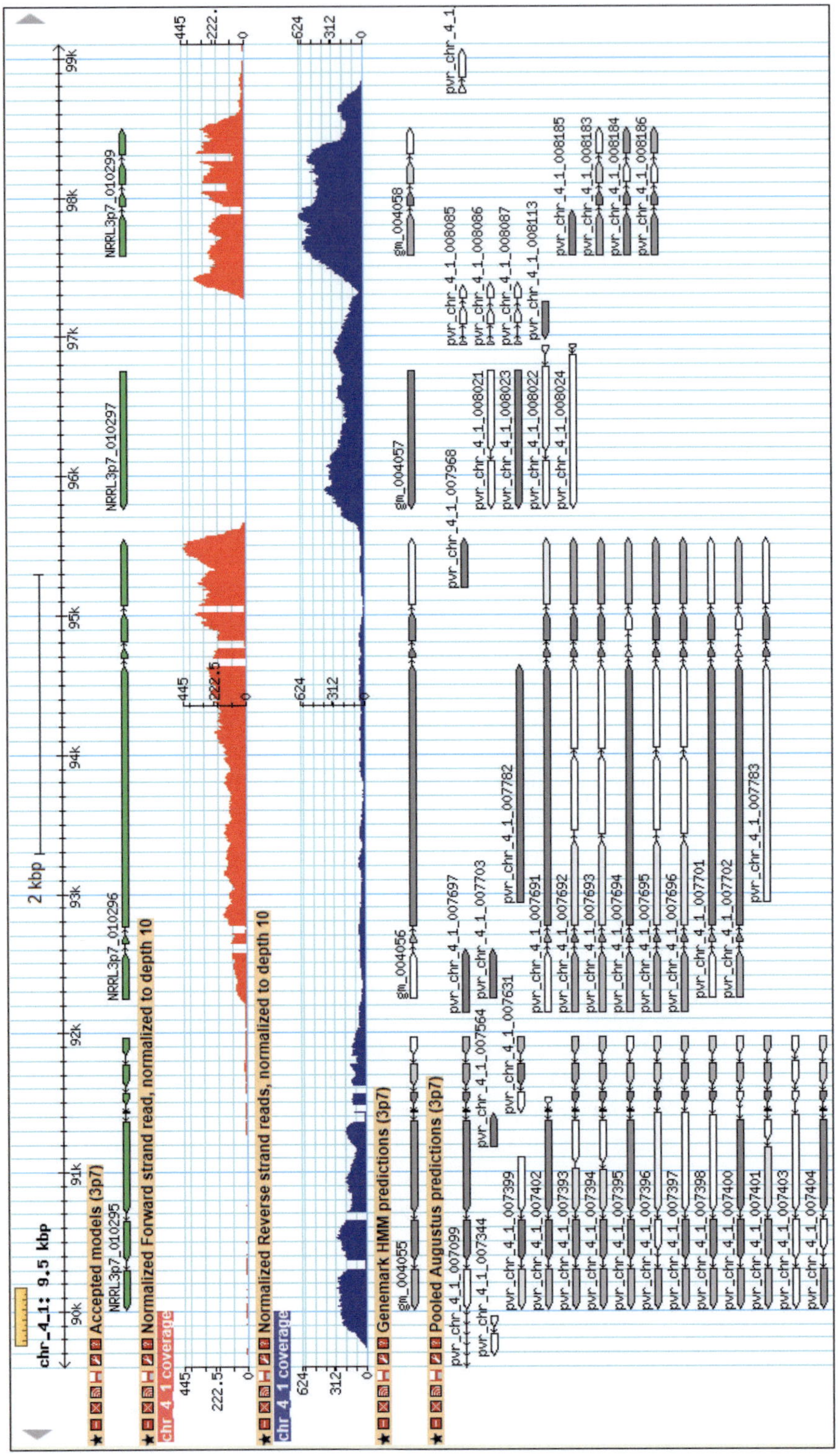

Fig. 9 Alternate gene models. Track 1: predicted gene models, track 2: forward strand RNA-seq reads, track 3: reverse strand RNA-seq reads, track 4: GeneMark predicted gene models [7], track 5: pool of alternate gene models generated and used by SnowyOwl for the selection of Accepted and Imperfect models

	A	B	C
1	ID	Decision (accept/demote/promote/change/flag)	Alternate model ID
2	NRRL3p7_000001		
3	NRRL3p7_000002		
4	NRRL3p7_000003		
5	NRRL3p7_000004		
6	NRRL3p7_000005		
7	NRRL3p7_000006		
8	NRRL3p7_000007		
9	NRRL3p7_000008		
10	NRRL3p7_000009		
11	NRRL3p7_000010		
12	NRRL3p7_000011		
13	NRRL3p7_000012		
14	NRRL3p7_000013		
15	NRRL3p7_000014		
16	NRRL3p7_000015		
17	NRRL3p7_000016		
18	NRRL3p7_000017		
19	NRRL3p7_000018		
20	NRRL3p7_000019		
21	NRRL3p7_000020		
22	NRRL3p7_000021		
23	NRRL3p7_000022		
24	NRRL3p7_000023		
25	NRRL3p7_000024		
26	NRRL3p7_000025		
27	NRRL3p7_000026		
28	NRRL3p7_000027		
29	NRRL3p7_000028		
30	NRRL3p7_000029		
31	NRRL3p7_000030		

Sheet1

Fig. 10 Excel spreadsheet for storing a curator's decisions on the gene models. Highlighting the gene models in the Imperfect set can speed up the process

models is available at http://genomearchitect.github.io/users-guide/ (*see* **Note 9**). Once the model is ready, assign it a unique gene model identifier and save it as a gff3 file. In this format, it can be used to replace the existing model(s) at the specified genomic location (*see* **Note 10**).

6. When the set of gene models is finalized, assign the gene models new chronological IDs. Typically, an acronym reflecting the species or strain is included followed by a number that indicates the gene order (*see* **Note 11**).

3.3 Manual Functional Annotation

1. Create a FASTA file with the protein sequences of all the gene models.
2. Perform a BLASTP [20] search (using the default parameters) to find homologous protein sequences in reviewed protein databases (*see* **Note 12**). The BLASTP output should be delivered in a tabular format.
3. Retain the top BLASTP match from each database (*see* **Note 13**).
4. Among the top BLASTP matches, find the one with the highest sequence identity score. This is the best BLASTP match.
5. Compile the BLASTP results in a Microsoft Excel worksheet. Column A contains the gene model ID numbers. The next column contains the functional descriptor of the best BLASTP match, followed by the name of the database, the database identifier, percent identity, percent query coverage, and percent target coverage in separate columns. Include these same fields for the top hit from each of the databases that were searched, and add them to succeeding columns of the spreadsheet (Fig. 11).
6. Insert four columns to the right of the gene model ID column. Assign the following headers to these columns: gene name, description, evidence, and evidence code. The functional predictions for the gene models and the supporting evidence will be added to these blank cells.
7. Scan the sequences in Pfam [12] and InterProScan [13] to locate the conserved protein domains.
8. Look for secondary, structural features by analyzing the sequences in: (1) Phobius [21] and TMHMM [22] to predict transmembrane topology; (2) SignalP [23] to detect signal peptides; (3) TargetP [24] and WoLF PSORT [25] to predict subcellular localization; (4) big-GPI [26] and KDEL pattern scan [27] to detect glycophospholipid anchors and endoplasmic retention signals, respectively.
9. Enter the conserved domain(s) and localization predictions in separate columns of the spreadsheet.

3.4 Assignment of Functional Annotation

1. Assigning functional annotation to genes that had been characterized experimentally. All of the information in the spreadsheet should be considered when assigning a functional annotation. Use the best BLASTP match as a starting point. If the gene model and its BLASTP match share ≥98% identity over their entire lengths, then the two proteins are considered to be functionally equivalent (*see* **Note 14**). If the reviewed protein has been experimentally characterized, then assign its gene name and functional descriptor to the gene model (*see* **Note 15**). The evidence for the annotation is the PubMed identifier of the reference article. If the activity was directly

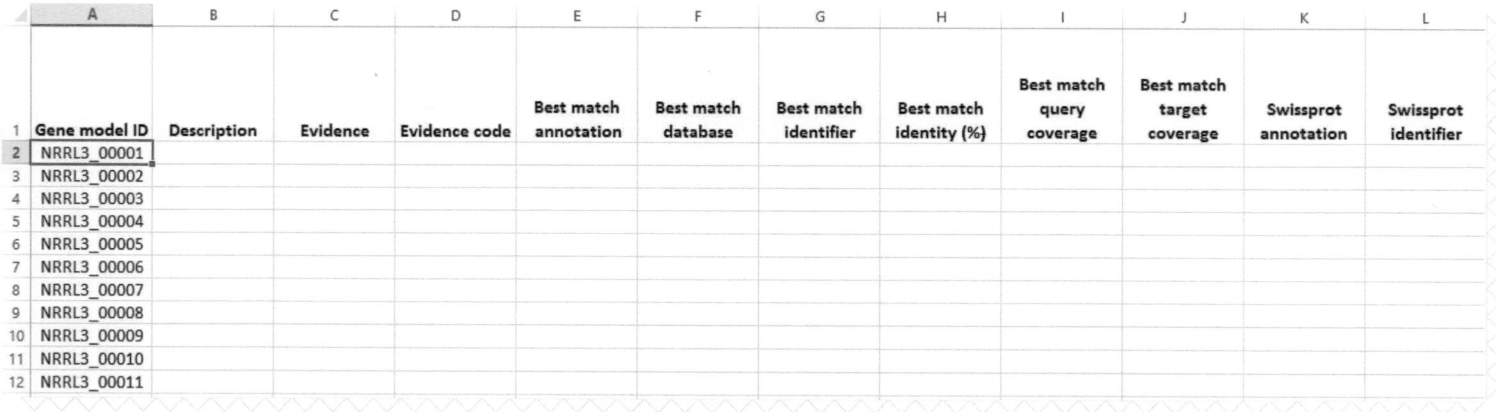

	A	B	C	D	E	F	G	H	I	J	K	L
1	Gene model ID	Description	Evidence	Evidence code	Best match annotation	Best match database	Best match identifier	Best match identity (%)	Best match query coverage	Best match target coverage	Swissprot annotation	Swissprot identifier
2	NRRL3_00001											
3	NRRL3_00002											
4	NRRL3_00003											
5	NRRL3_00004											
6	NRRL3_00005											
7	NRRL3_00006											
8	NRRL3_00007											
9	NRRL3_00008											
10	NRRL3_00009											
11	NRRL3_00010											
12	NRRL3_00011											

Fig. 11 Layout of the functional annotation spreadsheet. Gene model identifiers are listed in column A. Columns B, C, and D are used for the functional annotations and the supporting evidence

assayed using the purified protein or cell extracts, then use the Gene Ontology [28] evidence code IDA (inferred from direct assay). If the article describes a rescue experiment in which the gene restores function in another organism, use the evidence code IGI (inferred from genetic interaction) instead. If the function of the protein is deduced from a mutant or a knock-out strain then use the evidence code IMP (inferred from mutant phenotype).

2. The gene model and the reviewed protein are ≥70% identical over more than 70% of their lengths. If this applies, assign the functional descriptor of the reviewed protein to the gene model (*see* **Note 16**). Use the database identifier of the reviewed protein as evidence for the annotation in combination with the Gene Ontology evidence code "ISS" (inferred from sequence or structural similarity).
3. The proteins share 40–70% identity and ≥70% coverage. The two sequences are considered to be homologous if the conserved domains in the gene model support the function of the reviewed protein (*see* **Notes 17** and **18**). If this is the case, adopt the functional descriptor of the reviewed protein and use its database identifier and the Gene Ontology code ISS as evidence for the annotation. If this is not the case, then assign a more general function to the gene model based on its conserved domains (*see* case "IV" below).
4. The database match to the gene model is lower than 40%. Assign a function to the gene model based on its InterPro and/or Pfam domain(s). Add the words "domain-containing protein" or "family protein" or "-like protein" to the end of the functional descriptor (*see* **Note 19**). This also applies to gene models for which the BLASTP search did not return a match, and the only evidence available is the InterPro and/or Pfam domain profile (*see* **Notes 20** and **21**).
5. Gene models without a BLASTP match and without conserved domains are annotated as "hypothetical protein."
6. After the first pass, review the functional annotations to check for inconsistencies (*see* **Note 22**).
7. Display the functional annotations below the gene models in the genome browser.

4 Notes

1. SnowyOwl accurately predicted 87% of the manually curated gene models in *A. niger* NRRL 3 (manuscript in preparation).
2. No coverage does not necessarily mean that there is no gene (the gene might not have been expressed under the conditions

tested or at the time the sample was collected). The more the RNA-seq profiles that are available, the better. Different carbon sources or conditions may trigger the expression of different genes. Keep in mind that 5′ and 3′ UTRs may have coverage. Coverage on both strands is also sometimes observed and may indicate overlapping genes or antisense RNA.

3. Models with introns longer than 200 bp are suspicious.
4. Coverage of peptides is much lower than transcripts. Most, though not all, peptides map to highly expressed genes. Since peptides are mapped back to predicted models only, they do not provide evidence for models that were missed by the gene prediction pipeline.
5. Sometimes, only parts of a domain are detected in a gene model. This could indicate a pseudogene.
6. A window of 10 kilobase pairs typically provides an adequate resolution for reviewing and finding genes. Using a widescreen monitor may allow a larger region to be viewed and requires less scrolling and loading time.
7. It is highly recommended that each region of the genome be reviewed by at least two curators.
8. To speed up the process, difficult models that require more attention can be flagged and reviewed in detail after the first pass is complete.
9. Apollo clearly shows if a gene model is possible or not. What appears to be the correct gene model may not always be feasible. This could be due to errors in the genome sequence and, in these instances, the closest possible gene model that best fits the evidence is chosen. It could also indicate that the locus contains a pseudogene. The combined evidence from all of the tracks will help to determine if this is the case.
10. The genomic region that is opened and saved in Apollo will be used to replace everything previously contained in that region. Avoid including parts of neighboring models in the file as this will create issues.
11. It is wise to set up a numbering system that can accommodate future gene model additions or changes. Pfam and InterPro are frequently updated and new evidence (e.g., RNA-seq data from samples under different conditions) may provide information that can affect the models. Numbers such as .1 and .2 can be used to indicate changes to the gene models between release versions (e.g., gene model NRRL3_00009 was altered and reassigned the new ID NRRL3_00009.1).
12. Restricting the source of annotations to reliable records minimizes error in the functional predictions. BLASTP searches are, therefore, done only in manually reviewed protein-

sequence databases such as Swiss-Prot [29], the *Saccharomyces* genome database SGD [30], mycoCLAP [31, 32], and the Genozymes *Aspergillus niger* strain NRRL 3 database (http://gbrowse.fungalgenomics.ca/).

13. Exclude database sequences with <30% identity and/or <70% coverage.
14. If the differences are in the active site or other residues known to be important for enzymatic function, then the gene model should be flagged with a warning and/or assigned a more general annotation.
15. Gene names are assigned to gene models only if their function is supported by experimental evidence. If the reviewed protein has not been characterized, then only adopt its functional descriptor.
16. If the description of the best BLASTP match is a gene name only, then use a more descriptive annotation. For example, *A. niger* gene model NRRL3_11321 shares 89% amino acid sequence identity over its entire length with an *A. oryzae* protein described as "SNAP-25" in the AspGD database. SNAP-25 is a soluble NSF attachment protein receptor, also known as SNARE, involved in vesicular trafficking [33]. We annotated NRRL3_11321 as "Vesicular trafficking protein SNAP-25". Likewise, if the description of the best BLASTP match is ambiguous, such as "Putative glucan endo-1,3-beta-glucosidase," then use another reviewed protein as the source of annotation.
17. For example, the *A. niger* gene model NRRL3_04127 shares 44% amino acid sequence identity and 97% coverage with the *S. cerevisiae* alpha N-terminal protein methyltransferase Ntm1p (UniProt ID P38340). NRRL3_04127 contains the InterPro entry IPR008576:Alpha-N-methyltransferase NTM1. Since the InterPro classification of the gene model concurs with the description of the reviewed protein we annotated NRRL3_04127 as an "Alpha-N-methyltransferase." Here is another example. NRRL3_01132 is predicted to be a mitochondrial protein by TargetP and WoLF PSORT. It share 62% sequence identity with the Swiss-Prot entry Q9Y767. The proteins have the same multidomain organization in InterProScan including: DNA-directed DNA-polymerase, family A, mitochondria (IPR002297); Ribonuclease H-like domain (IPR012337); DNA-directed DNA polymerase, family A, palm domain (IPR001098); and DNA-directed DNA polymerase, family A, conserved site (IPR019760) (*see* Fig. 12). Q9Y767 is described in UniProt as a DNA polymerase gamma involved in the replication of mitochondrial DNA. Based on this evidence, NRRL3_01132 was annotated as "DNA polymerase gamma."

Fig. 12 InterPro domain architecture of (**a**) *A. niger* NRRL3_01132 and (**b**) UniProt entry Q9Y767

18. Multiple sequence alignments may also be used to decide whether proteins are homologous or nonhomologous.
19. As an example, the *A. niger* gene model NRRL3_02176 belongs to the InterPro family IPR008901:Ceramidase. It shares low (31%) sequence identity with a *S. cerevisiae* alkaline ceramidase involved in phytoceramide metabolism [34]. Based on this evidence, we annotated NRRL3_02176 as a "Ceramidase-like protein."
20. The *A. niger* gene model NRRL3_01780, for instance, contains only one SET domain (IPR001214). According to InterPro, this domain is evolutionarily conserved and found in proteins of diverse function (https://www.ebi.ac.uk/interpro/entry/IPR001214). We annotated NRRL3_01780 as a "SET domain-containing protein."
21. Gene models with InterPro and/or Pfam domains of unknown function (DUF) are annotated as "uncharacterized protein."
22. Gene models belonging to a particular group should be assigned the same functional annotation. For instance, *A. niger* NRRL 3 models containing the InterPro entry IPR011118:Tannase/feruloyl esterase were all annotated as "Tannase/feruloyl esterase family protein" (unless the gene model had a high-scoring BLASTP match). Likewise, the components of a complex should be given similar descriptions. *A. niger* gene models NRRL3_02728 and NRRL3_07071 are predicted to be part of the multimeric chaperone protein, prefoldin, and were annotated as "Prefoldin subunit 4" and "Prefoldin subunit 6," respectively.

Acknowledgment

This work was supported by Genome Canada and Génome Québec.

References

1. Reid I, O'Toole N, Zabaneh O, Nourzadeh R, Dahdouli M, Abdellateef M, Gordon PM, Soh J, Butler G, Sensen CW, Tsang A (2014) SnowyOwl: accurate prediction of fungal genes by using RNA-Seq and homology information to select among ab initio models. BMC Bioinformatics 15:229. https://doi.org/10.1186/1471-2105-15-229
2. Stein LD, Mungall C, Shu S, Caudy M, Mangone M, Day A, Nickerson E, Stajich JE, Harris TW, Arva A, Lewis S (2002) The generic genome browser: a building block for a model organism system database. Genome Res 12(10):1599–1610. https://doi.org/10.1101/gr.403602
3. Lee E, Helt GA, Reese JT, Munoz-Torres MC, Childers CP, Buels RM, Stein L, Holmes IH, Elsik CG, Lewis SE (2013) Web Apollo: a web-based genomic annotation editing platform. Genome Biol 14(8):R93. https://doi.org/10.1186/gb-2013-14-8-r93
4. Salamov AA, Solovyev VV (2000) Ab initio gene finding in Drosophila genomic DNA. Genome Res 10(4):516–522

5. Birney E, Clamp M, Durbin R (2004) GeneWise and Genomewise. Genome Res 14(5):988–995. https://doi.org/10.1101/gr.1865504
6. Grigoriev I, Martinez D, Salamov A (2006) Fungal genomic annotation. Appl Mycol Biotechnol 6:123–142. https://doi.org/10.1016/S1874-5334(06)80008-0
7. Ter-Hovhannisyan V, Lomsadze A, Chernoff YO, Borodovsky M (2008) Gene prediction in novel fungal genomes using an ab initio algorithm with unsupervised training. Genome Res 18(12):1979–1990. https://doi.org/10.1101/gr.081612.108
8. Cantarel BL, Korf I, Robb SM, Parra G, Ross E, Moore B, Holt C, Sanchez Alvarado A, Yandell M (2008) MAKER: an easy-to-use annotation pipeline designed for emerging model organism genomes. Genome Res 18(1):188–196. https://doi.org/10.1101/gr.6743907
9. Grabherr MG, Haas BJ, Yassour M, Levin JZ, Thompson DA, Amit I, Adiconis X, Fan L, Raychowdhury R, Zeng Q, Chen Z, Mauceli E, Hacohen N, Gnirke A, Rhind N, di Palma F, Birren BW, Nusbaum C, Lindblad-Toh K, Friedman N, Regev A (2011) Full-length transcriptome assembly from RNA-Seq data without a reference genome. Nat Biotechnol 29(7):644–652. https://doi.org/10.1038/nbt.1883
10. Li D, Liu CM, Luo R, Sadakane K, Lam TW (2015) MEGAHIT: an ultra-fast single-node solution for large and complex metagenomics assembly via succinct de Bruijn graph. Bioinformatics 31(10):1674–1676. https://doi.org/10.1093/bioinformatics/btv033
11. Johnson DS, Mortazavi A, Myers RM, Wold B (2007) Genome-wide mapping of in vivo protein-DNA interactions. Science 316(5830):1497–1502. https://doi.org/10.1126/science.1141319
12. Finn RD, Coggill P, Eberhardt RY, Eddy SR, Mistry J, Mitchell AL, Potter SC, Punta M, Qureshi M, Sangrador-Vegas A, Salazar GA, Tate J, Bateman A (2016) The Pfam protein families database: towards a more sustainable future. Nucleic Acids Res 44(D1):D279–D285. https://doi.org/10.1093/nar/gkv1344
13. Jones P, Binns D, Chang HY, Fraser M, Li W, McAnulla C, McWilliam H, Maslen J, Mitchell A, Nuka G, Pesseat S, Quinn AF, Sangrador-Vegas A, Scheremetjew M, Yong SY, Lopez R, Hunter S (2014) InterProScan 5: genome-scale protein function classification. Bioinformatics 30(9):1236–1240. https://doi.org/10.1093/bioinformatics/btu031
14. Cerqueira GC, Arnaud MB, Inglis DO, Skrzypek MS, Binkley G, Simison M, Miyasato SR, Binkley J, Orvis J, Shah P, Wymore F, Sherlock G, Wortman JR (2014) The Aspergillus genome database: multispecies curation and incorporation of RNA-Seq data to improve structural gene annotations. Nucleic Acids Res 42(Database issue):D705–D710. https://doi.org/10.1093/nar/gkt1029
15. Grigoriev IV, Nikitin R, Haridas S, Kuo A, Ohm R, Otillar R, Riley R, Salamov A, Zhao X, Korzeniewski F, Smirnova T, Nordberg H, Dubchak I, Shabalov I (2014) MycoCosm portal: gearing up for 1000 fungal genomes. Nucleic Acids Res 42(Database issue):D699–D704. https://doi.org/10.1093/nar/gkt1183
16. Stajich JE, Harris T, Brunk BP, Brestelli J, Fischer S, Harb OS, Kissinger JC, Li W, Nayak V, Pinney DF, Stoeckert CJ Jr, Roos DS (2012) FungiDB: an integrated functional genomics database for fungi. Nucleic Acids Res 40(Database issue):D675–D681. https://doi.org/10.1093/nar/gkr918
17. Andersen MR, Salazar MP, Schaap PJ, van de Vondervoort PJ, Culley D, Thykaer J, Frisvad JC, Nielsen KF, Albang R, Albermann K, Berka RM, Braus GH, Braus-Stromeyer SA, Corrochano LM, Dai Z, van Dijck PW, Hofmann G, Lasure LL, Magnuson JK, Menke H, Meijer M, Meijer SL, Nielsen JB, Nielsen ML, van Ooyen AJ, Pel HJ, Poulsen L, Samson RA, Stam H, Tsang A, van den Brink JM, Atkins A, Aerts A, Shapiro H, Pangilinan J, Salamov A, Lou Y, Lindquist E, Lucas S, Grimwood J, Grigoriev IV, Kubicek CP, Martinez D, van Peij NN, Roubos JA, Nielsen J, Baker SE (2011) Comparative genomics of citric-acid-producing Aspergillus niger ATCC 1015 versus enzyme-producing CBS 513.88. Genome Res 21(6):885–897. https://doi.org/10.1101/gr.112169.110
18. Pel HJ, de Winde JH, Archer DB, Dyer PS, Hofmann G, Schaap PJ, Turner G, de Vries RP, Albang R, Albermann K, Andersen MR, Bendtsen JD, Benen JA, van den Berg M, Breestraat S, Caddick MX, Contreras R, Cornell M, Coutinho PM, Danchin EG, Debets AJ, Dekker P, van Dijck PW, van Dijk A, Dijkhuizen L, Driessen AJ, d'Enfert C, Geysens S, Goosen C, Groot GS, de Groot PW, Guillemette T, Henrissat B, Herweijer M, van den Hombergh JP, van den Hondel CA, van der Heijden RT, van der Kaaij RM, Klis FM, Kools HJ, Kubicek CP, van Kuyk PA, Lauber J, Lu X, van der Maarel MJ, Meulenberg R, Menke H, Mortimer MA, Nielsen J, Oliver SG, Olsthoorn M, Pal K, van Peij NN, Ram AF, Rinas U, Roubos JA, Sagt CM, Schmoll M, Sun J, Ussery D, Varga J, Vervecken W, van de Vondervoort PJ, Wedler H, Wosten HA, Zeng AP, van Ooyen AJ, Visser J, Stam H (2007)

Genome sequencing and analysis of the versatile cell factory Aspergillus niger CBS 513.88. Nat Biotechnol 25(2):221–231. https://doi.org/10.1038/nbt1282

19. O'Leary NA, Wright MW, Brister JR, Ciufo S, Haddad D, McVeigh R, Rajput B, Robbertse B, Smith-White B, Ako-Adjei D, Astashyn A, Badretdin A, Bao Y, Blinkova O, Brover V, Chetvernin V, Choi J, Cox E, Ermolaeva O, Farrell CM, Goldfarb T, Gupta T, Haft D, Hatcher E, Hlavina W, Joardar VS, Kodali VK, Li W, Maglott D, Masterson P, McGarvey KM, Murphy MR, O'Neill K, Pujar S, Rangwala SH, Rausch D, Riddick LD, Schoch C, Shkeda A, Storz SS, Sun H, Thibaud-Nissen F, Tolstoy I, Tully RE, Vatsan AR, Wallin C, Webb D, Wu W, Landrum MJ, Kimchi A, Tatusova T, DiCuccio M, Kitts P, Murphy TD, Pruitt KD (2016) Reference sequence (RefSeq) database at NCBI: current status, taxonomic expansion, and functional annotation. Nucleic Acids Res 44(D1):D733–D745. https://doi.org/10.1093/nar/gkv1189
20. Altschul SF, Gish W, Miller W, Myers EW, Lipman DJ (1990) Basic local alignment search tool. J Mol Biol 215(3):403–410. https://doi.org/10.1016/S0022-2836(05)80360-2
21. Kall L, Krogh A, Sonnhammer EL (2007) Advantages of combined transmembrane topology and signal peptide prediction – the Phobius web server. Nucleic Acids Res 35(Web Server issue):W429–W432. https://doi.org/10.1093/nar/gkm256
22. Krogh A, Larsson B, von Heijne G, Sonnhammer EL (2001) Predicting transmembrane protein topology with a hidden Markov model: application to complete genomes. J Mol Biol 305(3):567–580. https://doi.org/10.1006/jmbi.2000.4315
23. Petersen TN, Brunak S, von Heijne G, Nielsen H (2011) SignalP 4.0: discriminating signal peptides from transmembrane regions. Nat Methods 8(10):785–786. https://doi.org/10.1038/nmeth.1701
24. Emanuelsson O, Nielsen H, Brunak S, von Heijne G (2000) Predicting subcellular localization of proteins based on their N-terminal amino acid sequence. J Mol Biol 300(4):1005–1016. https://doi.org/10.1006/jmbi.2000.3903
25. Horton P, Park KJ, Obayashi T, Fujita N, Harada H, Adams-Collier CJ, Nakai K (2007) WoLF PSORT: protein localization predictor. Nucleic Acids Res 35(Web Server issue):W585–W587. https://doi.org/10.1093/nar/gkm259
26. Eisenhaber B, Schneider G, Wildpaner M, Eisenhaber F (2004) A sensitive predictor for potential GPI lipid modification sites in fungal protein sequences and its application to genome-wide studies for Aspergillus nidulans, Candida albicans, Neurospora crassa, Saccharomyces cerevisiae and Schizosaccharomyces pombe. J Mol Biol 337(2):243–253. https://doi.org/10.1016/j.jmb.2004.01.025
27. Gattiker A, Gasteiger E, Bairoch A (2002) ScanProsite: a reference implementation of a PROSITE scanning tool. Appl Bioinforma 1(2):107–108
28. Ashburner M, Ball CA, Blake JA, Botstein D, Butler H, Cherry JM, Davis AP, Dolinski K, Dwight SS, Eppig JT, Harris MA, Hill DP, Issel-Tarver L, Kasarskis A, Lewis S, Matese JC, Richardson JE, Ringwald M, Rubin GM, Sherlock G (2000) Gene ontology: tool for the unification of biology. The gene ontology consortium. Nat Genet 25(1):25–29. https://doi.org/10.1038/75556
29. Magrane M, Consortium U (2011) UniProt Knowledgebase: a hub of integrated protein data. Database (Oxford) 2011:bar009. https://doi.org/10.1093/database/bar009
30. Cherry JM, Hong EL, Amundsen C, Balakrishnan R, Binkley G, Chan ET, Christie KR, Costanzo MC, Dwight SS, Engel SR, Fisk DG, Hirschman JE, Hitz BC, Karra K, Krieger CJ, Miyasato SR, Nash RS, Park J, Skrzypek MS, Simison M, Weng S, Wong ED (2012) Saccharomyces genome database: the genomics resource of budding yeast. Nucleic Acids Res 40(Database issue):D700–D705. https://doi.org/10.1093/nar/gkr1029
31. Murphy C, Powlowski J, Wu M, Butler G, Tsang A (2011) Curation of characterized glycoside hydrolases of fungal origin. Database (Oxford) 2011:bar020. https://doi.org/10.1093/database/bar020
32. Strasser K, McDonnell E, Nyaga C, Wu M, Wu S, Almeida H, Meurs MJ, Kosseim L, Powlowski J, Butler G, Tsang A (2015) mycoCLAP, the database for characterized lignocellulose-active proteins of fungal origin: resource and text mining curation support. Database (Oxford) 2015. https://doi.org/10.1093/database/bav008
33. Kuratsu M, Taura A, Shoji JY, Kikuchi S, Arioka M, Kitamoto K (2007) Systematic analysis of SNARE localization in the filamentous fungus Aspergillus oryzae. Fungal Genet Biol 44(12):1310–1323. https://doi.org/10.1016/j.fgb.2007.04.012
34. Mao C, Xu R, Bielawska A, Obeid LM (2000) Cloning of an alkaline ceramidase from Saccharomyces cerevisiae. An enzyme with reverse (CoA-independent) ceramide synthase activity. J Biol Chem 275(10):6876–6884

Chapter 17

Evaluating Programs for Predicting Genes and Transcripts with RNA-Seq Support in Fungal Genomes

Ian Reid

Abstract

The steps needed to computationally predict genes and transcripts in fungal genomes with support from RNA-Seq data are described in detail for three prediction programs: CodingQuarry, BRAKER1, and Harfang. These programs predicted from 86% to 92% (Harfang) of the genes in a manually curated reference set for *Aspergillus niger* strain NRRL3. Genes with little or no RNA-Seq read coverage were predicted less successfully than genes with adequate coverage.

Key words Gene prediction, Transcript prediction, RNA-Seq, Bioinformatics, Cleaning short sequence reads

1 Introduction

After a genome is sequenced and assembled, the next step is to identify the genes it contains and the proteins that they encode. Computational methods for predicting genes from a DNA sequence generally use probabilistic models of exon, intron, transcript, and intergenic sequences, such as Hidden Markov Models (HMMs), to discover likely genes ab initio [1]. Additional evidence, such as the sequences of known genes and proteins from related organisms or of messenger RNA from the target organism, can be used to guide the predictions or to evaluate the candidate gene models. The development of RNA-Seq technology has greatly increased the availability of messenger RNA sequences, albeit in short fragments [2]. The more recent development of methods for generating strand-specific reads from RNA increases the utility of RNA-Seq data by removing ambiguity about which strand of the DNA was transcribed [3]. When the strand-specific RNA-Seq reads are aligned to the genome they clearly delineate the exons and introns of well-expressed genes.

Ronald P. de Vries et al. (eds.), *Fungal Genomics: Methods and Protocols*, Methods in Molecular Biology, vol. 1775, https://doi.org/10.1007/978-1-4939-7804-5_17, © Springer Science+Business Media, LLC, part of Springer Nature 2018

Numerous programs for computational gene prediction have been developed [4], and the performance of some of them on animal genomes has been compared in organized competitions [5–7]. No comparisons of the success of different gene predictors on fungal genomes have been published, however. The genomes of fungi are somewhat more amenable to gene prediction than animal or plant genomes because they are smaller and have shorter introns and intergenic spaces. On the other hand, fungal genes may be so closely spaced that recognizing the gaps between them is difficult [8]. Also, sequence signals for initiation of transcription and translation are not well understood in fungal genomes [9]. Exon skipping is rare in fungal transcriptomes, but alternative splice donor and acceptor sites and intron retention result in alternatively spliced transcript isoforms in about 7% of the genes of filamentous fungi [10].

Evaluating transcript and genome predictions requires a reference set of correct models. Such a reference set has been lacking in the fungi, but now a complete set of carefully curated gene models from *Aspergillus niger* strain NRRL3 is available [11]. Here I describe the use of these models to evaluate the performance of three recently developed prediction programs on the NRRL3 genome with strand-specific RNA-Seq data. This chapter gives detailed instructions for processing RNA-Seq reads and running the three gene predictors.

All three of these predictors make use of RNA-Seq reads aligned to the genome, but in differing ways. BRAKER1 [12] uses the spliced RNA-Seq reads to produce hints about the positions and orientations of introns, and feeds the hints to Genemark-ET [13]. Then a filtered subset of the genes predicted by Genemark-ET is used to train Augustus [14]. The genes subsequently predicted by Augustus form the output of BRAKER1.

CodingQuarry [15] requires the aligned RNA-Seq reads to be assembled into transcript predictions by a program such as Cufflinks. CodingQuarry filters the predicted transcripts and uses the high-quality ones to train a Hidden Markov Model. The predictions of this HMM are combined with the filtered input transcripts in CodingQuarry's output.

Harfang is an unpublished derivative of SnowyOwl [16] that is designed to improve the output of other gene predictors such as BRAKER1. It uses Augustus trained by the precursor program along with both intron and read depth hints and relaxed model selection to generate a wide variety of candidate transcript models. The candidate models are scored by comparison to known proteins and to the RNA-Seq data, and transcripts scoring above a threshold are clustered into genes.

RNA-Seq reads received from a sequencing center may contain sequence errors, and may be contaminated with adapter sequences or reads from ribosomal RNA. For the most accurate gene predictions, the reads should be cleaned before use. There are numerous software tools and processing pipelines for read cleaning described in the literature; the instructions below describe the ones that I use.

2 Materials

In the following, text in monospaced font should be typed verbatim on the command line. "$" at the start of a line represents the command line prompt; do not include it. <*Italic text between angle brackets*> should be replaced by an appropriate value for your system.

2.1 Computer Workstation with Linux Operating System

Most bioinformatic software, including the programs described here, are designed to run with a command line interface on a Unix operating system, usually Linux. Any modern Linux distribution, including Ubuntu, should be suitable. The programs can run on a personal workstation or a compute server. Gene prediction programs are computation-intensive and benefit from using multiple processor cores. They are not very memory-hungry; 32 GB of RAM should be sufficient for fungal genomes. The examples in this chapter were run on a workstation with 16 processor cores and 48 GB of RAM.

2.2 A Sequenced and Assembled Genome

The genome file should be in FASTA format and include all the chromosomes, scaffolds, or contigs of the assembly.

First create a new directory to hold the gene prediction results.

```
$ PREDICTIONS=<Path to your gene prediction directory>
$ mkdir -p $PREDICTIONS
```

Make a copy of the genome sequence and save it as $PREDICTIONS/assembly.fasta:

```
$ cp <Genome sequence filepath>
$PREDICTIONS/assembly.fasta
```

2.3 RNA-Seq Reads

The reads should be from a strand-specific library. Paired-end reads are preferable but not essential, and the reads should be 100 bases or longer for best results. The reads should cover as much as possible of the transcriptome to a depth of ten or more. Pooling reads from cultures grown in different conditions helps to increase the coverage breadth. This chapter assumes Illumina reads from strand-specific libraries prepared by the dUTP method.

1. From your laboratory
   You likely have RNA-Seq reads produced in your own laboratory or by collaborators for gene expression studies or specifically for gene prediction.

2. From Sequence Read Archive
   Reads that have been placed in the public domain can be downloaded from the NCBI Sequence Read Archive (SRA). A convenient way to find available reads for your organism is through the NCBI Taxonomy Browser (https://www.ncbi.nlm.nih.gov/Taxonomy/taxonomyhome.html). The web pages for individual species with reads available have links to SRA Experiments that lead to lists of read sets. Choose RNA as the Source and follow the link "Send results to Run selector." The resulting table can be filtered by several criteria including LibraryLayout (single or paired), strand, and Load Date. Unfortunately strand specificity is not one of the criteria. To find strand-specific reads you can first select recently loaded runs and then follow their Experiment links to find details of the library preparation; if strand-specific is not mentioned, the reads are probably mixed-strand. After finding suitable runs, follow the Run link for download instructions.

2.4 Software (See Notes 1 and 2)

1. Bash shell
   The user interface to command line programs is provided by a shell program. Bash is the most common Linux shell, and the command lines in this chapter are designed for Bash. If Bash is not already available on your computer system, ask your system administrator to install it.
2. Perl and Python
   Many bioinformatic programs use either Perl or Python scripts. Both of these should be already installed on your system. Perl 5 is the current version. Python is in the midst of changing from version 2 to version 3, and your system may have both versions. The programs described here use the older Python version, specifically Python 2.7.
3. System utilities
   The gene prediction programs and their associated scripts depend on some utility programs, including sort, grep, awk, sed, wget, tar, make, and cut, that are normally available on Unix systems. A utility that might not be installed by default is parallel [17]. Packages to install parallel on many Linux distributions are available at https://www.gnu.org/software/parallel/. The compiler gcc will be needed to install software distributed as source code.
4. Read corrector
   Download and compile Rcorrector [18] to a folder in your home directory (*see* **Notes 2–4**).

```
$ mkdir ~/DL
$ cd ~/DL
$ wget
https://github.com/mourisl/Rcorrector/archive/
v1.0.2.tar.gz
```

```
$ tar -xzf v1.0.2.tar.gz
$ cd Rcorrector-1.0.2
$ make
```

5. Read trimmer

   Download and compile skewer [19].

```
$ cd ~/DL
$ wget
https://github.com/relipmoc/skewer/archive/
0.2.2.tar.gz
$ tar -xzf 0.2.2.tar.gz
$ cd skewer-0.2.2
$ make
```

   Skewer needs the sequences of the adapters that were used for your reads. The sequencing center can supply you with these sequences, and they can be conveniently put in a FASTA file in the skewer directory, e.g., adapters.fasta.

6. Download SortMeRNA [20] and its data files.

```
$ cd ~/DL
$ wget
http://bioinfo.lifl.fr/RNA/sortmerna/code/sortmerna-
2.1-linux-64.tar.gz
$ tar -xzf sortmerna-2.1-linux-64.tar.gz
$ cd sortmerna-2.1-linux-64
$ ./indexdb_rna -ref ./rRNA_databases/silva-bac-16s-
id90.fasta,./index/silva-bac-16s-db:\
./rRNA_databases/silva-bac-23s-id98.fasta,./index/
silva-bac-23s-db:\
./rRNA_databases/silva-arc-16s-id95.fasta,./index/
silva-arc-16s-db:\
./rRNA_databases/silva-arc-23s-id98.fasta,./index/
silva-arc-23s-db:\
./rRNA_databases/silva-euk-18s-id95.fasta,./index/
silva-euk-18s-db:\
./rRNA_databases/silva-euk-28s-id98.fasta,./index/
silva-euk-28s:\
./rRNA_databases/rfam-5s-database-id98.fasta,./in-
dex/rfam-5s-db:\
./rRNA_databases/rfam-5.8s-database-id98.fasta,./
index/rfam-5.8s-db
```

7. Read mapper

   Download precompiled STAR [21].

```
$ cd ~/DL
$ wget
https://github.com/alexdobin/STAR/archive/2.5.2b.tar.gz
$ tar -xzf 2.5.2b.tar.gz
$ ln -s ~/DL/STAR-
2.5.2b/bin/Linux_x86_64_static/STAR ~/bin
```

8. Install samtools, bgzip, and tabix (*see* **Note 5**).

```
$ cd ~/DL
$ wget
https://github.com/samtools/samtools/releases/down-
load/1.3.1/samtools-1.3.1.tar.bz2
$ tar -xjf samtools-1.3.1.tar.bz2
$ cd samtools-1.3.1
$ ./configure
$ make all all-htslib
$ ln -s $PWD/samtools ~/bin
$ ln -s $PWD/htslib-1.3.1/bgzip ~/bin
$ ln -s $PWD/htslib-1.3.1/tabix ~/bin
```

9. Install the bamtools toolkit required by BRAKER1 (*see* **Note 5**).

```
$ cd ~/DL
$ wget
https://cmake.org/files/v3.7/cmake-3.7.1-Linux-x86_64.
tar.gz
$ tar -xzf cmake-3.7.1-Linux-x86_64.tar.gz
$ wget https://github.com/pezmaster31/bamtools/ar-
chive/v2.4.1.tar.gz
$ tar -xzf v2.4.1.tar.gz
$ cd bamtools-2.4.1
$ mkdir build
$ cd build
$ ~/DL/cmake-3.7.1-Linux-x86_64/bin/cmake ..
$ make
$ cd ..
$ ln -s $PWD/bin/bamtools ~/bin
```

10. Install the seqtk toolkit (*see* **Note 5**).

```
$ cd ~/DL
$ wget
https://github.com/lh3/seqtk/archive/v1.2.tar.gz
$ tar -xzf v1.2.tar.gz
$ cd seqtk-1.2
$ make
$ ln -s $PWD/seqtk ~/bin
```

11. Download and prepare the latest version of Augustus [14]

```
$ cd ~/DL
$ wget
http://bioinf.uni-greifswald.de/augustus/binaries/
augustus.current.tar.gz
$ tar -xzf augustus.current.tar.gz
$ ln -s augustus-3.2.3 augustus
```

Compile bam2hints.

```
$ cd augustus/auxprogs/bam2hints
```

Open Makefile in a text editor and replace the two lines just below "# Variable definition" with:

```
BAMTOOLS = $(HOME)/DL/bamtools-2.4.1
INCLUDES = $(BAMTOOLS)/include
LIBS = $(BAMTOOLS)/lib/libbamtools.a -lz
```

Save the edited Makefile and enter

```
$ make clean
$ make
```

12. Install GeneMark-ET [13]
Browse to http://exon.gatech.edu/license_download.cgi. Select GeneMark-ES / ET LINUX64, fill in your name and address, and click "I agree to the terms of this license agreement." Note that this license forbids transferring or modifying the software.

    Click on the download link that appears. After the software downloads, decompress it with.

```
$ tar -xzf gm_et_linux_64
```

    GeneMark depends on some Perl modules that are not installed by default. If you do not have sudo privileges ask your system administrator to make sure that these are available. If you do have sudo privileges, you can install them from CPAN.

```
$ sudo cpan
cpan> install YAML
cpan> install Hash::Merge
cpan> install Logger::Simple
cpan> Parallel::ForkManager
cpan> exit
```

    At the end of each step, possibly after voluminous output, you should receive a message that the module was either already installed or has been successfully installed.

13. Download Braker [12]

```
$ cd ~/DL
$ wget
http://exon.gatech.edu/Braker/BRAKER1.tar.gz
$ tar -xzf BRAKER1.tar.gz
```

14. Download and compile CodingQuarry [15]

```
$ cd ~/DL
$ wget
https://sourceforge.net/projects/codingquarry/files/
latest/download
$ tar -xzf download
$ cd CodingQuarry_v2.0
$ make
```

Download the transcriptome assembler Stringtie [22].

```
$ cd ~/DL
$ wget
http://ccb.jhu.edu/software/stringtie/dl/stringtie-
1.3.1c.Linux_x86_64.tar.gz
```

```
$ tar -xzf stringtie-1.3.1c.Linux_x86_64.tar.gz
$ ln -s $PWD/stringtie-
1.3.1c.Linux_x86_64/stringtie ~/bin
```

15. Download Harfang and programs it depends on.

    (a) Program

    ```
    $ cd ~/DL
    wget
    https://sourceforge.net/projects/harfang/files/
    Harfang.v1.0.tar.gz
    tar -xzf Harfang.v1.0.tar.gz
    ```

    (b) Dependencies

    - Python modules

      ```
      $ sudo pip install biopython
      $ sudo pip install doit==0.29.0
      $ sudo pip install pysam
      ```

    - Blast+ [23]

      ```
      $ cd ~/DL
      $ wget ftp://ftp.ncbi.nlm.nih.gov/blast/
      executables/blast+/LATEST/ncbi-blast-
      2.6.0+-x64-linux.tar.gz
      $ tar -xzf ncbi-blast-2.6.0+-x64-linux.tar.gz
      $ export PATH=~/DL/blast-
      2.6.0+/bin:$PATH
      ```

    - Protein sequence database
      Download and format the latest release of Fungal RefSeq Proteins.

      ```
      $ mkdir protein_db
      $ cd protein_db
      $ wget ftp://ftp.ncbi.nlm.nih.gov/refseq/re-
      lease/RELEASE_NUMBER
      $ RELEASE=$(cat RELEASE_NUMBER)
      $ PROTEIN_FILENAME=RefSeq_Fungi.${RELEASE}.
      protein.faa
      $ wget ftp://ftp.ncbi.nlm.nih.gov/refseq/re-
      lease/fungi/fungi.*.protein.faa.gz
      $ FILE_COUNT=$(ls -l *.protein.faa.gz | wc -l)
      $ for d in $(seq 1 $FILE_COUNT) ; do zcat
      fungi.${d}.protein.faa.gz >> $PROTEIN_
      FILENAME ; done
      $ makeblastdb -dbtype prot -in $PROTEIN_
      FILENAME -out RefSeq_Fungi.protein -parse_
      seqids
      $ rm *.faa.gz
      ```

    - Cd-hit [24]

      ```
      $ cd ~/DL
      $ wget
      https://github.com/weizhongli/cdhit/releases/
      download/V4.6.6/cd-hit-v4.6.6-2016-0711.tar.gz
      ```

```
$ tar -xzf cd-hit-v4.6.6-2016-0711.tar.gz
$ cd cd-hit-v4.6.6-2016-0711
$ make
```

16. Genome browser

Download the Integrative Genomics Viewer [25, 26].

```
$ cd ~/DL
$ wget
http://data.broadinstitute.org/igv/projects/down-
loads/IGV_2.3.90.zip
$ unzip IGV_2.3.90.zip
```

17. Utilities

```
cd ~/DL
wget
https://sourceforge.net/projects/harfang/files/
Accessories.tar.gz
tar -xzf Accessories.tar.gz
cd ~/bin
wget
https://gist.githubusercontent.com/nathanhaigh/
3521724/raw/5d4cc310d65ce798c2b030756a2b855cf55ecbcd/
deinterleave_fastq.sh
chmod a+x deinterleave_fastq.sh
```

2.5 Reference Gene Models

```
$ cd $PREDICTIONS
$ mkdir curated
$ cp <Reference gene models>.gff3 curated/curated.
gff3
$ cd curated
$ grep -v '^##' curated.gff3 | sort -k1,1 -k4,4n -k5,5n |
bgzip -c > curated.gff3.gz
$ tabix -p gff curated.gff3.gz
```

3 Methods

3.1 Check RNA-Seq reads

See Subheading 2 for potential sources of RNA-Seq reads.

It is prudent to check that the reads really come from your target genome by seeing how well they align to that genome.

Index the genome.

```
$ mkdir -p index0
$ STAR --runMode genomeGenerate --run-
ThreadN 8 --genomeDir index0 --genomeFastaFiles
$PREDICTIONS/assembly.fasta
```

For each read file, map its first two million reads to the index and check that more than half of them are mapped.

```
$ mkdir -p test_map
$ STAR --runMode alignReads --runThreadN 12 --genomeDir
index0 --readFilesIn <Reads.fastq.gz> --readFilesCommand
zcat --readMapNumber 2000000000 --outFileNamePrefix test_
map/ --alignIntronMin 9 --alignIntronMax 2000 --outFilter-
```

```
ScoreMinOverLread 0 --outFilterMatchNminOverLread 0.5 --
outFilterMatchNmin 40 --alignEndsType EndToEnd
$ grep 'Uniquely mapped reads %' test_map/Log.final.out
```

If the number printed out is less than 50%, discard the read file.

3.2 Pool Strand-Specific RNA-Seq Reads

Choose a directory to hold all your reads and put its path in a bash variable for convenience.

```
$ READS=<Path to your read directory>
```

Make subdirectories for paired and single-end reads (*see* **Note 1**).

```
$ mkdir -p $READS/PE
$ mkdir -p $READS/SE
```

Move your read files into these subdirectories.

Pool all the available read pairs in interleaved format:

```
$ cd $READS
$ for read2 in PE/*2.fastq.gz ; do read1=${read2%2.
fastq.gz}1.fastq.gz ; seqtk mergepe $read1 $read2 |
gzip -c >> PEpool.fastq.gz ; done
```

Separately pool single-end reads:

```
$ zcat SE/*.fastq.gz | gzip -c > SEpool.fastq.gz
```

3.3 Correct Read Sequence Errors (See Note 6)

```
$ mkdir Rcorrected
$ perl ~/DL/Rcorrector-1.0.2/run_rcorrector.pl -i
PEpool.fastq.gz -s SEpool.fastq.gz -k 25 -od ./
Rcorrected -t 12 &> Rcorrector.log
$ cd Rcorrected
$ gunzip PEpool.cor.fq.gz
$ gunzip SEpool.cor.fq.gz
$ python ~/DL/Utilities/filter_fastq_by_tag.py -i
PEpool.cor.fq -t unfixable_error -a PEpool
$ python ~/DL/Utilities/filter_fastq_by_tag.py -s
SEpool.cor.fq -t unfixable_error -a SEpool
$ rm ../PEpool.fastq.gz
$ rm ../SEpool.fastq.gz
$ rm PEpool.cor.fq*
$ rm SEpool.cor.fq*
```

3.4 Trim Sequence Adapters (See Note 7)

```
$ mkdir trimmed
$ ~/DL/skewer-0.2.2/skewer -x ~/DL/skewer-0.2.2/
adapters.fasta -m tail -q 1 -l 80 -t 12 -o
trimmed/PEpool PEpool.correct.1.fastq PEpool.
correct.2.fastq
$ ~/DL/skewer-0.2.2/skewer -x ~/DL/skewer-0.2.2/
adapters.fasta -m tail -q 1 -l 80 -t 12 -o trimmed/
SEpool SEpool.correct.fastq
$ rm PEpool.correct.?.fastq
$ rm SEpool.correct.fastq
```

3.5 Remove Ribosomal RNA Reads (See Notes 8 and 9)

```
$ cd trimmed
$ mkdir rRNA
$ mkdir non_rRNA
```

```
$ run_sortmerna.sh PEpool-trimmed-pair1.fastq
PEpool-trimmed-pair2.fastq$PWD
$ run_sortmerna_single.sh SEpool-trimmed.fastq.gz
$PWD
$ mv non_rRNA/SEpool-trimmed_non_rRNA.READS1 non_
rRNA/SEpool-trimmed_non_rRNA.fastq
$ rm PEpool-trimmed-pair?.fastq.gz
$ rm SEpool-trimmed.fastq.gz
```

3.6 Map RNA-Seq Reads to the Genome Assembly (See Note 10)

```
$ cd non_rRNA
$ mkdir -p STAR
$ cd STAR
$ mkdir -p index0
$ STAR --runMode genomeGenerate --run ThreadN
8 --genomeDir index0 --genomeFas taFiles
$PREDICTIONS/assembly.fasta --sjd bOverhang 0
$ mkdir -p map 1_PE
$ STAR --runMode alignReads --runThreadN 12 --geno-
meDir index0 --readFilesIn ../PEpool-trimmed_non_rR-
NA-pair1.fastq.gz ../PEpool-trimmed_non_rRNA-pair2.
fastq.gz--readFilesCommand zcat --outFileNamePrefix
map 1_PE/ --alignIntronMin 9 --alignIntronMax
2000 --outFilterScoreMinOverLread 0 --outFilter-
MatchNminOverLread 0.5 --outFilterMatchNmin 40 --
outSAMstrandField intronMotif --outSJfilterOver-
hangMin 30 4 8 12 --outSJfilterCountUniqueMin 4 2
2 3 --outSJfilterCountTotalMin 8 4 4 6 --outSJfil-
terIntronMaxVsReadN 500 1000 2000 --alignEndsType
EndToEnd
$ mkdir -p map 1_SE
$ STAR --runMode alignReads --runThreadN 12 --geno-
meDir index0 --readFilesIn ../RNA-Seq/Rcorrected/
trimmed/non_rRNA/SEpool-trimmed_non_rRNA.fastq --
outFileNamePrefix map 1_SE/
$ mkdir -p index1
$ STAR --runMode genomeGenerate --runThreadN 8 --geno-
meDir index1 --genomeFastaFiles $PREDICTIONS/assembly.
fasta --sjdbOverhang 100 --sjdbFileChrStartEnd map
1_PE/SJ.out.tab map 1_SE/SJ.out.tab
$ mkdir -p map 2_PE
$ STAR --runMode alignReads --runThreadN 12 --geno-
meDir index1 --readFilesIn ../PEpool-trimmed_non_rR-
NA-pair1.fastq.gz ../PEpool-trimmed_non_rRNA-pair2.
fastq.gz --readFilesCommand zcat--outFileNamePrefix
map 2_PE/ --outSAMtype BAM SortedByCoordinate --
outFilterType BySJout --outReadsUnmapped Fastx --
alignIntronMin 9 --alignIntronMax 2000 --outFil-
terScoreMinOverLread 0 --outFilterMatchNminOverL-
read 0.5 --outFilterMatchNmin 40 --limitBAMsortRAM
30000000000 --outSAMstrandField intronMotif --out-
FilterIntronMotifs RemoveNoncanonicalUnannotated --
outSAMattributes All --outSJfilterOverhangMin 30 4 8
12 --outSJfilterCountUniqueMin 4 2 2 3 --outSJfilter-
```

```
CountTotalMin 8 4 4 6 --outSJfilterIntronMaxVsReadN
500 1000 2000 --alignEndsType EndToEnd
$ samtools merge sorted.bammap 2_PE/Aligned.sort-
edByCoord.out.bam map 2_SE/ Aligned.sortedByCoord.
out.bam
```

3.7 Run BRAKER1 (See Notes 11 and 12)

```
$ cd $PREDICTIONS
$ export PATH=~/DL/BRAKER1:~/DL/augustus/bin:$PATH
$ export AUGUSTUS_CONFIG_PATH=~/DL/augustus/config
$ export GENEMARK_PATH=~/DL/gm_et_linux_64/gmes_
petap
$ export BAMTOOLS_PATH=~/DL/bamtools-2.4.1/bin
$ export SAMTOOLS_PATH=~/bin
$ braker.pl --cores=12 --fungus --species=Aspni_brak-
er --genome=$PREDICTIONS/assembly.fasta --bam=$READS/
Rcorrected/trimmed/non_rRNA/STAR/sorted.bam
$ cd braker/Aspni_braker
$ python ~/DL/Utilities/gtf2gff3.py augustus.gtf
```

3.8 Run CodingQuarry (See Note 13)

```
$ cd $PREDICTIONS
$ mkdir CodingQuarry
$ ~/DL/stringtie-1.3.1c.Linux_x86_64/stringtie -f
0.25 -m 200 -o CodingQuarry/stringtie-transcripts.
gtf -j 3 -p 12 -x Mito $READS/Rcorrected/trimmed/
non_rRNA/STAR/sorted.bam
$ cd CodingQuarry
$ python ~/DL/Utilities/gtf2gff3.py stringtie-tran-
scripts.gtf
$ export QUARRY_PATH=~/DL/CodingQuarry_v2.0/
QuarryFiles
$ ~/DL/CodingQuarry_v2.0/CodingQuarry -f
$PREDICTIONS/assembly.fasta -t stringtie-tran-
scripts.gff3 -p12
$ cd out
$ python ~/DL/Utilities/fixCodingQuarryGFF3.py -i
PredictedPass.gff3 -o fixed.PredictedPass.gff3
```

3.9 Run Harfang (See Note 14)

```
$ cd $PREDICTIONS
$ mkdir Harfang
$ cp ~/DL/Harfang/CONFIG.template Harfang/CONFIG
```

Open Harfang/CONFIG in a text editor and change these entries to the values shown:

```
ProjectName = Harfang
ProjectDir = $PREDICTIONS/Harfang
Genome = $PREDICTIONS/assembly.fasta
MappedReads = $PREDICTIONS
label = Harfang
np = 12
ExternalPredictions = $PREDICTIONS/braker/Aspni_
braker/augustus.gff3
Species = Aspni_braker
blastp_db = ~/DL/protein_db/RefSeq_Fungi.protein
```

```
config_file = $PREDICTIONS/Harfang/CONFIG
```

Search and replace $PREDICTIONS with its value, save the file, and close the text editor.

```
$ bash ~/DL/Harfang/bin/scripts/strand-specific_
BAM_to_juncs_and_coverage.sh Rcorrected/trimmed/
non_rRNA/STAR/sorted.bam $PREDICTIONS/assembly.
fasta $PREDICTIONS
$ export PATH=~/DL/augustus/bin:~/DL/gm_et_
linux_64/gmes_petap:~/DL/cd-hit-v4.5.4-2011-03-
07:$PATH
$ python ~/DL/Harfang/Harfang -c Harfang/CONFIG &>
run_Harfang.log
```

3.10 Compare Predicted Genes and Transcripts to the Reference Annotation (See Notes 15 and 16)

```
$ cd $PREDICTIONS/braker/Aspni_braker
$ sort -k1,1 -k4,4n -k5,5n augustus.gff3 | bgzip -c
> augustus.gff3.gz
$ tabix -p gff augustus.gff3.gz
$ cd $PREDICTIONS
$ python ~/DL/Utilities/get_coincident_predictions.
py -i curated/curated.gff3 -q braker/Aspni_braker/
augustus.gff3.gz -m braker/Aspni_braker/curated_in_
augustus.gff3 | tee braker/Aspni_braker/curated_in_
augustus.log
$ python ~/DL/Utilities/get_coincident_predictions.
py -i braker/Aspni_braker/augustus.gff3 -q curated/
curated.gff3.gz -m braker/Aspni_braker/augustus_in_
curated.gff3
$ cd $PREDICTIONS/CodingQuarry/out
$ grep -v '^##' fixed.PredictedPass.gff3 |
sort -k1,1 -k4,4n -k5,5n | bgzip -c > fixed.
PredictedPass.gff3.gz
$ tabix -p gff fixed.PredictedPass.gff3.gz
$ cd $PREDICTIONS
$ python ~/DL/Utilities/get_coincident_predictions.
py -i curated/curated.gff3 -q CodingQuarry/out/
fixed.PredictedPass.gff3.gz -m CodingQuarry/out/cu-
rated_in_PredictedPass.gff3
$ python ~/DL/Utilities/get_coincident_predic-
tions.py -i CodingQuarry/out/fixed.PredictedPass.
gff3 -q curated/curated.gff3.gz -m CodingQuarry/out/
PredictedPass_in_curated.gff3
$ cd Harfang/Predictions
$ grep -v '^##' accepted.gff3 | sort -k1,1 -k4,4n -k5,5n
| bgzip -c > accepted.gff3.gz
$ tabix -p gff accepted.gff3.gz
$ cd $PREDICTIONS
$ python ~/DL/Utilities/get_coincident_predictions.
py -i curated/curated.gff3 -q Harfang/Predictions/
accepted.gff3.gz -m Harfang/Predictions/curated_in_
accepted.gff3 | tee Harfang/Predictions/curated_in_
accepted.log
$ python ~/DL/Utilities/get_coincident_predictions.
py -i Harfang/Predictions/accepted.gff3 -q curated/
```

```
curated.gff3.gz -m Harfang/Predictions/accepted_in_
curated.gff3 | tee Harfang/Predictions/accepted_in_
curated.log
```

3.11 Visualize Results

The Integrated Genome Viewer is a convenient tool for quickly visualizing the position and structure of predicted transcripts in comparison to the RNA-Seq evidence. To use it you must be on a workstation that allows graphics. You can launch the program with

```
~/DL/IGV_2.3.90/igv.sh
```

In the window that opens use the menus to load the genome sequence ($PREDICTIONS/assembly.fasta), the strand-specific RNA-seq read alignments ($READS/Rcorrected/trimmed/non_rRNA/STAR/sorted.R.bam, $READS/Rcorrected/trimmed/non_rRNA/STAR/sorted.F.bam), the reference gene models ($PREDICTIONS/curated/curated.gff3), the BRAKER1 predictions ($PREDICTIONS/braker/Aspni_braker/augustus.gff3), the CodingQuarry predictions ($PREDICTIONS/CodingQuarry/out/fixed.PredictedPass.gff3), and the Harfang predictions ($PREDICTIONS/Harfang/accepted.gff3).

4 Notes

1. The instructions assume that both paired and single-end RNA-Seq reads are available. If you have only one type of reads, ignore the instructions for the missing type.
2. The list of software to install is daunting, but many of these programs will be generally useful in bioinformatic work. All the programs are freely available to download and use. You will notice a simple pattern: download, decompress, and sometimes compile; this is a useful skill. URLs for downloading software often include the version number; these instructions specify the release version that is current at the time of writing. Much of the software we use is being actively maintained or developed, and new versions with bugs fixed and features added are released frequently. You should use the latest release and adjust the version numbers in URLs and in directory names to match.
3. You can put these programs in any directory for which you have write permissions. I find it convenient to have all downloaded software in one place; a folder named DL or Downloads in my home directory accomplishes this. These instructions use ~/DL/; "~" is an alias for your home directory.
4. After compiling programs with make, you usually have the option to install the executable programs in the system tree with the command "make install." This is not essential; it makes the program available to all users of the system, but it requires

root or sudo privileges. An alternative, if you are the sole user of a program, is to copy the executable or make a symbolic link to it in a common directory such as ~/bin. Then adding ~/bin to the system PATH (it may already be there) makes it possible to execute any of the programs without specifying its location.

5. The programs might already be installed. To check, use the "which" utility, e.g., "which samtools." If the computer responds with a file path, such as "/usr/local/bin/samtools", there is no need to install the program.
6. Rcorrector tags erroneous reads for which it cannot find a correction as "unfixable_error" but leaves them in the output. The filter_fastq_by_tag.py script removes these reads.
7. The command options tell skewer to look for adapter sequence only at the 3′ ends of the reads, trim terminal bases with quality values less than 1, drop any reads shorter than 80 bases after trimming, use 12 threads, and compress the output files.
8. SortMeRNA requires a tedious command listing all the rRNA libraries. The bash scripts run_sortmerna.sh and run_sortmerna_single.sh handle this and the conversion of paired read input to interleaved format. $PWD is a bash variable containing the path to the current working directory.
9. The precompiled version of SortMeRNA is single-threaded and can be slow. If you want faster execution, compile the program from source code and enable multithreading.
10. These commands run two passes of STAR. When STAR begins it has no information about splice junction positions and may miss junctions that are close to the ends of reads. The first pass collects reliable splice junction positions, allowing more accurate read mapping in the second pass. The resulting sorted.bam file is used by all three of the gene predictors.
11. BRAKER1 is designed to resume after being stopped by an error without repeating successful previous steps. If BRAKER1 does stop prematurely, look at the end of braker.log for information about the problem, most often a missing file. Correct the problem, delete any incomplete intermediate files, and restart braker.pl with the original command.
12. When BRAKER1 completes, it will put the gene predictions in $PREDICTIONS/Aspni_braker/augustus.gtf and predicted protein sequences in $PREDICTIONS/Aspni_braker/augustus.aa. To make the predictions comparable to the reference set, the script gtf2gff3.py converts them into GFF3 format.
13. The developers of CodingQuarry suggest using Cufflinks to assemble the mapped reads into transcripts; I have substituted the newer program Stringtie. The program to convert GTF to GFF3 format supplied with CodingQuarry does not work with

Stringtie transcripts, but the gtf2gff3.py script mentioned above does. The main predictions of CodingQuarry are in out/PredictedPass.gff3. These models do not include transcript records; the script fixCodingQuarryGFF3.py adds a transcript record to each gene to make them comparable to the reference set.

14. Harfang can be restarted after it halts, and will continue where it left off without repeating earlier work. After fixing the problem that caused the halt, delete any incomplete or corrupted files before restarting. The file Harfang/logs/Harfang.log lists the steps that have been started and finished; it will end with an error message whenever the program halts. When the program completes, the log will contain "Finished Publish_accepted_models". The gene predictions are in Harfang/accepted.gff3.
15. get_coincident_predictions.py looks in the transcripts of the query set for exact matches or matches that differ only in start codon position for each transcript in the master set. A gene is considered to have a match if any of its transcripts have a match.
16. To evaluate the performance of the gene prediction programs, I carried out the instructions of this chapter with a finished genome assembly for *Aspergillus* niger NRRL3 and eight sets of single-end strand-specific RNA-Seq reads from our laboratory (410 million reads; 326 million after cleaning). The curated reference models for *Aspergillus niger* NRRL3 contain only one transcript per gene. To allow assessment of transcript isoform predictions as well as gene predictions, variant transcripts containing each of the splice junctions found in the aligned RNA-Seq reads were generated, and their expected abundances were estimated by multiplying the observed frequencies of their splice junctions. Variant transcripts with relative expected abundances over 10% were added to the reference set. This added 5249 variant transcripts to the 11,861 genes in the reference set.

    The number of RNA-Seq reads mapping to each of the reference transcripts was estimated with Salmon [27] and used to calculate mean coverage depths. The reference transcripts were stratified by coverage depth: Fully covered if depth > 9.99; Partially covered if 3 < depth < = 9.99; Uncovered if depth < =3. The reads mapping to the transcripts of each gene were summed to calculate mean depth for the gene, and the genes were also stratified by coverage depth. Most (87%) of the reference genes had full read coverage.

    Table 1 summarizes the numbers of reference genes and transcripts that were predicted by each of the programs. All of them performed well; from 85.9% to 92.2% of the reference

Table 1
Prediction of reference genes and transcripts with high or low read coverage

Predictor	Curated	CodingQuarry	BRAKER1	Harfang
All genes	11,861	10,193 (85.9%)	10,334 (87.1%)	10,936 (92.2%)
Fully covered	10,331	9203 (89.1%)	9339 (90.4%)	9632 (93.2%)
Partly covered	554	383 (69.1%)	405 (73.1%)	497 (89.7%)
Uncovered	911	607 (66.6%)	590 (64.8%)	807 (88.6%)
All transcripts	17,110	9180 (53.7%)	9364 (54.7%)	12,251 (71.6%)
Fully covered	14,557	8353 (57.4%)	8358 (57.4%)	10,639 (73.1%)
Partly covered	972	317 (32.6%)	355 (36.5%)	638 (65.6%)
Uncovered	1581	510 (32.3%)	484 (30.6%)	974 (61.6%)

genes were predicted. Harfang predicted 5.1% more than BRAKER1, which predicted 1.2% more than CodingQuarry. The performance of all three degraded as the read coverage depth decreased; this shows the contribution that RNA-Seq data makes to gene prediction. The performance difference of Harfang relative to the other two increased at lower coverage depth, probably because Harfang uses additional information from protein homology to assess candidate models.

Smaller fractions of the reference transcripts than genes were predicted. Harfang did better than the other two because it is designed to find alternative transcripts. Note that the numbers of transcripts predicted by CodingQuarry and BRAKER1 are lower than the numbers of predicted genes because alternate start forms were counted as matches for genes but not for transcripts.

Acknowledgments

This work was supported by Genome Canada and Génome Québec.

References

1. Majoros WH (2007) Methods for computational gene prediction. Cambridge University Press, Cambridge
2. Hrdlickova R, Toloue M, Tian B (2017) RNA-Seq methods for transcriptome analysis. Wiley Interdiscip Rev RNA 8(1). https://doi.org/10.1002/wrna.1364
3. Levin JZ, Yassour M, Adiconis X, Nusbaum C, Thompson DA, Friedman N, Gnirke A, Regev A (2010) Comprehensive comparative analysis of strand-specific RNA sequencing methods. Nat Methods 7(9):709–715. https://doi.org/10.1038/nmeth.1491
4. Wikipedia (2017) List of gene prediction software. https://en.wikipedia.org/wiki/List_of_gene_prediction_software
5. Reese MG, Hartzell G, Harris NL, Ohler U, Abril JF, Lewis SE (2000) Genome annotation assessment in Drosophila melanogaster. Genome Res 10:483–501
6. Guigó R, Flicek P, Abril JF, Reymond A, Lagarde J, Denoeud F, Antonarakis S, Ashburner M, Bajic VB, Birney E, Castelo R, Eyras E, Ucla C, Gingeras TR, Harrow J, Hubbard T, Lewis SE, Reese MG (2006) EGASP: the human ENCODE genome annotation assessment project. Genome Biol 7(Suppl 1):S2.1–S231. https://doi.org/10.1186/gb-2006-7-s1-s2
7. Coghlan A, Fiedler TJ, SJ MK, Flicek P, Harris TW, Blasiar D, nGASP Consortium, Stein LD (2008) nGASP--the nematode genome annotation assessment project. BMC Bioinformatics 9:549. https://doi.org/10.1186/1471-2105-9-549
8. Galagan JE, Henn MR, Ma L, Cuomo CA, Birren B (2005) Genomics of the fungal kingdom: insights into eukaryotic biology. Genome Res 15:1620–1631
9. Nakagawa S, Niimura Y, Gojobori T, Tanaka H, Miura K (2008) Diversity of preferred nucleotide sequences around the translation initiation codon in eukaryote genomes. Nucleic Acids Res 36:861–871
10. Grützmann K, Szafranski K, Pohl M, Voigt K, Petzold A, Schuster S (2014) Fungal alternative splicing is associated with multicellular complexity and virulence: a genome-wide multi-species study. DNA Res 21(1):27–39. https://doi.org/10.1093/dnares/dst038
11. McDonnell E, Strasser K, Tsang A. (2018) Manual Gene Curation and Functional Annotation. This book
12. Hoff KJ, Lange S, Lomsadze A, Borodovsky M, Stanke M (2016) BRAKER1: Unsupervised RNA-Seq-based genome annotation with GeneMark-ET and AUGUSTUS. Bioinformatics 32:767–769. https://doi.org/10.1093/bioinformatics/btv661
13. Lomsadze A, Burns PD, Borodovsky M (2014) Integration of mapped RNA-Seq reads into automatic training of eukaryotic gene finding algorithm. Nucleic Acids Res 42:e119. https://doi.org/10.1093/nar/gku557
14. Stanke M, Diekhans M, Baertsch R, Haussler D (2008) Using native and syntenically mapped cDNA alignments to improve de novo gene finding. Bioinformatics 24:637–644
15. Testa AC, Hane JK, Ellwood SR, Oliver RP (2015) CodingQuarry: highly accurate hidden Markov model gene prediction in fungal genomes using RNA-seq transcripts. BMC Genomics 16:170. https://doi.org/10.1186/s12864-015-1344-4
16. Reid I, O'Toole N, Zabaneh O, Nourzadeh R, Dahdouli M, Abdellateef M, Gordon PM, Soh J, Butler G, Sensen CW, Tsang A (2014) SnowyOwl: accurate prediction of fungal genes by using RNA-Seq and homology information to select among ab initio models. BMC Bioinformatics 15:229. https://doi.org/10.1186/1471-2105-15-229
17. Tange O (2011) Gnu parallel – the command-line power tool. Login: The USENIX Magazine 36:42–47
18. Song L, Florea L (2015) Rcorrector: efficient and accurate error correction for Illumina RNA-seq reads. GigaScience 4(48). https://doi.org/10.1186/s13742-015-0089-y
19. Hongshang J, Lei R, Ding S-W, Zhu S (2014) Skewer: a fast and accurate adapter trimmer for next-generation sequencing paired-end reads. BMC Bioinformatics 15:1–12
20. Kopylova E, Noé L, Touzet H (2012) SortMeRNA: fast and accurate filtering of ribosomal RNAs in metatranscriptomic data. Bioinformatics 28:3211–3217. https://doi.org/10.1093/bioinformatics/bts611
21. Dobin A, Davis CA, Schlesinger F, Drenkow J, Zaleski C, Jha S, Batut P, Chaisson M, Gingeras TR (2013) STAR: ultrafast universal RNA-seq aligner. Bioinformatics 29:15–21. https://doi.org/10.1093/bioinformatics/bts635
22. Pertea M, Pertea GM, Antonescu CM, Chang TC, Mendell JT, Salzberg SL (2015) StringTie enables improved reconstruction of a transcriptome from

RNA-seq reads. Nat Biotechnol 33:290–295. https://doi.org/10.1038/nbt.3122

23. Camacho C, Coulouris G, Avagyan V, Ma N, Papadopoulos J, Bealer K, Madden TL (2008) BLAST+: architecture and applications. BMC Bioinformatics 10:421. https://doi.org/10.1186/1471-2105-10-421
24. Li W, Godzik A (2006) Cd-hit: a fast program for clustering and comparing large sets of protein or nucleotide sequences. Bioinformatics 22:1658–1659
25. Robinson JT, Helga Thorvaldsdóttir H, Winckler W, Guttman M, Lander ES, Getz G, Mesirov JP (2011) Integrative genomics viewer. Nat Biotechnol 29:24–26
26. Thorvaldsdóttir H, Robinson JT, Mesirov JP (2013) Integrative Genomics Viewer (IGV): high-performance genomics data visualization and exploration. Brief Bioinform 14:178–192
27. Patro R, Duggal G, Love MI, Irizarry RA, Kingsford C (2016) Salmon provides accurate, fast, and bias-aware transcript expression estimates using dual-phase inference. BioRxiv. https://doi.org/10.1101/021592

Chapter 18

Genomic Sequence Variation Analysis by Resequencing

Joel Martin, Wendy Schackwitz, and Anna Lipzen

Abstract

Whole-genome resequencing is a method for determining the differences between individuals and a reference genome. The experiments are performed by sequencing the individuals, aligning generated reads to a common reference and discovering variation within the data set by analysis of the alignment with software tools. When correlated with phenotypic information, sites of causative genomic variation may be putatively assigned.

While the analysis is generally straightforward, there are many nuances, and we aim to help you understand how to generate an initial result, sift through it to identify likely candidates for a phenotype of interest, and flag false positive calls.

Key words Resequencing, Genotyping, SNP, SNV, Indel

1 Introduction

Resequencing detects genomic differences between sequenced individuals and a reference genome. These include single nucleotide polymorphisms (SNPs), small insertions and deletions (indels) as well as larger structural rearrangements. Due to the short length of next generation sequencing reads, 100–150 bp, the individuals must be closely related to the reference sequence for mapping software to align the reads properly. When there are too many mismatches, reads can no longer align and there will be a gap in coverage wherein variation information cannot be acquired. We find 95% nucleotide identity is a reasonable target for comparing organisms with current technologies, though somewhat greater divergence can be accommodated. Below this threshold of identity, reads may fail to align, resulting in spotty data coverage (*see* **Note 1**). After reads have been aligned to the reference and sequence variation reported, gene annotation information is added, identifying genes containing changes as well as the codon and amino acid effects of those changes.

Ronald P. de Vries et al. (eds.), *Fungal Genomics: Methods and Protocols*, Methods in Molecular Biology, vol. 1775, https://doi.org/10.1007/978-1-4939-7804-5_18,

The following outlines the steps for reference based sequencing read analysis.

1. The reference fasta file is prepared by indexing with bwa [1] and samtools [2].
2. Sequenced reads, in fastq format, are aligned to the reference fasta with bwa to produce a bam file. The bam alignment file contains the details of each read and its mapping to the reference, along with metadata such as sample, library and run identifiers.
3. Variant discovery programs are run to identify SNPs and small indels by assessing the bam file to determine locations of the genome that vary from the reference and assign the locations a confidence value and report the results in a vcf (Variant Call Format) [3] file. Bcftools and GATK are two commonly used snp and small indel callers and we will cover the use of bcftools in Subheading 3.
4. Structural variation discovery programs such as BreakDancer [4] and Pindel [5] are run to discover larger scale variation such as inversions, large insertions and deletions (*see* **Note 2**), as well as inter and intra contig translocations. Pindel results will be converted to, and reported in, the same vcf format as snp and small indels.
5. Annotation of the variants identifies which variants affect protein sequence and what that effect is. We use snpEff [6] to apply annotation to the vcf files from the variant callers.

2 Materials

Sequencing libraries should be generated from samples in a haploid form if possible, and enough sequence run to provide 15× depth on average according to the reference. If only diploid DNA is available, then 30× depth should be attempted. Target depth in reads is calculated as total bp of the reference × target depth/read length. We recommend 100–150 bp paired end reads sequenced from 500 bp fragment Illumina libraries. The unsequenced region of the fragment allows the sequenced ends to align outside of potentially complex breakpoint regions and is required by some structural variation detection software.

Once the reads, in fastq format, come off the sequencer, a suite of programs are needed to (1) align the reads to the reference, (2) identify the differences in the reads when compared to the reference, (3) identify high confidence variants by comparing where differences in reads compared to reference are correlated amongst multiple reads, and (4) generate a report of variants with a score of confidence. Below is a list of the software needed to accomplish the steps listed above.

The software binaries used are not all available from Linux package distributions so here follow rudimentary instructions for acquiring and building the tools. More details and troubleshooting instructions are often available through their websites.

For each program built, either copy the binaries to a location on your path, or add its location to your path.

```
Example: export PATH=/home/jmartin/bwa-0.7.12:$PATH
```

Prerequisites for building the software, these will be available for installation from most Linux distributions.

gcc

cmake

make

gnu binutils

git

Read alignment:

download bwa from https://sourceforge.net/projects/bio-bwa/files/

Unpack with

```
tar xjf bwa-0.7.12.tar.bz2
```

then cd into the created directory and build with 'make'.

Snp and small indel identification, bam and vcf manipulation:

Download samtools, bcftools, and htslib from http://www.htslib.org/download into the same directory and unpack them as with bwa. They are then each built and installed with the same procedure, cd into each directory in turn.

```
cd samtools-1.3.1
make
make prefix=/where/to/install install
```

The executables will then be installed to /where/to/install/bin which can be added to your path with

```
export PATH=/where/to/install/bin:$PATH
```

Bam and vcf manipulations:Picard project at https://github.com/broadinstitute/picard/releases Download picard.jar.

Annotation of effects of variants on coding:

snpEff http://snpeff.sourceforge.net/

Download and uncompress.

Identification of structural variation:

Pindel http://gmt.genome.wustl.edu/packages/pindel

Pindel is cloned with the tool 'git' then provided with the path where htslib was built.

```
git clone git://github.com/genome/pindel.git
cd pindel && ./INSTALL /path/to/htslib
```

BreakDancer http://gmt.genome.wustl.edu/packages/breakdancer/install.html

BreakDancer has detailed instructions on the installation page where it is downloaded from.

3 Methods

1. Create indices for the reference fasta.

```
samtools faidx FUNGAL.fa
java -Xmx4g /path/to/picard.jar \
CreateSequenceDictionary R=FUNGAL.fa O=FUNGAL.dict
bwa index FUNGAL.fa
```

2. Use the bwa mem algorithm to align reads in gzip compressed fastq format (.fq.gz or .fastq.gz) against the reference and produce a bam file.

```
bwa mem -p -R '@RG\tID:123\tSM:456\tPL:illumina\
tLB:ABC' \
FUNGAL.fa reads.fq.gz |  samtools view -bt \
FUNGAL.fa.fai -o sample_456.bam -
```

The trailing '-'above is intentional and notifies samtools to read input from stdin.

The -p parameter informs bwa that your reads are paired end and interleaved.

A simple way to determine if your reads are interleaved or not is,

```
gzip -cd reads.fq.gz | head -5
```

If the first and fifth lines are identical, the file is interleaved. If not, then you should have two files for reads, read1.fq.gz and read2.fq.gz, and the bwa command will be:

```
bwa mem -R '@RG\tID:123\tSM:456\tPL:illumina\
tLB:ABC' \
FUNGAL.fa read1.fq.gz read2.fq.gz | samtools view
\
-bt FUNGAL.fa.fai -o sample_456.bam -
```

3. Fix mate information, which corrects read pairing information in the bam file.

```
picard FixMateInformation I=sample_456.bam \
O=sample_456.fd.bam VALIDATION_STRINGENCY=SILENT \
SO=coordinate
```

4. Mark duplicates and remove the intermediate stage bam file.

```
picard MarkDuplicates I=sample_456.fd.bam \
O=sample_456.bam M=sample_456.bam.dupeMetrics \
VALIDATION_STRINGENCY=SILENT \ MAX_FILE_HANDLES_
FOR_READ_ENDS_MAP=950 \
CREATE_INDEX=true && rm sample_456.fd.bam
```

5. Generate some mapping statistics and determine depth of coverage.

```
samtools stats sample_456.bam > sample_456.bam.
stats
```

Depth can be calculated by dividing the 'bases mapped (cigar):' entry in the stats file by the total genome size. Total genome size is the sum of contigs (or chromosomes, for our purposes here they are equivalent), which can be listed with:

```
samtools view -H *.bam | grep ^\@SQ
```

@SQ SN:ChrI LN:3470898
@SQ SN:ChrII LN:4070061
the number after LN: is the size of that contig.

Repeat **steps 2–5** for all samples that will be aligned to same reference.

6. Feed the pileup format view of a bam to bcftools for snp and indel calling.

```
samtools mpileup -ugf FUNGAL.fa sample_456.bam \
sample_567.bam sample_678.bam | bcftools call \
-vmO z -o experimentX.vcf.gz
```

7. Index the produced vcf with tabix.

```
tabix -pvcf experimentX.vcf.gz
```

Filter the vcf to flag low quality calls and calls at regions of excessive depth; the depth cutoff we recommend is 3 × the sum of average depth across all bams processed in the samtools mpileup step.

8. Filter the indexed vcf file to flag suspect variant calls.

```
bcftools filter -O z -o experimentX.filt.vcf.gz \
-s LOWQUAL -i'%QUAL>20' experimentX.vcf.gz
```

9. To run BreakDancer and determine potential structural variation breakpoint regions create a plain text config file, e.g., sample_456.bam.bd.config.

```
bam2cfg.pl sample_456.bam > sample_456.bam.bd.config
```

The breakdancer configuration file should appear similar to this example, tab delimited.

readgroup:123 platform:illumina map:sample_456.bam readlen:100 lib:ABC num:4000000 lower:393 upper:821 mean:504.10 std:154.08 SWnormality:minus infinityexe:samtools view

10. BreakDancer is then run and the output filtered for a minimum score of 90 and four contributing reads for any calls made.

```
breakdancer-max sample_456.bam.bd.config | \
awk '$9>90 && $10 > 4' > sample_456.bam.bd.max.xls
```

11. To run pindel and determine more exact breakpoint coordinates, first create a tab delimited file containing on one line, e.g., sample_456.pindel.config
bam file name, mean insert size, sample name
Example pindel config file:
sample_456.bam 504.10 sample_456
12. Launch pindel, with the config file and breakdancer output file as inputs.

```
pindel -T 4 -f FUNGAL.fa -b sample_456.filter.max.
xls \
-c ALL -o sample_456.pindel \
-I sample_456.pindel.BND.config > sample_456.pin-
del.out \
2> sample_456.pindel.err
```

-T # number of threads to run,
-f REF.fa reference in fasta format
-b file breakdancer output file

13. Convert the pindel output to vcf format with pindel2vcf

```
pindel2vcf -co 200 -P sample_456.pindel -r FUNGAL.
fa \
-R FUNGAL.fa.v1.0 -d '11-25-16'
```

-R is a string indicating version of the reference assembly

14. Compress and index the vcf

```
bgzip sample_456.pindel.vcf
tabix -pvcf sample_456.pindel.vcf.gz
```

15. Do this step for each final result vcf, 'GENOMEID', below, is an identifier for an annotation database available and downloaded from the snpEff website or built with 'snpEff.jar build ...' and the appropriate annotation file

```
java -Xmx4G -jar snpEff.jar eff -c snpEff.config
GENOMEID\ sample_456.vcf.gz
```

4 Interpreting Results

Variants are reported in VCF (Variant Call Format) files, a tab delimited text file with a descriptive header block. The header is marked with leading '#' characters and the actual calls follow with a standard 9 columns plus 1 column for each sample in the analysis as seen in Fig. 1.

#CHROM	POS	ID	REF	ALT	QUAL	FILTER	INFO	FORMAT	sample_456
ChrIII_A_nidulans_FGSC_A4	158	.	AG	A	.	PASS	END=159;SVLEN=-1;SVTYPE=DEL	GT:AD	0/0:66,1

Fig. 1 Deletion and column definitions from a vcf

##INFO=<ID=END,Number=1, Type=Integer, Description="End position of the variant described in this record">
##INFO=<ID=SVLEN,Number=1, Type=Integer, Description="Difference in length between REF and ALT alleles">
##INFO=<ID=SVTYPE,Number=1, Type=String, Description="Type of structural variant">
##FORMAT=<ID=PL,Number=G, Type=Integer, Description="List of phred-scaled genotype likelihoods">
##FORMAT=<ID=GT,Number=1, Type=String, Description="Genotype">

Fig. 2 INFO and FORMAT definitions from a vcf header

CHROM is the chromosome or contig, POS is the coordinate within that contig, ID is used when variants have a known identity in a database of variation such as dbsnp for humans and is generally blank (.) for fungal genomes, REF is the reference allele, ALT is the variant allele and where multiple variant flavors exist they will be listed here and concatenated with ',' (e.g., A,C), QUAL is the phred scaled likelihood of the call being true, FILTER is either '.' for unfiltered data, PASS for sites that have passed a filter or various strings for sites failing to pass a filter, INFO contains various values related to the site of this call, FORMAT defines the data in the per sample information and is followed by a corresponding block of information on the call for each sample.

Keys and values in the FILTER, INFO and FORMAT fields will vary depending on the tools used to produce or process a vcf but are always described in the header section at the top of a VCF. The definitions (Fig. 2) tell you which field, which key (ID) and at the end a description.

The FORMAT block is a colon delimited set of keys describing the data that follows in the sample block(s). For example, in GT:PL 0/1:129,0,255. GT:PL indicates the values will be a genotype call, 0/1, and a phred likelihood value of 129,0,255.

Genotypes are listed as 0 for the reference genotype and 1 for the alt genotype. In higher ploidies there are additional values for each copy of a chromosome. In a vcf with diploid calls the values are 0/0 for homozygous ref., 0/1 a heterozygous call and 1/1 is homozygous variant allele. The values for PL are the probability of each possible genotype being miscalled, and scaled relative to the most likely call which is 0. In the 129,0,255 that is the heterozygous call.

The effect of a variant located in a protein coding gene will be listed in the INFO column as the EFF = block, a set of '|' delimited values.

```
Effect ( Effect_Impact | Functional_Class | Co-
don_Change | Amino_Acid_Change| Amino_Acid_Length |
Gene_Name | Transcript_BioType | Gene_Coding | Tran-
script_ID | Exon_Rank | Genotype_Number [| ERRORS |
WARNINGS ] )
```

```
Example: EFF=NON_SYNONYMOUS_CODING(MODERATE
|MISSENSE|tCg/tTg|S1537 L|1571|182|protein_
coding|CODING|AN52000|4|1)
```

A quick way to find candidate severe mutations such as introduced stop codons and frameshifts is to filter the snpEff annotated vcf by searching for the term 'HIGH'.

Filtering the list of candidate SNPs beyond the initial 'bcftools filter' is usually necessary, to winnow down a large list of candidate sites, to quickly focus in on a set of potential causative mutations. A way to filter through "uninteresting differences" and potentially causative mutations is to sequence a parental strain or control strain which does not contain the phenotype of interest. Differences found in the parental strain are either true variants that existed in the original strain and are not associated with the causative mutation or are false positives, including reference errors, and sequencing artifacts from difficult to sequence spots such as long homopolymer tracks.

Each analysis has its own expected patterns of variation (*see* **Note 3**), and there are no simple rules to locate sample mishandling. The key to identifying signatures of mishandling is to understand the pattern of variation you expect in your data set and look for unexpected patterns, for example which samples should have variants in common, which should not. Some further signatures that could indicate sample mishandling include samples that were independently evolved sharing variants.

It is very useful to review candidate sites in a bam visualization tool that allows inspection of the alignment in detail. The Integrated Genome Viewer [7] from the Broad Institute is an excellent tool for this and has very thorough documentation available from their website.

http://software.broadinstitute.org/software/igv

Using the viewer SNPs can be confirmed (*see* Fig. 3) and more complex events such as inversion can be visualized (*see* Fig. 4).

5 Notes

1. Detection of false negative calls is difficult, but regions that likely could not be analyzed can be identified. Regions where a sample is too diverged from the reference for reads to align will appear as a gap in coverage similar to a large deletion but lacking signatures of a deletion such as read pairs spanning it. All positions with no depth can be listed with this samtools command that prints out the depth at every base, the awk part is selecting for depth values less than 0.

   samtools depth –aa –q 0 –Q 0 –reference FUNGAL.fa sample_456.bam | awk '$3 < 1' > gaps.in.coverage.txt

Fig. 3 Two homozygous SNPs viewed in IGV; the snps are the highlighted Ts, the grey lines represent mapped reads, and the white lines are reads mapped with 0 mapping quality as they could map equally well in multiple parts of the genome. The bars in a row along the top report the depth at each base

2. Within large deletion events some reads will often align with low depth and low to no map quality. These are likely merely reads from repetitive regions that occur both within and outside the deleted region in the reference and should therefore be discounted.

3. An easy way to identify sample mishandling is to look for unexpected patterns of variation. For example, that a haploid sample has a high percentage of multiallelic snps (genotype 0/1), especially if these sites are also found in additional individuals included in the analysis and sequenced or processed together, it could be contaminated with a second sample.

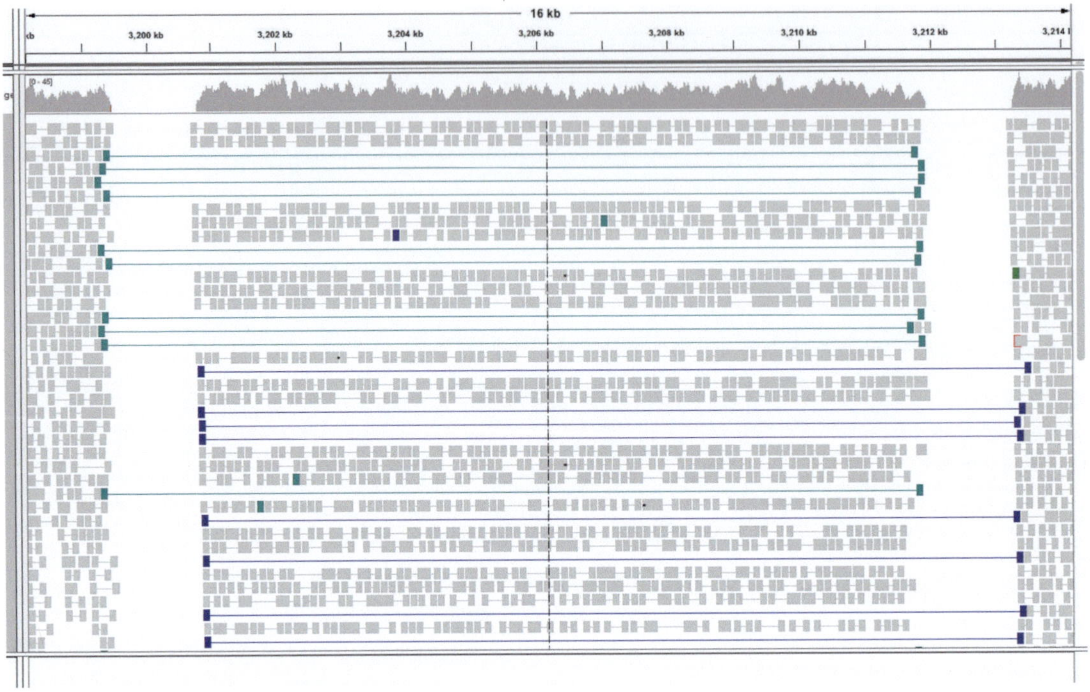

Fig. 4 An inversion viewed with IGV set to color alignments by pair orientation and view reads as pairs. When viewing reads as pair a line is drawn between each sequenced end of the template DNA fragment

Acknowledgment

The work conducted by the US Department of Energy Joint Genome Institute, a DOE Office of Science User Facility, was supported by the Office of Science of the US Department of Energy under Contract No. DE-AC02-05CH11231.

References

1. Li H, Durbin R (2009) Fast and accurate short read alignment with burrows-wheeler transform. Bioinformatics 25:1754–1760
2. Li H (2011) A statistical framework for SNP calling, mutation discovery, association mapping and population genetical parameter estimation from sequencing data. Bioinformatics 27(21):2987–2993
3. Samtools organization for next gen sequencing developers file format specifications (2016) https://samtools.github.io/hts-specs Accessed 15 Dec 2016
4. Chen K, Wallis JW, McLellan MD, Larson DE, Kalicki JM, Pohl CS, McGrath SD, Wendl MC, Zhang Q, Locke DP, Shi X, Fulton RS, Ley TJ, Wilson RK, Ding L, Mardis ER (2009) BreakDancer: an algorithm for high-resolution mapping of genomic structural variation. Nat Methods 6:677–681
5. Ye K, Schulz MH, Long Q, Apweiler R, Ning Z (2009) Pindel: a pattern growth approach to detect break points of large deletions and medium sized insertions from paired-end short reads. Bioinformatics 25(21):2865–2871

6. C Cingolani P, Platts A, Wang le L, Coon M, Nguyen T, Wang L, Land SJ, Lu X, Ruden DM (2012) A program for annotating and predicting the effects of single nucleotide polymorphisms, SnpEff: SNPs in the genome of Drosophila Melanogaster strain w1118; iso-2; iso-3. Fly 6(2):80–92
7. Robinson JT, Thorvaldsdóttir H, Winckler W, Guttman M, Lander ES, Getz G, Mesirov JP (2011) Integrative genomics viewer. Nat Biotechnol 29:24–26

Chapter 19

ChIP-Seq Analysis in *Neurospora crassa*

Aileen R. Ferraro and Zachary A. Lewis

Abstract

Chromatin immunoprecipitation paired with next-generation sequencing (ChIP-seq) can be used to determine genome-wide distribution of transcriptions factors, transcriptional machinery, or histone modifications. DNA–protein interactions are covalently cross-linked with the addition of formaldehyde. Chromatin is prepared and sheared, then immunoprecipitated with the appropriate antibody. After reversal of cross-linking and treating with protease, the resulting DNA fragments are sequenced and mapped to the reference genome to determine overall enrichment. Here we describe a method of ChIP-seq for investigating protein–DNA interactions in the filamentous fungus *Neurospora crassa*.

Key words Chromatin immunoprecipitation, Protein–DNA interactions, Histone modifications, Transcription factor binding

1 Introduction

Protein–DNA interactions regulate diverse nuclear processes such as gene expression, DNA repair and maintenance, chromosome segregation, and establishing and maintaining epigenetic modifications. The advent of chromatin immunoprecipitation (ChIP) has proven to be critical in the study of protein–DNA interactions and associated processes. ChIP followed by high throughput sequencing (ChIP-seq) has become a standard method in genome biology, allowing researchers to look at diverse DNA–protein interactions, including histone occupation, transcription factor binding, histone modifications, histone turnover, and other features of the genome-wide chromatin landscape including base modifications.

ChIP was first described by Gilmour and Lis [1] as a method to investigate localization of regulatory factors such as RNA polymerase II (Pol II) and histone occupation in *Drosophila* [2]. These original studies were performed with UV cross-linking followed by restriction digest and Southern blotting. Reversible formaldehyde cross-linking was introduced by Solomon et al. [3] to determine the association of Pol II with heat shock protein (*hsp*) genes in

Ronald P. de Vries et al. (eds.), *Fungal Genomics: Methods and Protocols*, Methods in Molecular Biology, vol. 1775, https://doi.org/10.1007/978-1-4939-7804-5_19,

Drosophila. Chromatin was fragmented via sonication or restriction digest followed by immunoprecipitation of covalently cross-linked protein–DNA complexes with the appropriate antibodies. After immunoprecipitation, cross-linking of immunoprecipitated protein–DNA complexes was reversed with heat, and remaining DNA fragments were analyzed by Southern blot. Reversible cross-linking has allowed for the advancement of ChIP applications, including ChIP followed by microarray (ChIP-chip), ChIP followed by quantitative polymerase chain reaction (ChIP-qPCR), and ChIP followed by next-generation sequencing (ChIP-seq).

Early application of ChIP in fungi was described in *Saccharomyces cerevisiae* [4] and *Schizosaccharomyces pombe* [5]. Here we describe a ChIP-seq method for use in the filamentous fungus *Neurospora crassa*. which is a derivation of the protocol originally developed by Tamaru and colleagues [6]. Chromatin fractions are prepared by covalent formaldehyde cross-linking and fragmentation by sonication. Fragmented chromatin is then immunoprecipitated with the appropriate antibody bound to agarose beads. Covalent cross-linking of protein–DNA complexes is then reversed by heat, and the chromatin fractions are treated with RNase and proteinase. Remaining DNA fragments are purified, and Illumina sequencing libraries are then prepared and sequenced (Fig. 1). Additionally, we provide a brief summary of available methods for downstream analysis of sequencing results and a sample pipeline for data analysis (Fig. 2).

2 Materials

2.1 Chromatin Immunoprecipitation

1. ChIP Lysis buffer without protease inhibitors: 50 mM HEPES (pH 7.5), 140 mM NaCl, 1 mM EDTA, 1% Triton X-100, 0.1% deoxycholate. Combine 160.6 mL sterile distilled water, 10 mL 1 M HEPES–KOH or HEPES–NaOH (pH 7.5), 7 mL 4 M NaCl, 400 μL 0.5 M EDTA, 20 mL 10% Triton X-100, 2 mL 10% DOC. Store at 4 °C.
2. ChIP Lysis buffer + 0.5 M NaCl: 50 mM HEPES (pH 7.5), 500 mM NaCl, 1 mM EDTA, 1% Triton X-100, 0.1% deoxycholate. Combine 142.6 mL sterile distilled water, 10 mL 1 M HEPES–KOH or HEPES-NaOH (pH 7.5), 25 mL 4 M NaCl, 400 μL 0.5 M EDTA, 20 mL 10% Triton X-100, 2 mL 10% DOC. Store at 4 °C.
3. ChIP LiCl Wash Buffer: 1 mM Tris–HCl, 250 mM LiCl, 0.5% NP-40, 0.5% deoxycholate, 1 mM EDTA. Combine 167.6 mL sterile distilled water, 2 mL 1 M Tris–HCl (pH 8.0), 10 mL 5 M LiCl, 10 mL 10% NP40, 10 mL 10% DOC, 400 μL 0.5 M EDTA, Store at 4 °C.
4. TE buffer: 10 mM Tris–HCl (pH 7.4), 1 mM EDTA.

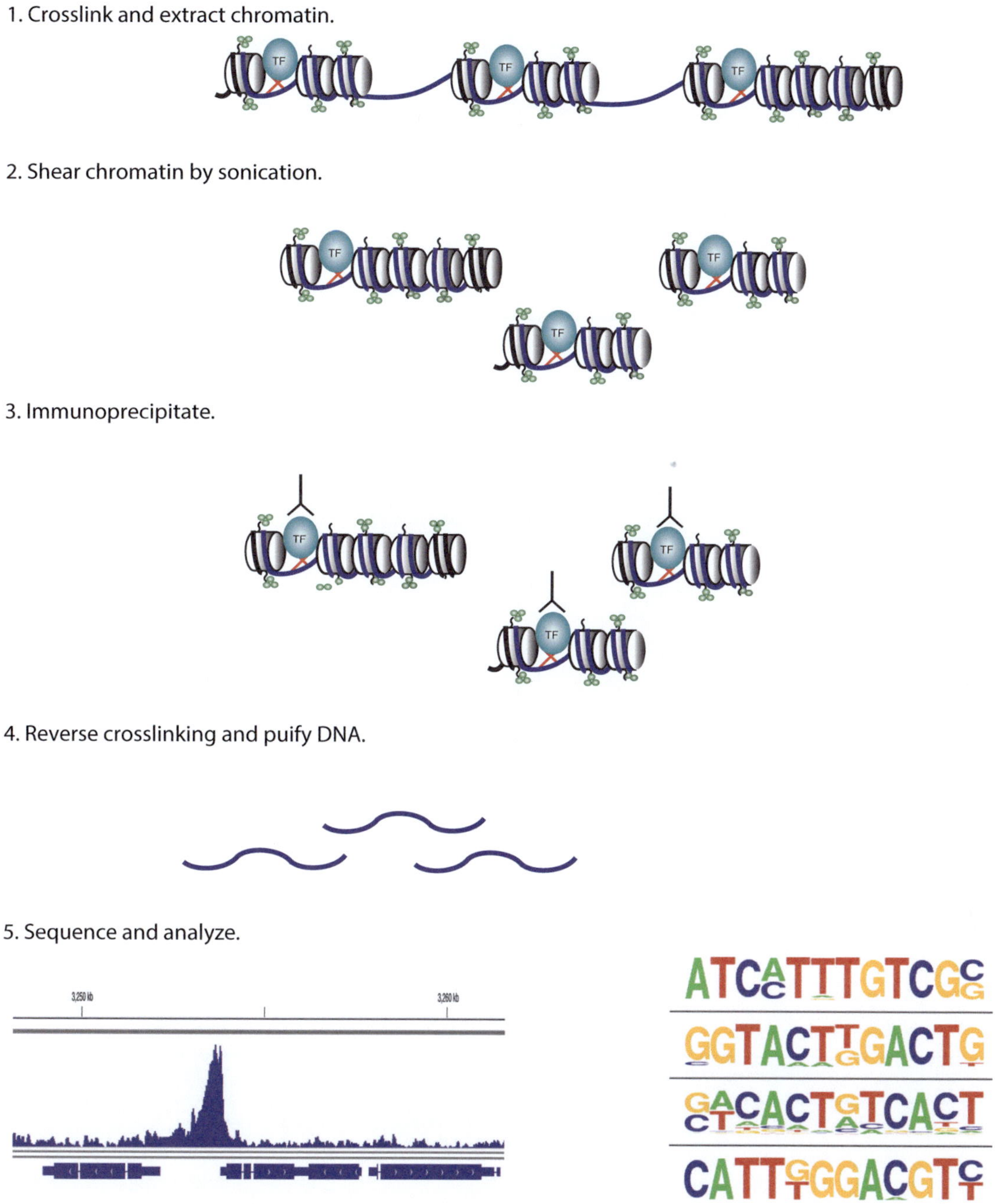

Fig. 1 Schematic diagram of a ChIP-seq experiment. DNA-binding proteins are covalently cross-linked to chromatin in vivo. The chromatin fiber is sheared by sonication into small fragments, which are subjected to immunoprecipitation using an antibody that binds a specific DNA-binding protein. Shown here as a transcription factor (TF). Following immunoprecipitation, the cross-links are reversed and DNA is purified, sequenced, and analyzed

5. ChIP TES Buffer: 50 mM Tris–HCl, 10 mM EDTA, 1% SDS. Combine 41.5 mL sterile distilled water, 2.5 mL 1 M Tris–HCl (pH 8.0), 1 mL 0.5 M EDTA, 5 mL 10% SDS. Store at room temperature.

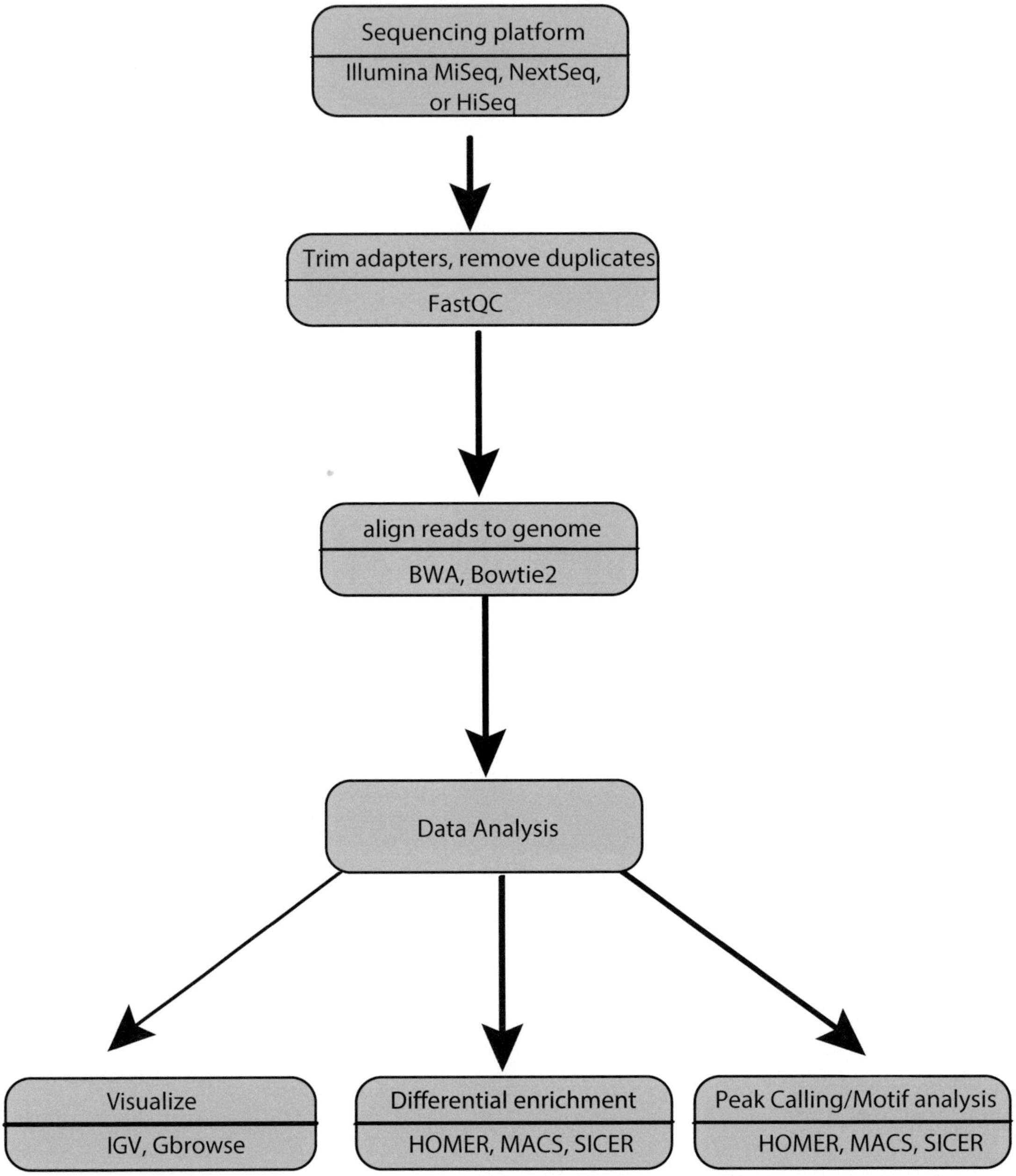

Fig. 2 General bioinformatics workflow for ChIP-seq analysis. Individual ChIP-seq analysis pipelines will vary based on the specific ChIP-seq application, but analyses workflows include several basic steps. Most ChIP-seq experiments in fungi will require a minimum of 1–4 million sequence reads generated using an Illumina sequencing instrument. Raw sequence reads should be preprocessed using a program such as FastQC [7], to remove Illumina adaptor sequences, and remove PCR and optical duplicates. Preprocessed reads are then aligned to a reference genome using a short read aligner such as bowtie2 [8] or the Burrows-Wheeler Aligner [9]. Aligned reads can be visualized using genome browser software, such as the Broad Integrative Genomics Viewer [10] or Gbrowse [11]. Aligned reads can then analyzed using a variety of software packages, depending on the specific goals of the ChIP-seq. For example, software such as HOMER [12], MACS [13], or SICER [14] can be used to call peaks, identify DNA sequence motifs, or perform differential enrichment analyses

6. Roche Complete Protease inhibitor cocktail tablets.
7. PMSF: 100 mM in isopropanol. Store at room temperature.
8. 37% formaldehyde.
9. 2.5 M glycine.
10. Santa Cruz Biotechnology A/G agarose beads.
11. 10 mg/mL RNase A.
12. 20 mg/mL Proteinase K.
13. Ambion 5 μM Glycogen.
14. 3 M sodium acetate (pH 5.2).
15. Phenol–chloroform–isoamyl alcohol (25:24:1).
16. Chloroform.
17. Phosphate buffered saline.

2.2 Library Preparation

1. Ampure XP PCR purification beads.
2. 10 mM Tris–HCl (pH 7.5).
3. Double strand adaptor for Illumina sequencing: (NEB or comparable supplier).
4. Dual index primers for library amplification (NEB or comparable supplier).
5. NEB Ultra II End Repair Module.
6. NEBNext Ultra II Q5 Hot Start HiFi PCR Master Mix.
7. T4 DNA ligase.

3 Methods

3.1 Chromatin Immunoprecipitation

1. Grow 5 mL overnight culture in liquid medium.
2. Collect mycelia by vacuum filtration using a Buchner funnel and wash mycelium with 100 mL of PBS.
3. Transfer mycelia to 10 mL PBS in a 125-mL Erlenmeyer flask.
4. Add 270 μL of 37% formaldehyde for a final concentration of 1%.
5. Incubate on rotating platform for 30 min at room temperature.
6. Add 500 μL 2.5 M glycine to each sample to quench the formaldehyde. Let samples sit at room temperature for 5 min.
7. Collect mycelia by filtration. Wash with PBS.
8. Transfer mycelia to a 1.5-mL microcentrifuge tube.
9. Add 100 μL PMSF and 1 Roche protease inhibitor tablet to 9.9 mL ChIP lysis buffer.

10. Resuspend mycelia in 500 μL ice-cold ChIP lysis buffer with PMSF and protease inhibitors.
11. Lyse mycelia by sonicating (*see* **Note 1**).
12. Shear chromatin by sonicating (*see* **Note 1**).
13. Centrifuge samples at maximum speed in benchtop centrifuge (e.g. 14,000 RPM or 20,000 × *g* in an Eppindorf 5430 centrifuge). 14k RPM for 5 min at 4 °C.
14. Transfer supernatant containing sheared chromatin to a new tube.
15. Save 20 μL of sheared chromatin extract in new tube and store at −20 °C. This will be your input. Use the remaining extract for immunoprecipitation.
16. Aliquot 20 μL agarose beads per reaction + 10% total volume into a 1.5-mL microcentrifuge tube.
17. Spin at 5000 RPM for 1 min. Discard supernatant.
18. Resuspend beads in 1 mL ChIP lysis buffer without protease inhibitors.
19. Repeat **steps 17** and **18**.
20. Resuspend beads in original volume (20 μL/sample + 10%) ChIP lysis buffer without protease inhibitors.
21. Add 20 μL equilibrated protein A/G beads to each sample. Add 1–3 μL desired antibody.
22. Incubate overnight at 4 °C on rotator to allow antibody binding.
23. Spin samples at 2600 × *g* (e.g. 5000 RPM in an Eppendorf 5430 microcentrifuge) 5000 RPM for 1 min to pellet beads. Discard the supernatant by pipetting. Be sure not to disrupt the pellet.
24. Add 1 mL ice-cold ChIP lysis buffer without protease inhibitors to each sample. Incubate for 10 min at 4 °C on a rotating platform.
25. Spin samples for 1 min at 5000 RPM at 4 °C. Discard the supernatant.
26. Repeat **steps 24** and **25**.
27. Wash (as in **steps 24** and **25**) with ice-cold ChIP lysis buffer + 0.5 M NaCl.
28. Wash (as in **steps 24** and **25**) with ice-cold LiCl wash buffer.
29. Wash (as in **steps 24** and **25**) with ice-cold TE buffer.
30. Collect immunoprecipitated chromatin by adding 62.5 μL TES buffer to each sample. Incubate at 65 °C for 10 min. Mix by inversion several times during incubation.
31. Spin at 5000 RPM for 1 min. Transfer supernatant to a new 1.5-mL microcentrifuge tube and save.

32. Repeat **steps 30** and **31**, saving the supernatant in the same microcentrifuge tube as **step 31**.
33. Remove input sample (from **step 15**) from −20 °C. Add 105 μL TES to each input sample.
34. De-cross-link samples by incubating overnight at 65 °C.
35. Add 125 μL sterile distilled water and 2.5 μL 10 mg/mL RNaseA to samples. Incubate for 2 h at 50 °C. Mix samples by vortexing multiple times during incubation.
36. Add 6.25 μL 20 mg/mL Proteinase K. Incubate for 2 h at 50 °C. Mix samples by vortexing multiple times during incubation.
37. Add 250 μL phenol–chloroform–isoamyl alcohol to each sample. Mix well by vortexing.
38. Spin at 14k RPM for 5 min. Transfer the aqueous layer to a new 1.5-mL microcentrifuge tube.
39. Add 250 μL chloroform. Mix well by vortexing.
40. Spin at 14k RPM for 5 min. Transfer the aqueous layer to a new 1.5-mL microcentrifuge tube.
41. Add 1 μL glycogen, 25 μL 3 M Na-acetate (pH 5.2), and 865 μL 100% ethanol to each sample. Precipitate overnight at −20 °C.
42. Retrieve samples from −20 °C.
43. Spin at 14k RPM for 10 min. Discard the supernatant.
44. Add 300 μL 70% ethanol to each sample.
45. Spin at 14k RPM for 5 min. Discard the supernatant.
46. Air-dry samples or dry in Speed Vac.
47. Resuspend samples in 25 μL TE.
48. Store at −20 °C.

3.2 Library Preparation

1. Thaw End Repair buffer on ice. Vortex thoroughly to make sure all buffer components are in solution.
2. In a low-bind PCR tube, mix 25.5 μL ChIP DNA, 3 μL 10× End Repair Reaction buffer, and 1.5 μL End Prep Enzyme Mix.
3. Incubate for 30 min at 20 °C, 30 min at 65 °C, hold at 4 °C.
4. Thaw 10× Adaptor Ligation buffer on ice. Vortex thoroughly to make sure all buffer components are in solution.
5. Dilute double stranded Illumina adaptor to 1.5 μM in 10 mM Tris.
6. Add the following directly to end repair mix: 4 μL of 10× Ligase Buffer with dATP, 2 μL of T4 DNA Ligase, 2 μL of double stranded adaptor, and 2 μL of water.

7. Incubate overnight at 16 °C.
8. Add 40 μL of AmpPure beads and mix by pipetting up and down ten times.
9. Incubate for 5 min at room temperature.
10. Place on magnet stand for 5 min to clear supernatant.
11. Carefully remove supernatant. Be sure to avoid removing beads.
12. Leaving the tubes on the magnet stand, add 200 μL freshly prepared 80% ethanol.
13. Incubate for 30 s and remove ethanol wash. Be sure to avoid removing beads.
14. Repeat **steps 12** and **13**.
15. Air-dry beads for 5 min on magnet stand with lid open. Be sure not to overdry, as this will make elution difficult.
16. Remove tubes from magnet and elute DNA in 22 μL of 10 mM Tris–HCl (pH 7.5–8.0). Mix solution up and down, incubating beads for 5 min at room temperature to elute DNA.
17. Place tubes on magnet stand and transfer 20 μL of supernatant to a new PCR tube.
18. In a low-bind PCR tube, combine 20 μL Adaptor ligated DNA fragments, 5 μL dual index primer mix containing 10 μM of each primer (use unique dual index combination for each sample you plan to multiplex), and 25 μL 2× Q5 Hot start polymerase master mix.
19. Amplify libraries: (a) denature at 98 °C for 30 s; (b) for 2–12 cycles (*see* **Note 2**), 98 °C for 10 s, 55 °C for 30 s, and 72 °C for 60 s; (c) final extension at 72 °C for 3 min; and (d) hold at 10 °C.
20. Add 50 μL of SeraPure beads (1:1 ratio) and mix by pipetting up and down ten times.
21. Incubate for 5 min at room temperature.
22. Place on magnet for 5 min to clear supernatant.
23. Remove supernatant.
24. Leaving the tubes on the magnet stand, add 200 μL freshly prepared 80% ethanol.
25. Incubate for 30 s and remove ethanol.
26. Repeat **steps 24** and **25**.
27. Air-dry beads for 5 min on magnet with lid open. Be sure not to overdry.
28. Remove tubes from magnet and elute DNA in 15 μL of 10 mM Tris–HCl (pH 7.5–8.0). Mix solution up and down,

incubating beads for 5 min at room temperature to elute DNA.

29. Transfer 13 μL of supernatant to a new tube. Be sure not to carry over beads, as they will inhibit downstream applications. If carryover occurs, add solution to magnet a second time.
30. Quantify using a bioanalyzer or Qubit fluorometer. If sufficient material is obtained, run 10–20 ng of library DNA on a 1.5% agarose gel to confirm correct size distribution and lack of primer dimers.
31. Dilute samples to a concentration of 10 nM. For the 40 Mb Neurospora genome, 50–80 individual ChIP-seq samples can be pooled and sequenced on a single flow cell of an Illumina Next-Seq or Hi-Seq instrument. Most ChIP-seq experiments in fungi will require a minimum of 1–4 million sequence reads generated using an Illumina sequencing instrument.

3.3 Data Analysis

Several analysis options exist for ChIP-seq data. While the specifics of these options may differ based on specific experimental details, the overall approach will require several key steps. Here we will present a general workflow, as well as a small sample of available analysis software.

1. Preprocess sequence reads: Duplicate reads should be removed and Illumina adaptor sequences should be trimmed from any reads that contain them. This is done using FastQC [7] or similar software.
2. Align reads to reference genome using a short read aligner such as bowtie2 [8] or the Burrows-Wheeler Aligner [9].
3. Visualize sequence alignments in a genome browser such as the Broad Integrative Genome Viewer [10] or Gbrowser [11].
4. Perform project specific analyses such as peak calling, analysis of differential enrichment, and Motif analysis. HOMER [12], MACS [13], or SICER [14] are commonly used software packages for ChIP-seq analyses.

4 Notes

1. Sonication conditions will need to be optimized to ensure proper tissue homogenization and chromatin shearing. Check efficiency by running sheared chromatin on a 1.5% agarose gel to ensure a fragment size of 500 bp.
2. *N. crassa* genomes contain A:T-rich domains, which can be underrepresented due to PCR bias [15]. Bias can be reduced by limiting the number of PCR cycles used to amplify libraries. Be sure to optimize the amplification step to determine the appropriate number of PCR cycles for your samples.

References

1. Gilmour DS, Lis JT (1985) In vivo interactions of RNA polymerase II with genes of Drosophila Melanogaster. Mol Cell Biol 5(8):2009–2018
2. Champlin DT, Frasch M, Saumweber H, Lis JT (1991) Characterization of a drosophila protein associated with boundaries of transcriptionally active chromatin. Genes Dev 5(9):1611–1621
3. Solomon MJ, Larsen PL, Varshavsky A (1988) Mapping protein-DNA interactions in vivo with formaldehyde: evidence that histone H4 is retained on a highly transcribed gene. Cell 53(6):937–947
4. Tanaka T, Knapp D, Nasmyth K (1997) Loading of an mcm protein onto DNA replication origins is regulated by Cdc6p and CDKs. Cell 90(4):649–660
5. Ekwall JP (1999) Fission yeast chromosome analysis: fluorescence in-situ hybridization (FISH) and chromatin immunoprecipitation (CHIP). In: Chromosome structural analysis: a practical approach. Oxford University Press, Oxford, UK
6. Tamaru H, Zhang X, McMillen D, Singh PB, Nakayama J-i, Grewal SI, Allis DC, Cheng X, Selker EU (2003) Trimethylated lysine 9 of histone H3 is a mark for DNA methylation in Neurospora crassa. Nat Genet 34(1):75–79. https://doi.org/10.1038/ng1143
7. Andrews S (2010) FastQC: a quality control tool for high throughput sequence data. Available online at http://www.bioinformatics.babraham.ac.uk/projects/fastqc
8. Langmead B, Salzberg SL (2012) Fast gapped-read alignment with bowtie 2. Nat Methods 9(4):357–359. https://doi.org/10.1038/nmeth.1923
9. Li H, Durbin R (2009) Fast and accurate short read alignment with burrows–wheeler transform. Bioinformatics 25(14):1754–1760. https://doi.org/10.1093/bioinformatics/btp324
10. Robinson JT, Thorvaldsdóttir H, Winckler W (2011) Integrative genomics viewer. Nat Biotechnol 29(1):24–26. https://doi.org/10.1038/nbt.1754
11. Stein LD (2002) The generic genome browser: a building block for a model organism system database. Genome Res 12(10):1599–1610. https://doi.org/10.1101/gr.403602
12. Heinz S, Benner C, Spann N, Bertolino E, Lin YC, Laslo P, Cheng JX, Murre C, Singh H, Glass CK (2010) Simple combinations of lineage-determining transcription factors prime cis-regulatory elements required for macrophage and B cell identities. Mol Cell 38(4):576–589
13. Feng J, Liu T, Zhang Y (2011) Using MACS to identify peaks from ChIP Seq data. Curr Protoc Bioinformatics. https://doi.org/10.1002/0471250953.bi0214s34
14. Xu S, Grullon S, Ge K, Peng W (2014) Spatial clustering for identification of ChIP-enriched regions (SICER) to map regions of histone methylation patterns in embryonic stem cells. Methods Mol Biol 1150:97–111
15. Ji L, Sasaki T, Sun X, Ma P, Lewis ZA, Schmitz RJ (2014) Methylated DNA is over-represented in whole-genome bisulfite sequencing data. Front Genet 5:341. https://doi.org/10.3389/fgene.2014.00341

Chapter 20

Fungal Phylogenomics

Robert Riley and Laszlo Nagy

Abstract

Phylogenomics aims to infer the evolutionary relationships of organisms, and their genomes, genes, and proteins, from genomic data. Understanding the evolution of these components can provide clues about their biological functions. Here we describe minimal protocols for inferring families of genes (and the proteins they encode), and using them in phylogenomic analyses to infer species trees.

Key words Phylogenomics, Genomics, Fungi, Phylogeny inference, Phylogenetic tree, MCL clustering, Protein families, Orthologs

1 Introduction

A key insight emerging from early genomics studies [1, 2] was that proteins occur in families: groups of proteins with a common evolutionary origin, inferred by their similar sequence, structure, and biological function. Protein families are groups of homologous proteins, that comprise both orthologs (sets of sequences separated from each other by speciations) and paralogs (resulting from duplications after speciation). Orthologs generally retain the function of the ancestral gene, whereas paralogs can evolve new functions (through neofunctionalization), or the two descendent paralogs might partition the ancestral function between each other (subfunctionalization), an important consideration for choosing gene families for phylogenomics.

Despite floods of genome sequence data the bioinformatic detection of protein families, given a set of predicted genes, and proteins they encode, remains a significant challenge [3–6]. Identifying groups of orthologous proteins is important because they usually have similar functions, and similarity to proteins of known function is often a basis for assigning functional annotations to the proteins predicted from newly sequenced genomes. Moreover, for protein families of known general function, observing the copy number variations (expansions and contractions) can

Ronald P. de Vries et al. (eds.), *Fungal Genomics: Methods and Protocols*, Methods in Molecular Biology, vol. 1775,
https://doi.org/10.1007/978-1-4939-7804-5_20, © Springer Science+Business Media, LLC, part of Springer Nature 2018

provide insights into the biology of organisms, as reported for the distribution of plant cell wall degrading enzyme genes (e.g., class II lignin peroxidases) in white rot versus brown rot fungi [7].

Markov clustering methods, e.g., OrthoMCL [5], are useful for assigning proteins into families given a matrix of all-vs-all pairwise sequence similarity in a set of proteins. Importantly for phylogenomic analyses, these protein sets can be from multiple organisms. OrthoMCL is an example of readily available software that applies the Markov Cluster algorithm [8], a clustering method for graphs, to the problem of assigning proteins to families. Other MCL-based methods, such as TRIBE-MCL [3], will produce comparable results for the analyses we describe here.

Phylogenetic tree inference [9–11] is generally based on multiple sequence alignments as input. A typical approach is to concatenate the sequences of a few conserved genes, resulting in a "super-sequence" from each organism. Genome sequencing, and the comprehensive sets of proteins they provide, enables us to use hundreds to thousands of orthologous genes and the combined historical evidence they make up for phylogenetic tree inference, thus taking full advantage of genome-wide data. Thus, the differences in the genealogies, evolutionary rates, potential biases and phylogenetic signal between individual gene families—if randomly distributed—are expected to average out and lead to robust support at multiple phylogenetic depths.

Genome-scale data provide a wealth of data for inferring phylogeny. These include ultraconserved genetic elements [12], conserved noncoding sequence regions, and whole transcriptome, genome, or proteome-based methods when complete genome sequences are available. Here, we will focus on the latter. We present a strategy for inferring a phylogenetic tree of a collection of organisms for which we have sequenced genomes and predicted protein-coding gene sets. The strategy is based on identifying single copy gene families (those families where each organism contributes one and only one gene) among MCL-inferred clusters, inferring multiple sequence alignments for the protein sequences, and performing phylogenetic analyses to infer species trees.

2 Materials

2.1 Data Download

We provide an example data based on the dataset from [7]. Download the "filtered" protein sets for the organisms and the downloads section of the given URLs in Table 1.

2.2 Software Download and Install

Download and install the software in Table 2 according to each program's documentation and your system's requirements.

Table 1
Organisms and URLs to access them

Organism	JGI URL	Abbreviation in Fig. 1
Aspergillus niger	http://genome.jgi.doe.gov/Aspergillusniger	asp
Auricularia Subglabra	http://genome.jgi.doe.gov/Auricularia	aur
Batrachochytrium dendrobatidis	http://genome.jgi.doe.gov/Batrachochytrium	bat
Coniophora puteana	http://genome.jgi.doe.gov/Coniophora	con
Coprinopsis cinerea	http://genome.jgi.doe.gov/Coprinopsis	cop
Cryptococcus neoformans	http://genome.jgi.doe.gov/Cryptococcus	cry
Cryphonectria parasitica	http://genome.jgi.doe.gov/Cparasitica	cryp
Dacryopinax primogenitus	http://genome.jgi.doe.gov/Dacryopinax	dac
Dichomitus squalens	http://genome.jgi.doe.gov/Dsqualens	dic
Fomitiporia mediterranea	http://genome.jgi.doe.gov/Fomitiporia	fom
Fomitopsis pinicola	http://genome.jgi.doe.gov/Fomitopsis	fomp
Gloeophyllum trabeum	http://genome.jgi.doe.gov/Gloeophyllum	glo
Heterobasidion annosum	http://genome.jgi.doe.gov/Heterobasidion	het
Laccaria bicolor	http://genome.jgi.doe.gov/Laccaria_bicolor	lac
Malassezia globosa	http://genome.jgi.doe.gov/Malassezia	mal
Melampsora laricis-populina	http://genome.jgi.doe.gov/MelampsoraLaricisPopulinaV2	mel
Phanerochaete chrysosporium	http://genome.jgi.doe.gov/Phanerochaete	phc
Phycomyces blakesleeanus	http://genome.jgi.doe.gov/Phycomyces	phy

(continued)

Table 1 (continued)

Organism	JGI URL	Abbreviation in Fig. 1
Pichia stipitis	http://genome.jgi.doe.gov/pichia	pic
Postia placenta	http://genome.jgi.doe.gov/PplacentaRSB12	pos
Punctularia strigosozonata	http://genome.jgi.doe.gov/Punctularia	pun
Schizophyllum commune	http://genome.jgi.doe.gov/Scommune3	sch
Serpula Lacrymans	http://genome.jgi.doe.gov/Serpula	ser
Sporobolomyces roseus	http://genome.jgi.doe.gov/Sroseus	spo
Stagonospora nodorum	http://genome.jgi.doe.gov/Stagonosporanodorum	sta
Stereum hirsutum	http://genome.jgi.doe.gov/Stereum	ste
Trametes versicolor	http://genome.jgi.doe.gov/Trametes	tra
Tremella mesenterica	http://genome.jgi.doe.gov/Tremella	tre
Trichoderma Reesei	http://genome.jgi.doe.gov/Treesei	tri
Ustilago maydis	http://genome.jgi.doe.gov/Ustilago	ust
Wolfiporia cocos	http://genome.jgi.doe.gov/Wolfiporia	Wol

Table 2
Required software

Program	Version	URL
OrthoMCL	2.0.9	http://orthomcl.org
Python	2.7.4	https://www.python.org
MAFFT	7.221	http://mafft.cbrc.jp/alignment/software/
Gblocks	0.91b	http://molevol.cmima.csic.es/castresana/Gblocks.html
ClustalW	2.1	http://www.clustal.org/clustal2/
RAxML	7.6.3	https://github.com/stamatak/standard-RAxML
FastTree	2.1.9	http://www.microbesonline.org/fasttree/#Install
ETE toolkit	3.0.0b36	http://etetoolkit.org

3 Methods

3.1 Assign Proteins to Families Using OrthoMCL

Run OrthoMCL on your protein set as in http://orthomcl.org/common/downloads/software/v2.0/UserGuide.txt.

Using OrthoMCL v2.0.9, we performed all-vs.-all BLAST with an E-value threshold of 10^{-5} and percent identity threshold of 50% on 383,910 protein sequences predicted from 31 fungal genomes used in [7] (note that several of the genome annotations were updated since 2012, so our numbers might not exactly match that study). The BLAST results were processed using OrthoMCL's scripts and clustering was performed with an inflation parameter of 2.0 (consult current version of software's documentation for details).

Running OrthoMCL on the above dataset resulted in some 32,372 nonsingleton clusters. OrthoMCL also identified some 2,149,920 pairs of orthologous genes (best-hit pairs of proteins that are across two species), 695,564 pairs of proteins whose best hit is within a species, and 634,134 co-orthologs (pairs of proteins across two species where the proteins are connected through both orthology and inparalogy).

3.2 Identify a Single-Copy Gene Set

Next, we identify the set of conserved single-copy clusters, in which each organism contributes only one protein (*see* **Note 1**). Such clusters could be extracted from the OrthoMCL results (e.g., called groups.txt file) using a Python script like the one in Text Box 1.

Text Box 1: Python Script to Extract Single-Copy Clusters from an OrthoMCL Run

```
# Extract single-copy clusters from OrthoMCL run, e.g. file 'groups.txt'
#
# File has one cluster per line, lines look like:

# cluster16469: asp|1176580 cryp|42756 sta|8442 tri|75434
# cluster16470: sch|2664171 dic|139591 fomp|1166292 glo|128976
# cluster16471: asp|1177110 cryp|357688 sta|3771 tri|22774

# Assume your proteins are named e.g. 'asp|1176580' where 'asp' is a species
# name and '1176580' is a unique identifier for that protein in asp, etc.

# Write out those clusters that have one and only one protein from each species

import sys
import re

TRUE = 1
FALSE = 0

try:
  f = open(sys.argv[1], 'r')
except:
  print "Supply a file, e.g. 'groups.txt' from an OrthoMCL run"
  sys.exit()

clusters = {}

# Read in the clusters
for line in f.readlines():
  m = re.match("^(\w+):", line)
  if m:
    cluster_id = m.group(1)
    clusters[cluster_id] = re.findall('(\w+)\|(\d+)', line)

# Determine what species are in the clusters
species = set([s for c in clusters.itervalues() for s, p in c])


# Find the cluster, where one and only one protein is contributed by each species
single_copy_clusters = {}
N = len(species)
for k, c in clusters.iteritems():
  if len(c) == N:
    species_in_cluster = set([s for s, p in c])  # Give the list of species in a cluster
    if len(species_in_cluster) == N:
      single_copy_clusters[k] = c

sys.stderr.write("%d single-copy clusters obtained\n" % len(single_copy_clusters))

for k, c in single_copy_clusters.iteritems():
  sys.stdout.write("%s:" % k)  # Write the cluster id
  for i in range(N):  # Write each protein in the cluster
    s, p = c[i]
    sys.stdout.write(" %s|%s" % (s, p))
  sys.stdout.write("\n")
```

Such a script could be run using the UNIX command line:

```
python orthomcl_single_copy.py groups.txt > single_copy.txt
```

Using this script, we extracted 510 single-copy clusters from the OrthoMCL results file groups.txt.

3.3 Concatenate Protein Sequences from the Single-Copy Clusters

To produce a FASTA-format input file for multiple sequence alignment, we concatenate each organism's protein sequences from the single-copy clusters, using a Python script as in Text Box 2. *See* also **Note 2**. on concatenation after multiple sequence alignment.

Text Box 2: Python Script to Produce a FASTA File with Concatenated Sequences from the Single-Copy OrthoMCL Clusters

```
# Read in a list of single-copy clusters from some MCL-based clustering method
# (e.g. OrthoMCL), and concatenate the sequences for each organism.  For each
# organism, write out one FASTA sequence containing the concatenated sequences,
# in the same cluster order for each organism.

import sys
import re
from Bio import SeqIO
from Bio.Seq import Seq
from Bio.SeqRecord import SeqRecord
from Bio.Alphabet import IUPAC

all_fastas = {}
clusters = []

try:
  try:
    fasta_sequences = SeqIO.parse(open(sys.argv[1]),'fasta')
    for fasta in fasta_sequences:
      name, sequence = fasta.id, str(fasta.seq)
      all_fastas[name] = fasta  # fasta is a SeqRecord object.  See http://biopython.org
  except:
    sys.stderr.write("ERROR: supply a valid FASTA file\n")
    sys.exit()
  try:
    f = open(sys.argv[2], 'r')
    for line in f.readlines():
      c = re.findall('(\w+)\|(\d+)', line)
      if c:
        clusters.append(c)
  except:
    sys.stderr.write("ERROR: supply a cluster file\n")
    sys.exit()
  try:
    out_file = open(sys.argv[3], 'w')
  except:
    sys.stderr.write("ERROR opening %s\n" % sys.argv[3])
    sys.exit()
except:
  print "ARGUMENTS: <input fasta file> <single copy clusters file> <output fasta file name>"
  sys.exit()

all_species = set([s for c in clusters for s, p in c])

concat_fastas = []  # Initialize dictionary

for species in all_species:
  concat_seq = ''
  for c in clusters:
    for s, p in c:
      if s == species:
        concat_seq += str(all_fastas[s + '|' + p].seq)
  concat_fastas.append(SeqRecord(Seq(concat_seq, IUPAC.protein), id=species, description=""))

SeqIO.write(concat_fastas, out_file, "fasta")
```

We would run this script as, for example:

```
python concatenate_seq.py allprot.fasta single_
copy.txt concat.fasta
```

3.4 Multiple-Align Sequences with MAFFT

Use MAFFT [13], to align the concatenated sequences (*see* **Note 2**). We generally run MAFFT with automatically determined run options as follows:

```
mafft --auto concat.fasta > concat.mafft
```

The –auto option determines an optimal tradeoff between speed and accuracy—for larger datasets it will choose one of the speed-oriented built-in algorithms of the software. On our data set, MAFFT v7.182 was able to align the 31 concatenated sequences, totaling about 2.7 Mb of data, in a few hours on a fairly typical Linux machine (compute time will vary depending on hardware and other considerations). Note that accuracy can be prioritized using other options of MAFFT (e.g., L-INS-I, G-INS-I, *see* also notes below) unless run times become prohibitive.

3.5 Extracting Well-Aligned Regions

If we manually inspect the resulting MAFFT alignment, we see that it contains a significant number of regions of low alignment quality. This can be caused by multiple factors, including protein sequence divergence [14], insertions and deletions, potential gene fragments and inaccuracies in the alignment. Gapped and highly variable alignment regions may introduce noise into the analyses, which can result in low signal-to-noise ratios during phylogenetic inference. Therefore it is desirable to extract only the well-aligned portions of the alignment. Gblocks [15] is the most widely used method for this, although more recent tools, such as Trim-al [16] and Aliscore [17] also exist. Gblocks searches for contiguous stretches of well-aligned regions of a minimum length flanked by regions containing gaps or low overall alignment score. To extract the well-aligned regions from the MAFFT alignment, run Gblocks as follows:

```
Gblocks concat.mafft -t=p -e=-gb1 -b4=5
```

The Gblocks output indicates that the original alignment consisted of 517,466 positions, whereas in the reduced Gblocks alignment, consisting of some 4736 conserved blocks, there are now 121,960 positions (23%).

3.6 Compute Neighbor Joining Tree with ClustalW

Use ClustalW to compute a tree using the neighbor-joining algorithm [18]:

```
clustalw2 -tree -infile=concat.mafft-gb1
```

This tree should take seconds to compute on a typical Linux machine. The tree file produced is `concat.ph`.

3.7 Convert Gblocks Output to Phylip Format

To prepare to run RAxML, use ClustalW to convert the Gblocks output to Phylip format:

```
clustalw2 -convert -infile=concat.mafft-gb1 -
output=phylip \
-outfile=concat.mafft-gb1.phy
```

This step is necessary because, depending on version, RAxML may require that the alignment is input in Phylip format, not the FASTA format output by Gblocks.

3.8 Compute Maximum Likelihood Species Phylogeny Using RAxML

Run the multithreading-enabled version of RAxML (a multicore computer with 16 cores is assumed for this example—specified with the parameter "–T") as follows:

Variant 1. Maximum likelihood (ML) search without bootstrap:

```
raxmlHPC-PTHREADS-SSE3 -T 16 -s concat.mafft-gb1.
phy -n \
concat.mafft-gb1.phy.RAxML -o bat -p 12345 -m
PROTGAMMAWAG
```

Variant 2. ML tree search with 100 rapid bootstrap replicates:

```
raxmlHPC-PTHREADS-SSE3 -T 16 -s concat.mafft-gb1.
phy -n \
concat.mafft-gb1.phy.RAxML -o bat -f a -x 12345 -p
12345 \
-# 100 -m PROTGAMMAWAG
```

Variant 3. ML tree search with more thorough bootstrap analysis:

```
raxmlHPC-PTHREADS-SSE3 -T 16 -s concat.mafft-gb1.
phy -n \
concat.mafft-gb1.phy.500bootstrap.RAxML -b 12345 -p
12345 \
-# 500 -m PROTGAMMAWAG
```

If bootstrapping and ML tree inference are separated in time, bootstrap frequencies will need to be mapped on the ML tree (or any other tree of interest), which can be done using the SumTrees script of the Dendropy package [19]:

```
sumtrees.py --decimals=0 --percentages \
--output-tree-filepath=ML_tree_annotated500boot-
strap.tre \
--target=concat.mafft-gb1.phy.RAxML \
concat.mafft-gb1.phy.500bootstrap.RAxML
```

Partitioned models can provide much better fit to the data (see below) and thus their use is recommended for careful tree searches. A partition table for RAxML has the definition and model for each gene on a separate line:

```
WAG, Cluster5679 = 1 - 194
WAG, Cluster4143 = 195 - 732
WAG, Cluster5655 = 733 - 887
...
```

We can perform partitioned tree search and bootstrap analysis as follows:

```
raxmlHPC-PTHREADS-SSE3 -T 16 -s concat.mafft-gb1.phy -n \
concat.mafft-gb1.phy.RAxML -o bat -p 12345 -q partition.table \
-m PROTGAMMAWAG
raxmlHPC-PTHREADS-SSE3 -T 16 -s concat.mafft-gb1.phy -n \
concat.mafft-gb1.phy.500bootstrap.RAxML -b 12345 -p 12345 \
-# 500 -q partition.table -m PROTGAMMAWAG
```

3.9 Compute Species Tree Using FastTree

Alternatively, we can use FastTree (which, as its name suggests, generally runs faster) to compute an approximately maximum likelihood tree as follows:

```
FastTreeMP -wag < concat.mafft-gb1.phy > concat.mafft-gb1.phy.wag.ft
```

3.10 Compare the Trees Obtained with Different Methods

How do the trees generated with RAxML, FastTree, and ClustalW compare with the published tree from [7]? We compare the trees using the ETE Toolkit [20] as follows:

```
ete3 compare -t RAxML_bestTree.concat.mafft-gb1.phy.RAxML \
concat.mafft-gb1.phy.wag.ft concat.ph -r floudas_2012.ph \
--unrooted
```

The `-r` option sets the published tree as a reference tree to compare the others to. We see in Table 3 that the trees generated with the more computationally expensive methods (maximum likelihood RAxML and approximately maximum likelihood FastTree), are somewhat more similar to the published tree than the NJ tree, as indicated by a shorter Robinson–Foulds distance [21] and greater percentage of shared edges. Although in this case the difference between the NJ tree and the ML trees are small, the this difference can be significant for larger and/or more challenging datasets, e.g., trees with short internal branches, long branches (long branch attraction) or in the presence of rate variation across genes or branches of the tree (*see* **Note 3**). Notice that while the tree topologies from the NJ and ML analyses are mostly identical, there are some differences (*see* **Note 4**).

Table 3
Comparison of trees using ETE Toolkit

Source target tree used	RAxML	FastTree	NJ
Effective tree size used for comparisons (after pruning not shared items)	31	31	31
Normalized Robinson–Foulds distance (RF/maxRF)	0.11	0.11	0.14
Robinson–Foulds symmetric distance	6	6	8
Maximum Robinson–Foulds value for this comparison	56	56	56
Frequency of edges in target tree found in the reference (1.00 = 100% of branches are found)	0.95	0.95	0.93
Frequency of edges in the reference tree found in target (1.00 = 100% of branches are found)	0.95	0.95	0.93

4 Notes

1. The quality of genome-scale datasets for phylogenomic inference largely determines the outcome of the analyses. The aim of dataset assembly is to maximize the amount of reliable phylogenetic information but minimize noise in the datasets. Some considerations for assembling maximally informative datasets follow.

   The number of single copy genes, and thus the amount of universally available information, naturally decreases as the number of species increases, due to rarely occurring gene duplications, even in housekeeping gene families. When analyzing a large number of species (>30) gene tree-based methods for identifying suitable marker genes may yield more genes that can be used for phylogenomic reconstruction (*see* methods in [22] for details). In this case, gene trees are used to distinguish deep paralogs (genes which were duplicated prior to the last speciation events) from inparalogs (paralogs that arise in terminal nodes of the species tree, i.e., species-specific). Note that deep paralogs interfere with species tree estimation, whereas inparalogs do not—the choice of which inparalog to retain for phylogenetic analysis can be arbitrary, or can be based on their distance from the root of the tree. Collections of gene trees can be screened for deep paralogs using the scripts published in [22].

Another factor that should be considered during the assembly of phylogenomic datasets is contamination by highly divergent genes (e.g., due to sequencing errors or pseudogenes). Excessively long branches in individual gene trees can are usually signs of such contamination: a general rule of thumb (but quite liberal cutoff) is to exclude genes whose branch length accounts for >60% of the sum of all branch lengths in the gene tree [23]. Lower cutoff values will result in less contamination by divergent genes, but the cutoff should depend on the number of species in the dataset too (e.g., in a 4-species tree, if all branches are of equal length then each branch accounts for 20% of the total tree length).

The identified single-copy clusters usually show a decreasing trend in taxon occupancy: some clusters contain sequences for all species, whereas from most clusters few or more species will be missing due to functional constraints or the incompleteness of genomic assemblies and annotations [24]. Incomplete clusters are still useful for phylogenomic inference, although nonrandomly distributed missing data can compromise results [25]. Some authors use only orthologous genes in which all the species are represented, while others apply taxon occupancy cut-off. Considering there is a tradeoff between taxon occupancy and combined alignment length and that concatenated phylogenetic analyses are generally robust to even high amounts (50–80%) of missing data, when the distribution of missing data is random, we recommend a taxon occupancy cutoff of 50%—only gene families that contain genes for >50% of the total number of species will be included in the analyses. Naturally, this cutoff can be adjusted to the specific phylogenetic exercise, dataset size and availability of genomic data.

2. As an alternative to a priori concatenation of sequences, single gene alignments are often inferred first, followed by the concatenation of quality-filtered alignments. This strategy preserves gene boundaries, allows for both concatenation and summary-based phylogenetic methods (which combine data from individual gene trees to infer a species tree) to be applied to the data, and may be substantially less computationally intensive. The implementation of this alternate approach is left as an exercise to the reader. Inferring alignments for each gene can be done by MAFFT (*see* above), or by alternative approaches such as the probabilistic method PRANK [26], which is among the most accurate multiple sequence alignment software available (which comes at a cost of longer run time). Note that PRANK produces more fragmented, but generally more accurate alignments.

3. In Maximum Likelihood or Bayesian methods, the evolutionary model used to model nucleotide or amino acid substitutions is an important parameter to consider. This is particularly true for phylogenomic analyses, where biases, such as long branch attraction, can become pronounced due to poor model fit in larger datasets [25] under some circumstances. Software like jModeltest [27] and Partitionfinder [28] can be used to identify best-fit models for each gene in the dataset, although run times often limit the use of these methods and constrain the analyses to be performed with ad hoc selected models. In these cases exploring the sensitivity of the results to alternative commonly used evolutionary models is recommended. Advanced models that can account for incongruence among gene genealogies (e.g., incomplete lineage sorting) or differences in the rate of evolution across sites or in time (heterotachy) are now available for the analysis of genome-scale data [25, 29]. One very straightforward way to improve the fit of the model to the data is the use of partitioned models. Partitioning is an important factor in accounting for data heterogeneity and different evolutionary rates. Most commonly, datasets are partitioned by gene, however, other partitioning schemes, e.g., binning genes by evolutionary rate are also commonly used. Finally, while concatenation-based methods can be sensitive to certain biases, summary-based methods that combine information from individual gene trees into a species tree hold promise to evade some of these caveats [30, 31].

   The promise of phylogenomics has been to eliminate uncertainty from the reconstructions of evolutionary relationships [32]. However, this turned out to be an optimistic expectation [33] as a number of case studies reporting spurious, but strongly supported, relationships came to light. It is therefore very important to assess the robustness of the inferred relationships under multiple parameter combinations. These can include phylogeny reconstruction under multiple partitioning schemes (e.g., partitioned vs. unpartitioned), methods (ML vs. Bayesian), evolutionary models or data selection strategies. Although a detailed review of these strategies is beyond the scope of this chapter, we have highlighted several possibilities for exploring the robustness of results attained using the outlined protocols.

4. The tree topologies from the various analyses are mostly in agreement, but as is to be expected with phylogenomic analyses, there are some differences (Fig. 1). The NJ and ML trees differ in their relative placement of the three clades shown in Fig. 1. In the NJ tree (Fig. 1, panel A) clade 1, containing the orders Polyporales (*Postia placenta*, *Wolfiporia cocos*, *Fomitopsis pinicola*, *Trametes versicolor*, *Dichomitus squalens*, and

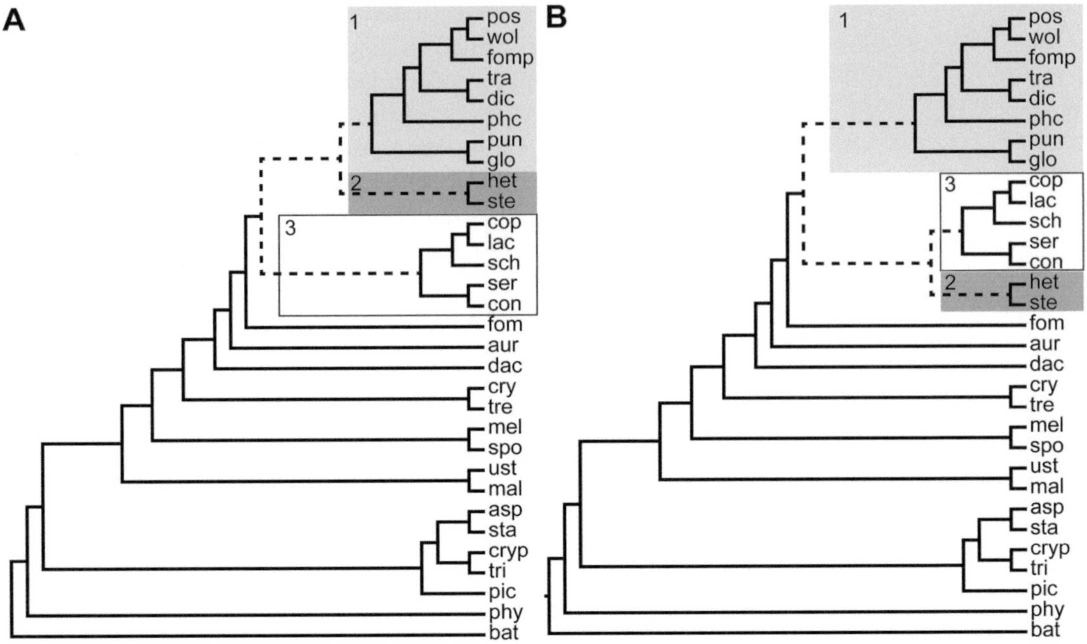

Fig. 1 Difference in tree topologies of ML and NJ trees. The branches that differ between the trees are indicated in dashed grey lines. The NJ tree (panel **a**) places clade 1 as a sister to clade 2, and clade 3 sister to them. The ML tree (panel **b**) places clade 2 as sister to clade 3, with clade 1 sister to them

Phanerochaete chrysosporium), Corticiales (*Punctularia strigosozonata*), and Gloeophyllales (*Gloeophyllum trabeum*), is a sister group (meaning that they have the same parent node) to clade 2, which consists of fungi of the order Russulales (*Heterobasidion irregulare* and *Stereum hirsutum*). Clade 3, containing the orders Agaricales (*Coprinopsis cinerea*, *Laccaria bicolor*, and *Schizophyllum commune*) and Boletales (*Serpula lacrymans* and *Coniophora puteana*) is a sister group to the clade made up of clades 1 and 2. However, in the ML analysis (Fig. 1, panel b), clades 2 and 3 are sister groups to each other, and clade 1 is sister to their combined clade. The published tree in [7], which used BEAST, a Bayesian method [34], is again slightly different, and places *P. strigosozonata* and *G. trabeum* in a separate clade sister to a clade containing the Agaricales, Boletales, Russulales, and Polyporales. The correct answer is an open research question, and the phylogeny of the fungi is continually being revised as new genomic data become available [7, 14, 22, 35, 36]. We thus see that varying the phylogenetic inference method used, data sets, and data selection strategies, can yield slightly different results. The wise researcher is advised to try them all.

Acknowledgments

The work conducted by the US Department of Energy Joint Genome Institute, a DOE Office of Science User Facility, is supported under Contract No. DE-AC02-05CH11231. LGN was supported by the "Momentum" Program LP 2014/12.

References

1. Goffeau A, Barrell BG, Bussey H et al (1996) Life with 6000 genes. Science 274. 546, 563-547
2. Lander ES, Linton LM, Birren B et al (2001) Initial sequencing and analysis of the human genome. Nature 409:860–921
3. Enright AJ, Van Dongen S, Ouzounis CA (2002) An efficient algorithm for large-scale detection of protein families. Nucleic Acids Res 30:1575–1584
4. Finn RD, Coggill P, Eberhardt RY et al (2016) The Pfam protein families database: towards a more sustainable future. Nucleic Acids Res 44:D279–D285
5. Li L, Stoeckert CJ Jr, Roos DS (2003) OrthoMCL: identification of ortholog groups for eukaryotic genomes. Genome Res 13:2178–2189
6. Tatusov RL, Fedorova ND, Jackson JD et al (2003) The COG database: an updated version includes eukaryotes. BMC Bioinformatics 4:41
7. Floudas D, Binder M, Riley R et al (2012) The Paleozoic origin of enzymatic lignin decomposition reconstructed from 31 fungal genomes. Science 336:1715–1719
8. Van Dongen S (2000) A cluster algorithm for graphs. Report-information systems:1–40
9. Thompson JD, Gibson T, Higgins DG (2002) Multiple sequence alignment using ClustalW and ClustalX. Curr Protoc Bioinformatics. https://doi.org/10.1002/0471250953.bi0203s00. 2.3. 1-2.3. 22
10. Price MN, Dehal PS, Arkin AP (2009) FastTree: computing large minimum evolution trees with profiles instead of a distance matrix. Mol Biol Evol 26:1641–1650
11. Stamatakis A (2014) RAxML version 8: a tool for phylogenetic analysis and post-analysis of large phylogenies. Bioinformatics 30: 1312–1313
12. Faircloth BC, McCormack JE, Crawford NG, Harvey MG, Brumfield RT, Glenn TC (2012) Ultraconserved elements anchor thousands of genetic markers spanning multiple evolutionary timescales. Syst Biol 61:717–726
13. Katoh K, Standley DM (2013) MAFFT multiple sequence alignment software version 7: improvements in performance and usability. Mol Biol Evol 30:772–780
14. James TY, Kauff F, Schoch CL et al (2006) Reconstructing the early evolution of fungi using a six-gene phylogeny. Nature 443:818–822
15. Castresana J (2000) Selection of conserved blocks from multiple alignments for their use in phylogenetic analysis. Mol Biol Evol 17:540–552
16. Capella-Gutierrez S, Silla-Martinez JM, Gabaldon T (2009) trimAl: a tool for automated alignment trimming in large-scale phylogenetic analyses. Bioinformatics 25:1972–1973
17. Misof B, Misof K (2009) A Monte Carlo approach successfully identifies randomness in multiple sequence alignments: a more objective means of data exclusion. Syst Biol 58:21–34
18. Saitou N, Nei M (1987) The neighbor-joining method: a new method for reconstructing phylogenetic trees. Mol Biol Evol 4:406–425
19. Sukumaran J, Holder MT (2010) DendroPy: a python library for phylogenetic computing. Bioinformatics 26:1569–1571
20. Huerta-Cepas J, Serra F, Bork P (2016) ETE 3: reconstruction, analysis, and visualization of phylogenomic data. Mol Biol Evol 33:1635–1638
21. Robinson DF, Foulds LR (1981) Comparison of phylogenetic trees. Math Biosci 53:131–147
22. Nagy LG, Riley R, Tritt A et al (2016) Comparative genomics of early-diverging mushroom-forming fungi provides insights into the origins of lignocellulose decay capabilities. Mol Biol Evol 33:959–970
23. dos Reis M, Inoue J, Hasegawa M, Asher RJ, Donoghue PC, Yang Z (2012) Phylogenomic datasets provide both precision and accuracy in estimating the timescale of placental mammal phylogeny. Proc Biol Sci 279:3491–3500
24. Parra G, Bradnam K, Korf I (2007) CEGMA: a pipeline to accurately annotate core genes in eukaryotic genomes. Bioinformatics 23:1061–1067

25. Philippe H, Brinkmann H, Lavrov DV et al (2011) Resolving difficult phylogenetic questions: why more sequences are not enough. PLoS Biol 9:e1000602
26. Loytynoja A (2014) Phylogeny-aware alignment with PRANK. Methods Mol Biol 1079:155–170
27. Posada D (2008) jModelTest: phylogenetic model averaging. Mol Biol Evol 25:1253–1256
28. Lanfear R, Calcott B, Ho SY, Guindon S (2012) PartitionFinder: combined selection of partitioning schemes and substitution models for phylogenetic analyses. Mol Biol Evol 29:1695–1701
29. Lartillot N, Philippe H (2004) A Bayesian mixture model for across-site heterogeneities in the amino-acid replacement process. Mol Biol Evol 21:1095–1109
30. Mirarab S, Reaz R, Bayzid MS, Zimmermann T, Swenson MS, Warnow T (2014) ASTRAL: genome-scale coalescent-based species tree estimation. Bioinformatics 30:i541–i548
31. Szollosi GJ, Tannier E, Daubin V, Boussau B (2015) The inference of gene trees with species trees. Syst Biol 64:e42–e62
32. Gee H (2003) Evolution: ending incongruence. Nature 425:782
33. Jeffroy O, Brinkmann H, Delsuc F, Philippe H (2006) Phylogenomics: the beginning of incongruence? Trends Genet 22:225–231
34. Drummond AJ, Suchard MA, Xie D, Rambaut A (2012) Bayesian phylogenetics with BEAUti and the BEAST 1.7. Mol Biol Evol 29:1969–1973
35. Binder M, Justo A, Riley R et al (2013) Phylogenetic and phylogenomic overview of the Polyporales. Mycologia 105:1350–1373
36. Hibbett DS, Binder M, Bischoff JF et al (2007) A higher-level phylogenetic classification of the fungi. Mycol Res 111:509–547

Chapter 21

Phylogenetic Analysis of Protein Family

Letian Song, Sherry Wu, and Adrian Tsang

Abstract

With the number of sequenced genomes increasing rapidly, it is impractical to perform functional and structural analyses on all individual proteins. Phylogenetic analysis employs a combination of molecular and statistical approaches to infer or estimate relationships among individuals. It provides a credible method to explore the relationship between sequence similarity and function of proteins belonging to the same family. This chapter describes a standardized framework of phylogenetic analysis to study large protein families. Bioinformatic approaches and online tools used in phylogenetic analyses are presented.

Key words Phylogenetic analysis, Protein family, Pfam domain, Protein sequence searching, Multiple sequence alignment, Phylogenetic tree, Subfamily cluster

1 Introduction

Molecular phylogeny examines evolutionary relationships of biomolecules such as genes and proteins. In phylogenetic analyses, changes occurred in molecular sequences are evaluated and their correlations are intuitively illustrated in tree-like diagrams. Based on phylogenetic tree topologies, sequences belonging to the same family can be further divided into subfamilies which provide information on evolutionary relationship and functional diversity within the family. In addition, by mapping existing experimental data of biochemically characterized proteins onto the family phylogenetic tree, one can explore the correlation between sequence similarity and biochemical properties to gain insight on the structure and the function of each subfamily [1]. It is hoped that, by incorporating experimental data, phylogenetic tree can be used as a prediction tool to assign function unambiguously to uncharacterized members of a protein family. In addition, by mapping existing experimental data of biochemically characterized proteins onto the family phylogenetic tree, one can easily identify subfamilies that lack biochemically characterized members [2]. In this case, phylogenetic analysis can be used as a screening tool to select target proteins

Ronald P. de Vries et al. (eds.), *Fungal Genomics: Methods and Protocols*, Methods in Molecular Biology, vol. 1775, https://doi.org/10.1007/978-1-4939-7804-5_21, © Springer Science+Business Media, LLC, part of Springer Nature 2018

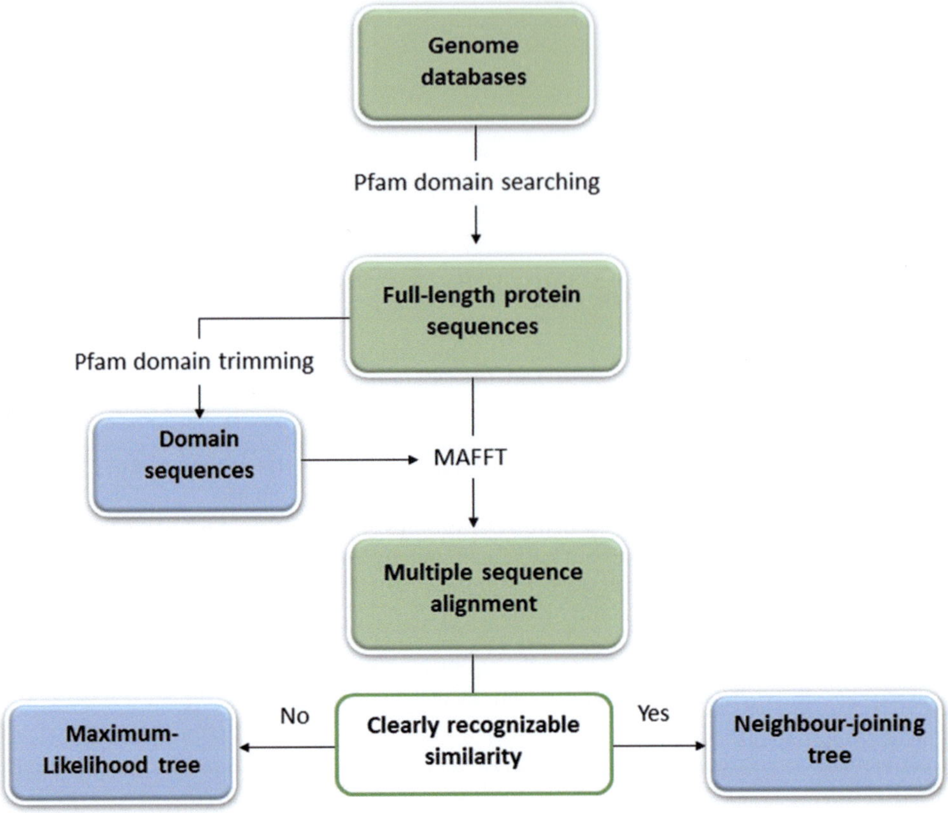

Fig. 1 Processing pipeline of phylogenetic analysis of protein family

from uncharacterized subfamilies for further study. In this chapter, we provide a standardized framework for phylogenetic analysis of protein families (Fig. 1).

2 Materials

2.1 Open-Source Databases for Protein Sequence Searching

1. MycoCosm (http://jgi.doe.gov/fungi) [3] and NCBI (https://www.ncbi.nlm.nih.gov/gene) databases are used to collect protein data from sequenced fungal genomes.
2. MycoCLAP database (https://mycoclap.fungalgenomics.ca/mycoCLAP) [1] is used to collect protein data of fungal carbohydrate-active enzymes (CAZyme) that have been biochemically characterized based on experimental evidence reported in published literature.

2.2 Automated Tools for Domain Identifying and Trimming

The PfamScan (http://www.ebi.ac.uk/Tools/pfa/pfamscan) is used to identify protein Pfam domain of protein sequences [4]. On the basis of PfamScan output, a perl script can be created for trimming the domain sequences associated with a specific Pfam ID.

2.3 Tools for Multiple Sequence Alignment

1. MAFFT (Multiple Alignment using Fast Fourier Transform, http://www.ebi.ac.uk/Tools/msa/mafft) is used for multiple sequence alignment. The advantages of MAFFT method are described in **Note 1**.
2. The free software of Jalview Desktop (download at http://www.jalview.org/download) is used to manually edit the multiple sequence alignment [5].

2.4 Phylogenetic Tree Construction and Analysis

1. RAxML (Randomized Axelerated Maximum Likelihood) [6] is used for maximum-likelihood phylogeny estimation (*see* **Note 2**).
2. The phylogeny module integrated in MEGA6 software [7] is used to construct neighbor-joining phylogenetic trees (*see* **Note 3**).
3. FigTree (http://tree.bio.ed.ac.uk/software/figtree) is used to graphically analyze phylogenetic trees.

3 Methods

As shown in Fig. 1, a standardized framework was established for the phylogenetic analysis of protein families. In brief, protein sequences are retrieved from annotated genomes in MycoCosm and NCBI databases using Pfam domain for searching. In addition, sequences of characterized fungal proteins are obtained from MycoCLAP database. Once all the sequences are retrieved, they are trimmed to their domain limit. The method of MAFFT is used for multiple sequence alignment as it has shown to be more accurate and less time-consuming [8]. Multiple sequence alignment profile is examined manually and sequences missing conserved residues or motifs are removed to improve the quality of the dataset. Finally, two methods used to construct phylogenetic trees are presented.

3.1 Sequence Search of MycoCosm Database

1. Pfam domain search on MycoCosm. Open MycoCosm portal, click the word "Fungi" shown in the fungal taxonomy tree to include all available fungal genomes in MycoCosm database (*see* **Note 4**), then click "Search" in the pop-up dialog box. To perform Pfam domain search (*see* **Note 5**), as shown in Fig. 2, enter the Pfam domain ID in the "Search" box (*see* **Note 6**), select "PFAM Terms" from the drop-down list of "Search By," and set other parameters as default.
2. Protein sequence retrieval from MycoCosm. On the result page select "As protein FASTA" from the drop-down list of "Download" to export all resulted protein sequences.

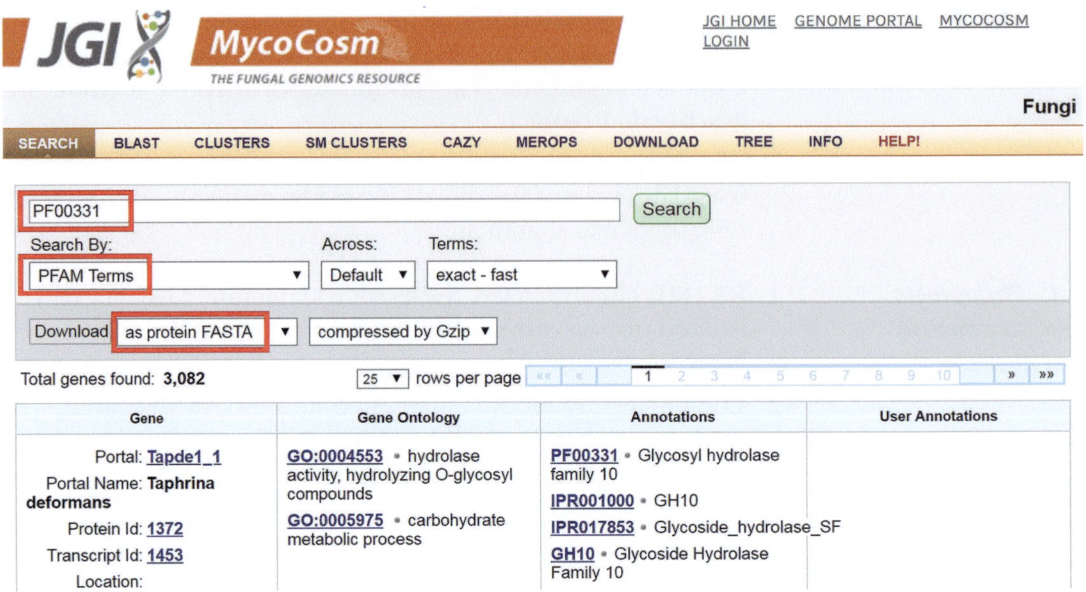

Fig. 2 Page of Pfam domain search and protein sequence download on MycoCosm website, using PF00331 as example

3.2 Sequence Search of NCBI Database

1. Pfam domain search on NCBI database using a web browser. On the home page of NCBI Gene database, enter the Pfam domain ID of interest in the search box and click on "Search." All sequences containing the queried Pfam domain are listed in the result page. To access fungal genes that have been annotated from sequenced genomes, as shown in Fig. 3, in the left of page check "Genomic" and "Annotated genes" under the filters of "Gene sources" and "Categories" respectively, and in the right of page click "Fungi" presented in the list of "Taxonomic Groups."
2. Protein sequence retrieval from NCBI database. To obtain protein sequences of genes retrieved from the above step, in the right corner of the page (Fig. 3) set the "Database" drop-down menu to "Protein" and "Refseq Proteins" in the drop-down list of "Option." By clicking on "Find items," the related protein sequences recorded in NCBI Refseq database are listed in the resulting page. To download the complete set in FASTA format, click "Send to" at the upper left: select "File," and "FASTA" in the drop-down list of "Format."

3.3 Sequence Search of MycoCLAP Database

1. To search a specific set of characterized proteins from entire database, keyword searching could be carried out by entering query in the "Term" filed on search toolbar, and click "Search" to display the data table. The search results can be further customized by selecting the features provided in drop-down menu of table "Entries."

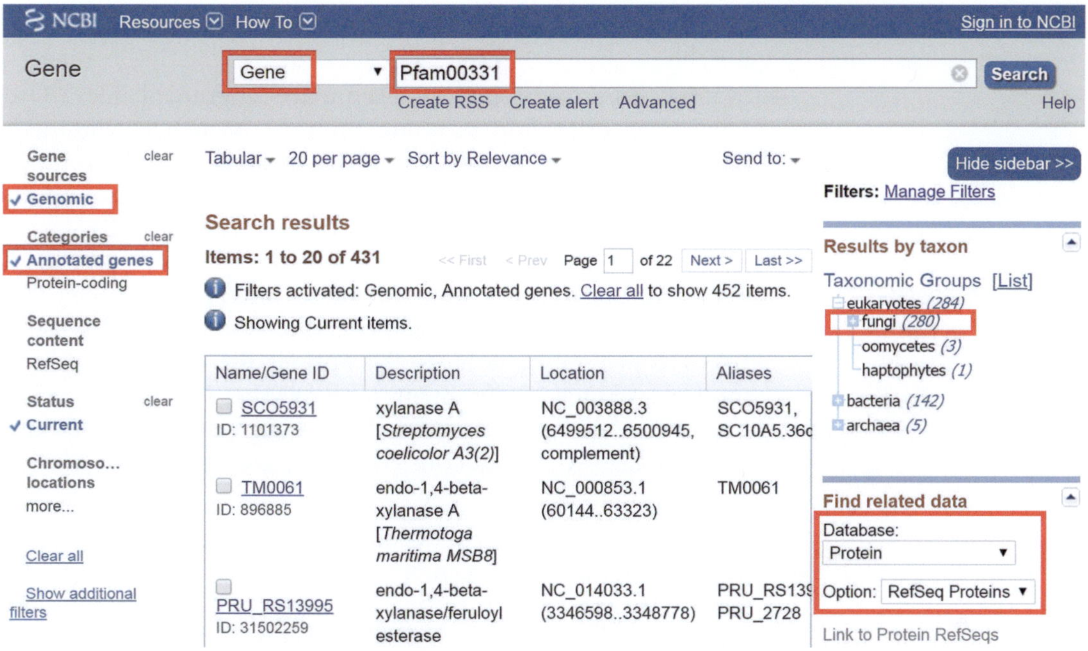

Fig. 3 Page of Pfam domain search on NCBI website, using Pfam00331 as example

2. Retrieve the sequence of selected proteins by clicking on "FASTA" button at the bottom of data table.

3.4 Multiple Sequence Alignment

1. Combine sequences retrieved from each database into one file. Before performing multiple sequence alignment, rename proteins in a standardized way and remove duplicate sequences and/or sequences with duplicate names (*see* **Note 7**).
2. Domain sequence extraction. This step can be ignored if using full-length protein sequences for multiple sequence alignment. Nevertheless, multiple sequence alignment generated using protein domains is more accurate than using entire protein sequences as less ambiguously aligned sites are produced. Domain limits can be identified by using the PfamScan tool. On the webpage of PfamScan, enter the dataset of full protein sequences in FASTA format, choose "Plain Text" as output format and click "Run" to start analysis. In the output result, the data of "alignment start" and "alignment end" would be used as domain limit to trim domain sequences according to the Pfam domain ID of interest. Save the trimmed domain sequences as a FASTA format file.
3. Multiple sequence alignment by MAFFT. MAFFT is used for generating multiple sequence alignment (*see* **Note 1**). Open MAFFT website, enter either complete protein sequences or trimmed domain sequences, select "Personal/FASTA" as the

output format, and save the alignment results in a FASTA format file.

4. Manual editing of multiple sequence alignment files. Use Jalview to open and examine multiple sequence alignment files. To improve the quality of the dataset, carefully remove some gap positions and ambiguously aligned regions, and do not change conserved residues and motifs (*see* **Note 8**).

3.5 Phylogenetic Tree Construction

1. Construction of maximum-likelihood phylogenetic tree. The maximum-likelihood phylogenetic tree is generated by using RAxML program (*see* **Note 9**). Submit the dataset of edited multiple sequence alignment to RAxML. The parameter of "BLOSUM62" model and a "bootstrap value of 1000" are used to carry out tree construction. Save the output tree in NWK format file.
2. Construction of neighbor-joining phylogenetic tree. The neighbor-joining phylogenetic tree is constructed by using software MEGA6. Upload the file containing edited multiple sequence alignment in MEGA6, and input as "Protein Sequences" for analysis. Click "Construct/Test Neighbour-Joining Tree" on the menu of "Phylogeny", then in the pop-up window, set parameters as "Bootstrap" method with "1000 replications," "*p*-distance" model, and "Pairwise deletion" of gaps to generate the tree. Export the tree file in NWK format.
3. Tree visualization and manipulation. Open tree file in FigTree. Format the tree in midpoint rooting (*see* **Note 10**) by checking the box of "Root Tree" on the submenu of "Trees" in the left menus of the page. The clusters or subfamilies could be initially determined by examining branch bootstrap values and the corresponding multiple sequence alignment profiles. For example, in the case of analyzing a tree including 800 xylanases of Glycoside Hydrolase 10 (GH10) family [2], a subfamily is assigned if it includes three or more sequences and is supported by 55% or more of the bootstrap replicates. Figtree also affords useful capabilities including scale, check node labeling and coloured appearance, etc. Finally, the publication-quality figures could be exported as a pdf or any other graphic format.

4 Notes

1. MAFFT is a popular program employing iterative approach to refine and improve the quality of alignment results from initial progressive alignment [9]. The advantages and disadvantages of MAFFT to other widely used multiple sequence alignment tools are summarized in Table 1.

Table 1
Comparison of different multiple sequence alignment tools. The ability of aligning the three conserved residues of the globin family was used to evaluate the accuracy

Program	Algorithm approach	Advantages	Disadvantages
ClustalW	Progressive	Fast	Unable to make a correction once a misalignment is introduced; does not guarantee optimal alignment; only works well for closely related sequences
MUSCLE	Iterative	Fast; able to correct misaligned position through iterative refinement steps	Less accurate; unable to align conserved residues of distantly related sequences
MAFFT	Iterative	Fast; accurate; able to correct misaligned position through iterative refinement steps; external sequences are included to obtain a more accurate alignment; refinement step also includes consistency-based score	None
T-coffee	Consistency based	Accurate; pairwise alignment score is supported by evidence from multiple sequences; both global and local alignment are assessed	Slow

2. Maximum-likelihood and neighbor-joining are two of the most common methods to build phylogenetic trees. Maximum-likelihood is a character-based method which examines character (nucleotide for DNA sequences and amino acid for protein sequences) at every single site of the multiple sequence alignment to assess the reliability of each position on the basis of all other positions. Maximum-likelihood method compares alternative tree phylogenies based on a predefined criterion to search for the optimal tree topology under that criterion. Maximum-likelihood method is informative and gives high confidence scores. However, it is a slow method that requires intensive computational calculation.

3. Neighbor-joining is a distance-based method which uses pairwise distance that the pair yields the smallest sum of branch lengths is evaluated as the closest neighbor and joined together. Neighbor-joining method is faster than maximum-likelihood method; however, it does not fully utilize the detailed alignment especially for distantly related sequences that might yield a biased tree. Under some conditions, for a small datasets neighbor-joining may give a better performance of local topology than maximum-likelihood. It is recommended to try multiple methods to build a tree. The selection of algorithm

depends on time, accuracy and interpretability. More information of different algorithms could be found in the paper by Ogden et al. [10].

4. Comparing to NCBI genome database, MycoCosm is a comprehensive repository for fungal genomes and that it has the advantage that functional annotation is predicted using the same annotation pipeline.
5. The advantage of Pfam domain search over BLAST search is that relatively diverged paralogs within each genome could be detected and included in the analysis [11]. On the other hand, dataset collected from BLAST search may differ depending on the query used. Moreover, a wide taxa spectrum could be easily covered by searching genome databases using Pfam domains.
6. To search in MycoCosm, the Pfam domain ID is written as PFxxxxx (x represents the number).
7. Duplicate IDs can be detected by an online tool (http://www.somacon.com/p568.php). Paste the sequence file in the form, and sort counts by line. The output value for duplicate or multiplicate is ≥2.
8. The deletion of gaps depends on how similar the sequences that are analyzing. It is recommended to delete columns with >90% of gap. However, retain the gap positions if the gap-related characters are phylogenetically informative. Some automated programs (e.g., Gblocks and trimAl) would help, but they are not always correct.
9. The code of RAxML is available at https://github.com/stamatak/standard-RAxML [6].
10. The midpoint rooting places the root of the tree at the midpoint of the longest path between any two tips and represents the ancestral point.

References

1. Murphy C, Powlowski J, Wu M, Butler G, Tsang A (2011) Curation of characterized glycoside hydrolases of fungal origin. Database 2011:bar020
2. Wu S (2015) Comprehensive bioinformatic analysis of glycoside hydrolase family 10 proteins. Dissertation, Concordia University
3. Cuomo CA, Birren BW (2010) Chapter 34–the fungal genome initiative and lessons learned from genome sequencing. In: Methods in enzymology. Academic Press, Cambridge, MA, pp 833–855
4. Mistry J, Bateman A, Finn RD (2007) Predicting active site residue annotations in the Pfam database. BMC Bioinformatics 8:298
5. Waterhouse AM, Procter JB, Martin DMA, Clamp M, Barton GJ (2009) Jalview version 2—a multiple sequence alignment editor and analysis workbench. Bioinformatics 25:1189–1191
6. Stamatakis A (2014) RAxML version 8: a tool for phylogenetic analysis and post-analysis of large phylogenies. Bioinformatics 30:1312–1313
7. Tamura K, Stecher G, Peterson D, Filipski A, Kumar S (2013) MEGA6: molecular evolutionary genetics analysis version 6.0. Mol Biol Evol 30:2725–2729
8. Katoh K, Misawa K, Kuma K, Miyata T (2002) MAFFT: a novel method for rapid

multiple sequence alignment based on fast Fourier transform. Nucleic Acids Res 30:3059–3066

9. Notredame C (2002) Recent progress in multiple sequence alignment: a survey. Pharmacogenomics 3:131–144

10. Ogden TH, Rosenberg MS (2006) Multiple sequence alignment accuracy and phylogenetic inference. Syst Biol 55:314–328

11. Punta M, Coggill PC, Eberhardt RY et al (2012) The Pfam protein families database. Nucleic Acids Res 40:D290–D301

Chapter 22

Parasexual Crossings for Bulk Segregant Analysis in *Aspergillus niger* to Facilitate Mutant Identification Via Whole Genome Sequencing

Mark Arentshorst and Arthur F. J. Ram

Abstract

The industrially important fungus *Aspergillus niger* is known to reproduce only asexually. The parasexual cycle of fungi can be used for crossing two different strains to produce segregants or progeny with combined mutations even in fungi without a known sexual cycle. In *A. niger*, the parasexual cycle has been extensively used to establish linkage groups and to generate genetic maps. With the advent of whole genome sequencing, the parasexual cycle has received renewed attention as a method to create segregants for bulk segregant analysis. Bulk segregant analysis is a genetic technique used to link and ultimately identify the mutation associated with a particular phenotype. In this chapter we describe the procedure for setting up parasexual crossings in *A. niger*. The segregants obtained with this method can be used in combination with next-generation sequencing to map mutations in the organism.

Key words Asexual reproduction, Mitotic recombination, Filamentous fungi, Parasexual cycle, Haploidization, Diploid, Auxotrophic markers, Color markers

1 Introduction

Much of our understanding of the molecular mechanisms that allow filamentous fungi—including *Aspergillus nidulans* and *Neurospora crassa*—to grow and develop is derived from classical mutant screens. Examples in fungal biology include the discovery of tubulin genes [1], cell cycle-related genes [2] and developmental genes [3]. The genetic characterization of a mutant requires complementation assays and the construction of high-quality genomic libraries which are time-consuming. Another limitation of the traditional complementation approach is that the restoration of the original phenotype has to be selectable, since thousands of transformants need to be analyzed. Whole genome sequencing offers a new and faster way to directly identify the mutation in a mutant strain.

Ronald P. de Vries et al. (eds.), *Fungal Genomics: Methods and Protocols*, Methods in Molecular Biology, vol. 1775,
https://doi.org/10.1007/978-1-4939-7804-5_22,

Most mutant screens use UV radiation or chemical mutagens (e.g. ethyl methane sulfonate) to introduce mutations. However, these mutagens have the drawback of introducing additional mutations that are not related to the mutant phenotype increasing the difficulty of analysis. To determine which mutation is responsible for the phenotype, different approaches can be used, including (1) performing a genetic linkage group analysis to locate the mutation, followed by inspection of mutations within this region [4], (2) repetitive backcrossing of mutants to outcross unrelated mutations [5], (3) sequencing of several mutants belonging to the same complementation group [6], or (4) bulk segregant analysis [7, 8]. In bulk segregant analysis, the mutant of interest is crossed with a wild-type strain to form a diploid strain. The diploid is subsequently induced to undergo meiosis (suitable for fungi with a known sexual cycle) or forced to lose one set of chromosomes via the parasexual cycle (for fungi without a sexual cycle). Both approaches result in haploid segregants which are then phenotypically screened for the phenotype to obtain a pool of segregants with a wild-type phenotype or a mutant phenotype. Genomic DNA from the pool of wild-type segregants and the pool of mutants is isolated and separately sequenced using deep sequencing techniques. The two parental strains used for the cross are also sequenced, and their genomic DNA sequences are compared to identify mutations between the parents. The DNA of the wild-type and mutant segregant pools is subsequently analyzed for the presence and ratio of the mutations identified between the parents. The mutation related to the phenotype will be present in all of the progeny (homozygous), while mutations that are not related to the phenotype have a 50% chance of being present in the genomic DNA of the pools (heterozygous). Mutations that are located close to the mutation of interest will cosegregate and separate only via mitotic recombination. These mutations are expected to be highly conserved in the pool of segregants but not necessarily completely conserved since mitotic cross-over events may allow their exchange.

Aspergillus niger is a biotechnologically important filamentous fungus known for its production of organic acids and enzymes [9, 10]. *A. niger* reproduces solely asexually through the high level production of melanized black conidia. The biosynthetic pathway for melanin production has been partially elucidated, and color mutants in three complementation groups have been described [11]. These complementation groups include *fwnA*, which encodes a polyketide synthase, *olvA* that codes for a hydrolyase involved in heptaketide processing, and *brnA* whose product is a multicopper oxidase thought to be involved in the polymerization of the polyketide precursors [11]. Targeted deletion or loss of function mutations in the *fwnA*, *olvA* or *brnA* genes give rise to fawn-, olive-, or brown-colored conidia, respectively [11].

Despite the lack of a sexual cycle, a genetic linkage analysis in *A. niger* has been developed initially by Pontecorvo [12] and subsequently by Bos and coworkers [13–15] using the parasexual cycle to generate genetic maps. The parasexual cycle is an alternative, nonsexual mechanism that allows recombination without meiosis in compatible strains or the formation of specialized developmental structures. The first step in setting-up parasexual crossings is to fuse hyphae from two different strains to form a heterokaryotic mycelium (Fig. 1a, b). Spontaneous fusion of the two different nuclei produces a diploid (Fig. 1c). The diploid is relatively stable, but haploidization can be induced by adding low concentrations of tubulin destabilizing agents such as benomyl (Fig. 1d). During haploidization, mitotic crossing-over can lead to the exchange of genes on homologous chromosomes. Like the sexual cycle, parasexuality affords the organism the chance to recombine its genome and produce offspring with novel genotypes (Fig. 1e). Unlike a sexual cycle, recombination is exclusively mitotic.

In this chapter, we describe a method for parasexual crossings in *A. niger*. The procedure generates diploid strains that can be used to obtain segregants. These segregants can be used in bulk segregant approaches to identify the genetic basis of the phenotype of interesting mutants.

2 Materials

Prepare all solutions and buffers in demineralized water and analytical grade reagents unless indicated otherwise.

2.1 Media

1. ASPA+N (50×): For 1 L, add 500 mL water to a 1-L graduated cylinder and start mixing. Weigh 297.5 g $NaNO_3$, 26.1 g KCl, and 74.8 g KH_2PO_4 and transfer slowly to the cylinder. Stir until fully dissolved. Adjust the pH to 5.5 with 5 M KOH. Add water up to 1 L, transfer to a 1 L bottle and autoclave.
2. 1 M $MgSO_4$ (500×): For 1 L, add 500 mL water to a 1-L graduated cylinder and start mixing. Weigh 246.5 g $MgSO_4 \cdot 7H_2O$ and transfer slowly to the cylinder. Stir until fully dissolved. Add water up to 1 L, transfer to a 1 L bottle and autoclave.
3. Trace element solution (1000×): For 1 L, add 800 mL water to a 1-L graduated cylinder and start mixing. Weigh 10.0 g EDTA, 4.4 g $ZnSO_4 \cdot 7H_2O$, 1.01 g $MnCl_2 \cdot 4H_2O$, 0.32 g $CoCl_2 \cdot 6H_2O$, 0.32 g $CuSO_4 \cdot 5H_2O$, 0.22 g $(NH_4)_6Mo_7O_{24} \cdot 4H_2O$, 1.1 g $CaCl_2$, and 1.0 g $FeSO_4 \cdot 7H_2O$ and transfer to the cylinder. Stir until fully dissolved. Adjust the pH to 4.0 with 1 M HCl and 1 M NaOH. Add water up to 1 L, aliquot

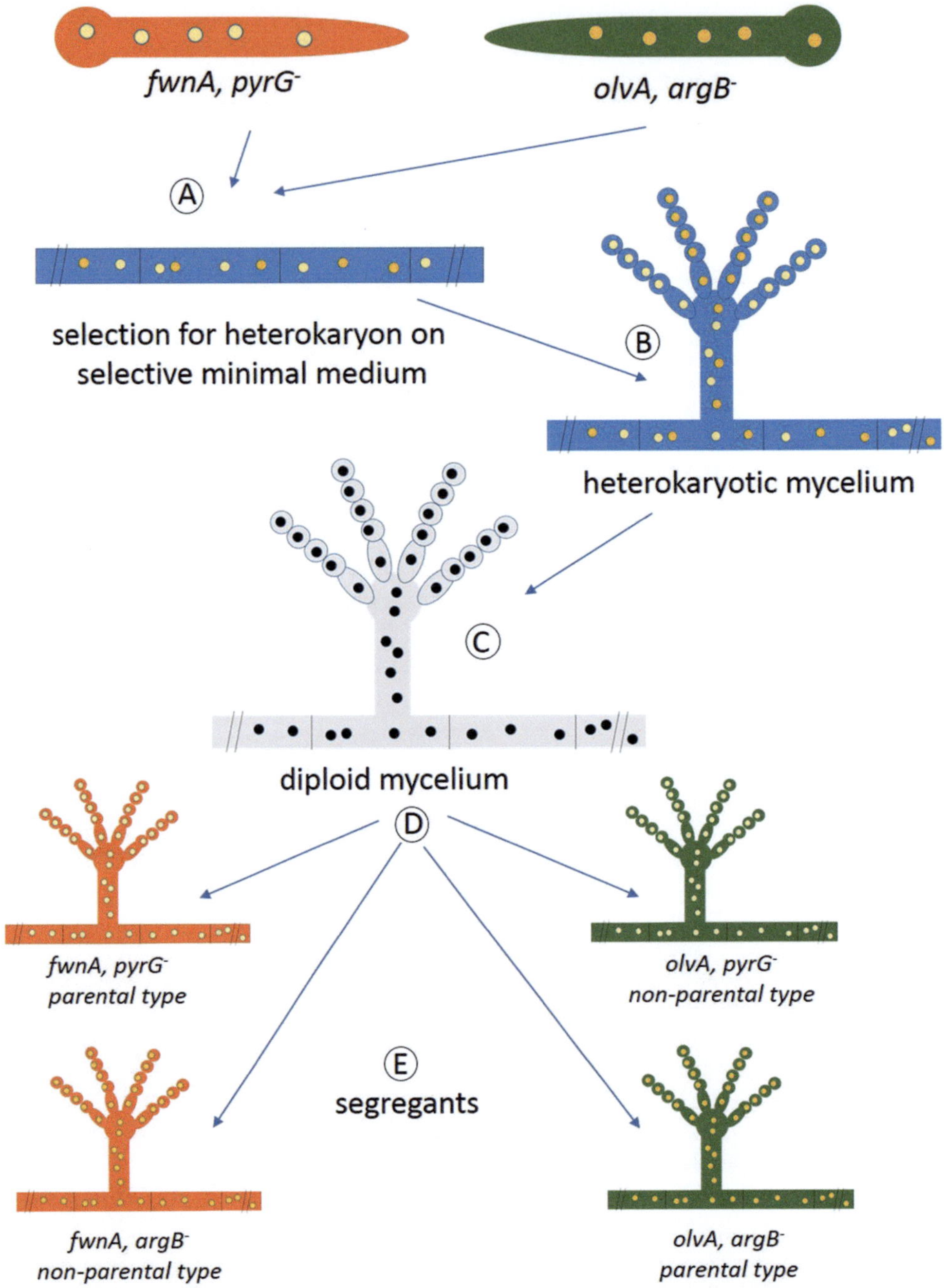

Fig. 1 Schematic overview of the parasexual cycle of *A. niger*. (**a** and **b**) Complementary strains (both for the color markers as well as for auxotrophies) are grown together on MM which selects for the formation of a heterokaryotic mycelium. (**c**) Spores from a heterokaryotic mycelium are inoculated on MM to select for diploids which produce only black spores. (**d**) Spores from a diploid strain can be forced to become haploid again by growing the diploid in the presence of benomyl. (**e**) Unlinked markers will be randomly distributed among the segregants, giving parental and nonparental types of segregants

into 100 mL bottles and autoclave. The color of the solution turns from green into purple within 2 weeks.

4. 1 M uridine (100×): For 250 mL, add 150 mL warm (~50 °C) water to a 250-mL graduated cylinder. Weigh 61.05 g uridine and transfer slowly to the cylinder. Stir until fully dissolved. Add water up to 250 mL, sterilize the solution by filtration (0.22 μm filter) and store at 4 °C.
5. Arginine (2%) (100×): For 250 mL, add 150 mL water to a 250-mL graduated cylinder. Weigh 5.0 g L-arginine and transfer to the cylinder. Stir until fully dissolved. Add water up to 250 mL, sterilize the solution by filtration (0.22 μm filter) and store at 4 °C.
6. Benomyl (10 mg/mL): For 10 mL, weigh 100 mg benomyl (methyl 1-(butylcarbamoyl)-2-benzimidazolecarbamate, 381586, Aldrich) and dissolve in 10 mL of 96% ethanol. Make aliquots of 250 μL and store at −20 °C.
7. Complete medium (CM): For 500 mL, add 400 mL water to a 500 mL graduated cylinder. Weigh 5.0 g D-glucose, 0.5 g Bacto casamino acids, and 2.5 g yeast extract and transfer to the cylinder. Stir until fully dissolved. Add 10 mL ASPA+N, 1 mL 1 M $MgSO_4$ and 0.5 mL trace element solution (1000×). Adjust the pH to 5.8 with 1 M HCl and 1 M NaOH. Add water up to 500 mL, transfer to a 500-mL bottle and autoclave. When required, add 5 mL of 1 M uridine and/or 5 mL of L-arginine (2%) after autoclaving.
8. Minimal medium (MM) + agar: For 400 mL, add 300 mL water to a 500 mL graduated cylinder. Weigh 4.0 g D-glucose and transfer to the cylinder. Stir until fully dissolved. Add 8 mL ASPA+N, 0.8 mL 1 M $MgSO_4$, and 0.4 mL trace element solution (1000×). Adjust the pH to 5.8 with 1 M HCl or 1 M NaOH. Add water up to 400 mL and transfer to a 500-mL bottle. Add 6.0 g of bacteriological agar (Scharlau, 07-004-500) and autoclave. When required, add 4 mL of 1 M uridine and/or 4 mL of arginine (2%) after autoclaving.
9. Saline solution (0.9% NaCl): For 1 L, add 800 mL water to a 1 L graduated cylinder and start mixing. Weigh 9.0 g NaCl and transfer to the cylinder. Stir until fully dissolved. Add water up to 1 L, transfer to a 1-L bottle and autoclave.
10. Complete medium (CM) + agar + benomyl: For 400 mL, add 300 mL water to a 500 mL graduated cylinder. Weigh 4.0 g D (+)-glucose, 0.4 g Bacto casamino acids, and 2.0 g yeast extract and transfer to the cylinder. Stir until fully dissolved. Add 8 mL ASPA+N, 0.8 mL 1 M $MgSO_4$ and 0.4 mL trace element solution (1000×). Adjust the pH to 5.8 with 1 M HCl or 1 M NaOH. Add water up to 400 mL and transfer to a 500-mL bottle. Add 6.0 g of bacteriological agar (Scharlau,

07-004-500) and autoclave. After autoclaving, add 24 μL benomyl (10 mg/mL) (final concentration of benomyl is 0.6 μg/mL). When required, add 4 mL of 1 M uridine and/or 4 mL of arginine (2%).

11. RNAse (10 mg/mL): For 10 mL, dissolve 100 mg RNAse A in 10 mL water, make aliquots of 500 μL and store at −20 °C.
12. DNA extraction buffer + RNAse: For 1 L, add 800 mL water to a 1-L graduated cylinder. Weigh 5.0 g SDS (sodium dodecyl sulfate), 1.21 g Tris and 37.2 g EDTA and transfer to the cylinder. Stir until fully dissolved. Adjust the pH to 8.0 with 1 M HCl or 1 M NaOH. Add water up to 1 L, transfer to a 1 L bottle and autoclave. Before use, add 2 μL RNAse (10 mg/mL) per mL of DNA extraction buffer.
13. Ultrapure phenol–chloroform–isoamyl alcohol (25:24:1, v/v).
14. Sodium acetate (3 M, pH 6.0): For 250 mL, add 150 mL warm (~50 °C) water to a 250-mL graduated cylinder. Weigh 102.06 g $CH_3COONa \cdot 3H_2O$ (sodium acetate trihydrate) and transfer slowly to the cylinder. Stir until fully dissolved. Adjust the pH to 6.0 with 1 M HCl and 1 M NaOH. Add water up to 250 mL, transfer to a 250-mL bottle and autoclave.
15. Isopropanol.
16. Ethanol (70%): For 100 mL, mix 73 mL of 96% ethanol with 27 mL of water.

3 Methods

3.1 Selection of Heterokaryons

1. Add 500 μL of CM plus the required supplements to a sterile Eppendorf tube. Inoculate equal amounts of spores (~10^5 spores) of the two strains used to set up the cross (Fig. 2a, b). Allow the strains to grow for 24 h at 30 °C (*see* **Note 1**).
2. After growth a mycelial mat will have formed on the surface of the culture medium. With sterile toothpicks, transfer the mycelial mat from the Eppendorf tube to a sterile surface such as the lid of a plastic petri dish. Using two sterile toothpicks, tear the mycelial mat into nine small pieces and transfer them to an MM + agar plate. The nine pieces can be placed onto a single plate, divided into a three-by-three grid. Incubate at 30 °C for 5–7 days until growing sectors are visible (Fig. 2c).
3. Isolate spores from a heterokaryotic mycelium using a wet cotton swab and transfer the spores to 10 mL saline solution by rotating the cotton swab in the saline solution to release the spores. Repeat until most spores have been harvested from the plate (*see* **Note 2**).

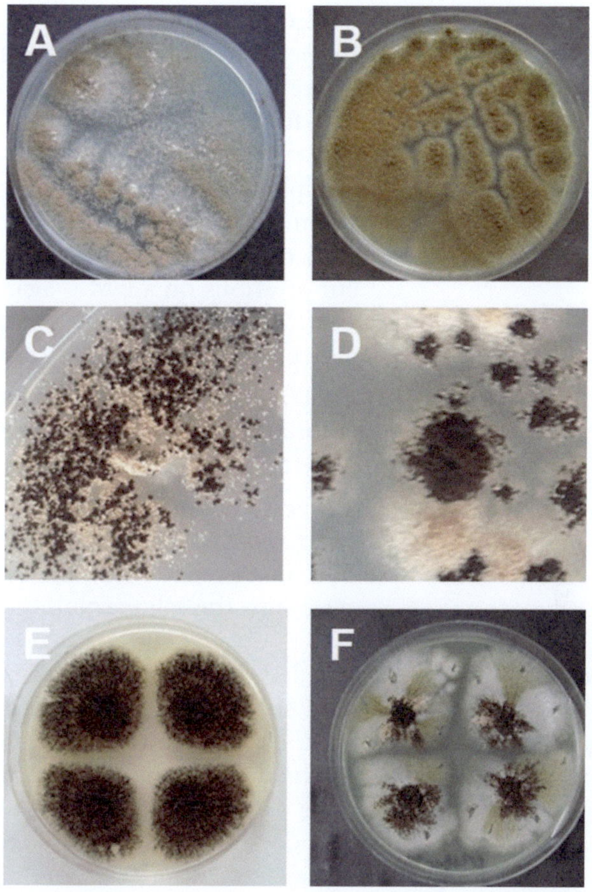

Fig. 2 The parasexual cycle of *A. niger*. (**a**) parental strain 1 (*fwnA, pyrG*⁻), (**b**) parental strain 2 (*olvA, argB*⁻), (**c**) heterokaryotic mycelia selected on MM + agar, (**d**) spontaneous diploid after plating out spores from a heterokaryon, (**e**) diploid point inoculated on CM + agar, (**f**) diploid point inoculated on CM + agar + benomyl (0.6 μg/mL)

3.2 Selection of Diploid Strain

1. Plate different amounts (5, 20, 100, and 250 μL) of the isolated spores onto separate MM + agar plates (*see* **Note 3**). Incubate the plates at 30 °C for 5–7 days until sporulating colonies are visible. Look for colonies that form only black conidia; these are likely to be a diploid (Fig. 2d). The plate will also contain many other colonies which still show the color phenotype, and these colonies are still heterokaryons, possibly derived from heterokaryotic mycelial fragments present in the spore solution (*see* **Note 4**).
2. To avoid contamination with other spores, carefully pick spores from the middle of a black colony using a sterile toothpick. Make a spore plate of the diploid strain by transferring the spores to a fresh MM + agar plate and incubating at 30 °C (Fig. 2e).

3. Collect the diploid spores by adding 10 mL of saline solution to the plate and rubbing the spores with a sterile cotton swab. Spores (10^6 spores/mL) can be stored at 4 °C in saline solution for at least 4 weeks.

3.3 Haploidization

1. A diploid can be forced to become haploid by growing the strain in the presence of low concentrations of benomyl. Spot 5 μL of the spore solution (1×10^6 spores/mL) at four, different positions on a CM + agar plate containing the appropriate amino acids or nucleobases and benomyl (0.6 μg/mL). Allow the cells to grow at 30 °C for 4–6 days (Fig. 2f).
2. Haploid segregants are recognizable by their fawn- or olive-colored sectors. Purify the haploid segregants on a MM + agar plate containing supplements (*see* **Note 5**).
3. Harvest the spores from a single colony using a wet cotton swab and transfer the spores to an Eppendorf tube containing 500 μL of saline solution. Release the spores from the cotton swab into the saline solution by gently mixing.

3.4 Genotypic Analysis of the Segregants

1. Purified segregants are analyzed for their auxotrophies and for their specific phenotype. Prepare selective MM + agar with and without supplements and use large Petri dishes (14 cm Ø). Inoculate 5 μL spore solution on one spot, 24 segregants per plate (*see* **Note 6**). Allow the strains to grow at 30 °C for 4–6 days and determine auxotrophies and phenotypes of the segregants.
2. Linkage analysis of the segregants. For each marker, it is expected that the occurrence of the marker is equally distributed among the segregants. This means that about half of the total number of segregants have the wild type gene while the other half of the segregants contain the auxotrophic marker. Unlinked markers are expected to be equally distributed among the segregants as long as these markers are not physically linked to the same chromosome. The distribution of the markers among the segregants and possible linkage can be used to determine the genetic map in *A. niger*.
3. For the bulk segregant analysis with the aim to facilitate mutant identification via whole genome sequencing, it is important that all segregants are carefully checked for their relevant phenotypes. We have used the method successfully to identify the mutation responsible for the nonacidifying phenotype in *A. niger* [8]. In this study, all 140 segregants obtained after a cross between the nonacidifying mutant and a "wild-type" strain were screened for the nonacidifying phenotype and subsequently collected.

3.5 Culturing Segregants for Genomic DNA Isolation

1. After performing the phenotypic analysis, two groups of segregants are created. The first group contains "wild-type" segregants and the second group consists of "mutant" segregants which display the desired phenotype. In bulk segregant analysis, the DNA of all segregants belonging to the first or second group is usually isolated, pooled and sequenced to identify mutations that are conserved in the pool. In our example, however, we sequenced only the pool of mutant segregants and looked for mutations that were conserved among the pool of segregants [8].
2. For genomic DNA isolation, each segregant is grown individually in a shaken suspension (*see* **Note 7**). In this case, about 80 segregants displaying the nonacidifying phenotype were grown separately in CM [8]. Then, 200 mg fresh weight mycelium is collected from each culture. The mycelia from 20 cultures (4 g total) were mixed together and ground, and the genomic DNA was isolated.

3.6 Preparation of Genomic DNA Pools for Sequencing

1. Grind the mycelia (4 g total) under liquid nitrogen into a fine powder.
2. Using a spatula, transfer 3–4 scoops of powder (approximately 500 μL of volume) to a 2 mL Eppendorf tube. Fill as many tubes as possible, until all of the powder has been collected.
3. Add 800 μL DNA extraction buffer + RNAse and shake or vortex to resuspend the powder.
4. Incubate at 37 °C for 30 min in a thermomixer; shake at maximum speed.
5. Add 800 μL phenol–chloroform–isoamyl alcohol and shake vigorously for 15 s by hand.
6. Centrifuge at maximum speed for 15 min.
7. Transfer 700 μL of the upper layer to a new 1.5 mL Eppendorf tube, add 700 μL phenol–chloroform–isoamyl alcohol and shake vigorously for 15 s by hand.
8. Centrifuge at maximum speed for 15 min.
9. Transfer 500 μL of the upper layer to a new 1.5 mL Eppendorf tube. Add 50 μL sodium acetate (3 M, pH 6.0) and 500 μL isopropanol and allow the DNA to precipitate for 10 min at room temperature.
10. Vortex and centrifuge at maximum speed for 15 min.
11. Remove the supernatant and add 250 μL 70% ethanol.
12. Centrifuge at maximum speed for 5 min.
13. Remove supernatant and dry the pellet for ~30 min at 37 °C, with the lid of the Eppendorf tube open.
14. Dissolve the DNA by adding 50 μL Milli-Q water to the DNA pellet and shaking at maximum speed for 30 min in a thermomixer set to 37 °C.

15. When the genomic DNA is completely dissolved, determine the genomic DNA concentration and purity by measuring OD_{260}/OD_{280} using a NanoDrop.
16. Equal amounts of DNA of each of the four pools is combined to obtain the genomic DNA pool for sequencing. Genomic DNA from parental strains and the pools is further purified using NucleoSpin Plant II columns (Macherey-Nagel, 740770.50) according to the supplier's instructions and used for DNA sequencing.

4 Notes

1. The two strains used to set up the cross should have different color markers and different auxotrophies to select for heterokaryons and diploids. Near isogenic *A. niger* strains carrying various color markers in combination with auxotrophies are available [16].
2. Because asexual *A. niger* spores contain a single nucleus, the heterokaryotic stage is broken during sporulation. These spores contain one of the parental nuclei which will not grow on MM + agar because of the auxotrophies.
3. Since the number of colonies growing as background is difficult to predict, different amounts of spores are plated. This will increase the chance of obtaining the right number of colonies on the plate. Bos and coworkers noted that the frequency of obtaining heterozygous diploid spores from a heterokaryon is 10^{-5}–10^{-6} [14].
4. Omit additional rounds of purification of the heterozygous diploid as mitotic recombination might occur which could lead to linkage bias due to recombination [14].
5. The fawn marker is located on the left arm of chromosome 1 and the olive and brown markers are next to each other on the right arm of chromosome 1 [11]. Therefore, haploid segregants from a cross between a fawn and an olive/brown parental strain are either fawn or olive/brown, unless a mitotic recombination has taken place between the color markers. As the frequency of mitotic recombination in *A. niger* is low, the occurrence of black-colored haploid segregants is rare.
6. Whereas replicate plating is optional, we prefer to make spore solutions of the segregants and spot spores on selective growth media to determine the genotype of each segregant.
7. An alternative is to inoculate spores from multiple strains in the same culture flask. However, there may be growth differences among the segregants. Therefore, we prefer to culture them individually to make sure that all segregants are equally represented in the pool.

Acknowledgments

We thank Felix Seelinger, Jing Niu, and Ebru Alazi for their help in developing the methods described in this chapter. We thank Fons Debets and Peter van der Vondervoort for helpful advice.

References

1. Oakley BR (2004) Tubulins in *Aspergillus nidulans*. Fungal Genet Biol 41:420–427
2. Morris NR, Enos AP (1992) Mitotic gold in a mold: *Aspergillus* genetics and the biology of mitosis. Trends Genet 8:32–37
3. Boylan MT, Mirabito PM, Willett CE et al (1987) Isolation and physical characterization of three essential conidiation genes from *Aspergillus nidulans*. Mol Cell Biol 7:3113–3118
4. McCluskey K, Wiest AE, Grigoriev IV et al (2011) Rediscovery by whole genome sequencing: classical mutations and genome polymorphisms in *Neurospora crassa*. G3 (Bethesda) 1:303–316. https://doi.org/10.1534/g3.111.000307
5. Yu Z, Armant O, Fischer R (2016) Fungi use the SakA (HogA) pathway for phytochrome-dependent light signalling. Nat Microbiol 29:16019. https://doi.org/10.1038/nmicrobiol.2016.19
6. Niu J, Alazi E, Reid ID et al (2017) An evolutionarily conserved transcriptional activator-repressor module controls expression of genes for D-galacturonic acid utilization in *Aspergillus niger*. Genetics 205:169–183. https://doi.org/10.1534/genetics.116.194050
7. Pomraning KR, Smith KM, Freitag M (2011) Bulk segregant analysis followed by high-throughput sequencing reveals the *Neurospora* cell cycle gene, *ndc-1*, to be allelic with the gene for ornithine decarboxylase, *spe-1*. Eukaryot Cell 10:724–733. https://doi.org/10.1128/EC.00016-11
8. Niu J, Arentshorst M, Nair PDS et al (2016) Identification of a classical mutant in the industrial host *Aspergillus niger* by systems genetics: LaeA is required for citric acid production and regulates the formation of some secondary metabolites. G3 (Bethesda) 6:193–204. https://doi.org/10.1534/g3.115.024067
9. Pel HJ, de Winde JH, Archer DB et al (2007) Genome sequencing and analysis of the versatile cell factory *Aspergillus niger* CBS 513.88. Nat Biotechnol 25:221–231. https://doi.org/10.1038/nbt1282
10. Andersen MR, Salazar MP, Schaap PJ et al (2011) Comparative genomics of citric-acid-producing *Aspergillus niger* ATCC 1015 versus enzyme-producing CBS 513.88. Genome Res 21:885–897. https://doi.org/10.1101/gr.112169.110
11. Jorgensen TR, Park J, Arentshorst M et al (2011) The molecular and genetic basis of conidial pigmentation in *Aspergillus niger*. Fungal Genet Biol 48:544–553. https://doi.org/10.1016/j.fgb.2011.01.005
12. Pontecorvo G, Roper JA, Forbes E (1953) Genetic recombination without sexual reproduction in *Aspergillus niger*. J Gen Microbiol 8:198–210
13. Bos CJ, Debets AJ, Swart K et al (1988) Genetic analysis and the construction of master strains for assignment of genes to six linkage groups in *Aspergillus niger*. Curr Genet 14:437–443
14. Swart K, Debets AJ, Bos CJ et al (2001) Genetic analysis in the asexual fungus *Aspergillus niger*. Acta Biol Hung 52:335–343
15. Debets AJ, Swart K, Hoekstra RF et al (1993) Genetic maps of eight linkage groups of *Aspergillus niger* based on mitotic mapping. Curr Genet 23:47–53
16. Niu J, Arentshorst M, Seelinger F et al (2016) A set of isogenic auxotrophic strains for constructing multiple gene deletion mutants and parasexual crossings in *Aspergillus niger*. Arch Microbiol 198:861–868. https://doi.org/10.1007/s00203-016-1240-6

Index

Ronald P. de Vries et al. (eds.), *Fungal Genomics: Methods and Protocols*, Methods in Molecular Biology, vol. 1775,
https://doi.org/10.1007/978-1-4939-7804-5, © Springer Science+Business Media, LLC, part of Springer Nature 2018

MIX
Papier aus verantwortungsvollen Quellen
Paper from responsible sources
FSC® C105338

If you have any concerns about our products,
you can contact us on
ProductSafety@springernature.com

In case Publisher is established outside the EU,
the EU authorized representative is:
Springer Nature Customer Service Center GmbH
Europaplatz 3, 69115 Heidelberg, Germany

Printed by Libri Plureos GmbH
in Hamburg, Germany